TH 2392 .A43 1994

Ambrose, James

NEW ENGLAND INSTITUTE
OF TECHNOLOGY
LEARNING RESOURCES CENTER

DESIGN OF BUILDING TRUSSES

DESIGN OF BUILDING TRUSSES

JAMES AMBROSE
Professor of Architecture
University of Southern California

NEW ENGLAND INSTITUTE
OF TECHNOLOGY
LEARNING RESOURCES CENTER

JOHN WILEY & SONS, INC.
New York · Chichester · Brisbane · Toronto · Singapore

This text is printed on acid-free paper.

Copyright © 1994 by John Wiley & Sons, Inc.

All rights reserved. Published simultaneously in Canada.

Reproduction or translation of any part of this work beyond that permitted by Section 107 or 108 of the 1976 United States Copyright Act without the permission of the copyright owner is unlawful. Requests for permission or further information should be addressed to the Permissions Department, John Wiley & Sons, Inc., 605 Third Avenue, New York, NY 10158-0012.

This publication is designed to provide accurate and authoritative information in regard to the subject matter covered. It is sold with the understanding that the publisher is not engaged in rendering legal, accounting, or other professional services. If legal advice or other expert assistance is required, the services of a competent professional person should be sought.

Library of Congress Cataloging in Publication Data:
Ambrose, James E.
   Design of building trusses / James Ambrose.
     p.   cm.
   Includes index.
   ISBN 0-471-55842-7
   1. Roof trusses—Design and construction.   I. Title.
TH2392.A43   1994
695—dc20   94-9043

Printed in the United States of America

10 9 8 7 6 5 4 3

CONTENTS

PREFACE ix

INTRODUCTION xi

PART ONE   HISTORICAL DEVELOPMENT 1

  1   Historical Development 1

      1.1   Origins and Basic Concepts, 3
      1.2   Applications to Buildings, 5
      1.3   Other Trussed Structures, 6
      1.4   Early Building Examples, 9

  2   The Nineteenth Century 13

      2.1   New Materials and Forms, 13
      2.2   Origins of Modern Trussed Construction and Use, 15

  3   Early Twentieth Century Construction 18

      3.1   Generic Truss Construction, 18
      3.2   Construction with Spanning Trusses, 19
      3.3   Trussed Bracing for High-rise Structures, 19
      3.4   Trusses as Industrial Products, 22

PART TWO   MODERN CONSTRUCTION WITH TRUSSES 25

  4   New Design Concerns 27

      4.1   Competing Long-span Systems, 27

4.2 Use of Interstitial Space, 28
4.3 Longer Spans, 28

5 New Truss Systems and Details · 31

5.1 New Fabrication Methods, 31
5.2 New Forms and Materials, 32
5.3 Space Frames, 34
5.4 Braced Frames for Lateral Stability, 35

PART THREE INVESTIGATION AND DESIGN OF PLANAR TRUSSES · 37

6 Analysis for Static Forces · 39

6.1 Properties of Forces, 39
6.2 Components of Forces, 42
6.3 Composition and Resolution of Forces, 43
6.4 Equilibrium, 44
6.5 Analysis of Concurrent, Coplanar Forces, 45
6.6 Graphical Analysis for Internal Forces in Planar Trusses, 50
6.7 Algebraic Analysis for Internal Forces in Planar Trusses, 55
6.8 Visualization of the Sense of Internal Forces, 62
6.9 Analysis of Nonconcurrent Forces, 65
6.10 Beams, 76
6.11 Graphic Representation of Beam Behavior, 79
6.12 Analysis of Statically Determinate Beams, 81
6.13 Behavior of Indeterminate Beams, 88
6.14 Analysis of Trusses by the Method of Sections, 93
6.15 Analysis of Trusses by the Beam Analogy Method, 96

7 Analysis of Planar Trusses: General Concerns · 99

7.1 Loads on Trusses, 99
7.2 Truss Reactions, 100
7.3 Stability and Determinacy, 100
7.4 Analysis for Gravity Loads, 104
7.5 Analysis for Wind Loads, 110
7.6 Design Forces for Truss Members, 112
7.7 Combined Stress in Truss Members, 114

8 General Considerations for Trussed Construction · 116

8.1 Planar Spanning Trusses, 116
8.2 Truss Forms, 118
8.3 Materials for Trusses, 125
8.4 Bracing for Trusses, 129
8.5 Truss Joints, 129
8.6 Manufactured Trusses, 133
8.7 Trussed Bracing for Lateral Forces, 135

9 Design of Wood Trusses — 145

9.1 Basic Considerations for Wood, 145
9.2 Joints in Wood Trusses, 165
9.3 Light Wood Trusses with Single Members, 185
9.4 Heavy Timber Trusses, 192
9.5 Wood Trusses with Multiple-Element Members, 196
9.6 Combination Trusses: Wood and Steel, 199
9.7 Trussed Bracing for Wood Structures, 200

10 Design of Steel Trusses — 203

10.1 Basic Considerations for Steel, 203
10.2 Joints in Steel Trusses, 221
10.3 Design of a Light Steel Truss with Bolted Joints, 225
10.4 Design of a Light Steel Truss with Welded Joints, 238
10.5 Design of a Steel Truss with Tubular Members, 243
10.6 Trussed Bracing for Steel Frame Structures, 247

11 Special Problems — 253

11.1 Deflection of Trusses, 253
11.2 Indeterminate Trusses, 257
11.3 Secondary Stresses, 258
11.4 Movements at Supports, 258

12 Special Truss Structures — 261

12.1 Space Frames, 261
12.2 Design Example: Spatial Truss, 273
12.3 Delta Trusses, 287
12.4 Trussed Columns and Towers, 290
12.5 Trussed Mullions, 294
12.6 Trussed Arches and Bents, 296

PART FOUR CONSTRUCTION PLANNING AND DESIGN — 301

13 Details for Construction — 303

13.1 Joint Details, 303
13.2 Control Joints in Trussed Structures, 304
13.3 Jointing for Erection and Assemblage, 305
13.4 Development of the General Construction, 306
13.5 The Revealed Structure, 307

14 General Construction Concerns — 313

14.1 Erection Problems, 313
14.2 Temporary Bracing, 315
14.3 Building Code and Standards, 317

PART FIVE USE OF GRAPHICAL METHODS — 329

15 Graphical Methods for Investigation of Forces, 331

15.1 Fundamentals of Force Investigation, 331
15.2 The Funicular Polygon, 332
15.3 Determination of Truss Reactions, 338
15.4 Applications of Graphical Methods, 346

16 Finding Efficient Forms for Trusses — 378

16.1 The Need for Efficient Truss Forms, 378
16.2 A Numerical Method for Finding an Optimal Truss Form, 378
16.3 A Graphical Method for Finding an Optimal Truss Form, 381
16.4 Finding a Truss Form That Has Uniform Force in the Curving Chord, 383
16.5 A Bottom-Loaded Constant Force Truss, 385
16.6 A Lenticular Truss, 386
16.7 A Constant Force Gable Truss, 387
16.8 Cantilever Trusses of Optimal Form, 388
16.9 Optimal Truss Forms: The General Solution, 389
16.10 Suboptimal Truss Forms: The Camelback Truss, 392
16.11 Visualizing Form Improvements in Trusses of Nonoptimal Shape, 393

APPENDIX A COEFFICIENTS FOR MEMBER FORCES IN SIMPLE TRUSSES — 397

APPENDIX B PROPERTIES OF SECTIONS — 400

APPENDIX C VALUES FOR TYPICAL BEAM LOADINGS — 414

APPENDIX D USE OF COMPUTERS — 416

GLOSSARY — 419

BIBLIOGRAPHY — 425

INDEX — 427

PREFACE

This book deals with the general use of trussing (basically triangulated framing) for building structures. It treats the subject broadly in terms of historic and modern construction, the variety of applications for buildings, and aspects of investigation and design of truss systems. It is a reference and a study source for all persons interested in the general topics of design and construction of building structures.

For persons with no background in the study of structural investigation and design, the truss offers a unique opportunity to learn the most elementary principles of applied mechanics and proceed to truly practical applications of design. That quality has made it a favorite of teachers for many generations. Readers desiring to pursue such a study will find the complete development of this work in Part 4 of this text.

A special form of study is that which employs graphical constructions for structural investigation. Using both historic developments and current applications, this area of study is presented in Part 5. While some real and practical benefits can be gained from this work, the biggest reward is in the sharpening of visualization skills by direct manipulations of vector analysis methods. This is truly an exercise in "seeing" mathematics at work.

Trussing has developed through many cultures as a pragmatic means for producing bracing of frames and the general development of very light structures for large constructions. With the advent of steel construction in the nineteenth century, it was used for some still impressive structures, such as those for the Eiffel Tower, the Firth of Forth bridge, and the large arched roofs of many exhibition halls and train sheds. In the twentieth century, some notable developments have been those of two-way spanning space frames and the truss bracing for high-rise buildings.

Still, what probably counts the most is the highly practical use of trussing for the light, long-span systems using simple open-web steel joists and combination wood and steel joists. These demonstrate most effectively the practical usefulness of the basic system.

Whether for highly practical applications and modest buildings, for dramatically monumental sized works, or for dazzling sculptural assemblages, trussing continues to be chosen by many engineers and architects as a favorite structural method. It is surely one of the most fundamental structural systems that we have inherited and will pass on to future builders.

My enthusiasm for trusses is shared by many people, and I have received support for this work from many others. The most direct assistance came from my friend, Edward Allen, who provided a very stimulating chapter (Chapter 16) for the part of the book devoted to the use of graphical methods for investigation and design. Others who contributed directly to this work through review of early drafts or other sharing were Anthony Schnarsky, Goetz Schierle, and Wilbur Yoder.

I have inherited a great deal personally from teachers, early writers, and professional mentors and colleagues, including Harry Parker, Cyrus Palmer, Edmund Toth, Fhazlur Kahn, and Stephen Tang. I am happy to be able to pass much of that heritage along through this book.

Having lived all of my academic and professional life in a dual environment of architecture and structural engineering, it has become increasingly difficult—and I believe artificial—for me to separate the issues in design between professional camps. I no longer think of myself as either an architect or an engineer, but rather as a person involved in the design of building structures with no preestablished boundaries or divisions of the design necessities. This book is for persons interested in the design of trusses for buildings; what they do with the materials is their own business.

Producing this publication would not have been possible without the support of my publisher, Peggy Burns, my editor, Everett Smethurst, and the hard-working production staff at John Wiley and Sons. I am especially indebted to the constant vigilant quest for excellence and the competent management of the production process by Robert Fletcher IV and Milagros Torres.

I am grateful to various sources for the permission to use materials from their publications, including the National Forest Products Association, the American Institute of Steel Construction, and the International Conference of Building Officials (publishers of the *Uniform Building Code*). I am also grateful to my publishers, John Wiley and Sons, for permission to reproduce materials from several early works.

Working largely in a home office, I must express my appreciation for the tolerance, support, and frequent assistance provided by my family, particularly by my wife, Peggy.

For what it may provide in assistance and inspiration, this book is dedicated to the readers. Hopefully, they will cherish and pass on any lasting value it contains.

JAMES AMBROSE

Westlake Village, California

INTRODUCTION

This book treats the subject of trusses with regard to their use for building structures. Wherever framed structures have been built, the use of triangulation of the frame has been utilized as a device for stabilizing the framework. This was discovered by primitive people by experimentation and continues as an option for the latest methods of lateral bracing for wind and seismic forces. For spanning structures, the truss offers an efficiency of materials unattainable by any other framed system. In an ever-increasing array of systems and methods, trussing continues to be a highly useful part of the inventory of available means for achieving building structures.

Apart from its practical applications, the study of behavior of simple trusses has long been a favorite means of involving the beginning student of structures. Resolution of internal forces in a simple truss is so elementary and easily visualized, that it has often been used as a first encounter with the visualization and computation of force mechanisms. Since the force mechanisms of the truss members and their fastening at connections is also simple and direct, it allows for a relatively easy progression from investigation of loads, to visualization and computation of internal forces, to design determination of parts; and finally, to the total definition of the complete, interactive system. In other words, it permits a person with limited training to cover the whole span from definition of a structural task to complete determination of the finished structure.

The work in this book is presented for both its practical applications to current designs and its potential for a learning experience. Individual readers may have needs related to one or the other of these uses. Or, they may have both as eventual goals—starting with learning. If fundamental learning about structures is a need, the materials in Part Three present a possibility for starting at the ground level and proceeding to some practical design work.

Readers with some amount of background in structural investigation and design may be more selective about the parts of the book they may find more useful. Most of the basic uses of trussing are presented here, many with example computations and illustrations of construction details.

A method used some years ago for investigation of trusses was that of graphical devices using the vector properties of forces and the simple equilibrium of force systems. With the gradual removal of training in graphics from engineering curricula, and its drastic reduction in most architectural curricula, the usefulness of these methods to students became less significant. However, some of the simple graphical methods offer support for visualization of structural behaviors and are a potential supplement to algebraic methods. Exploration of more complex graphical methods requires some skill in drafting, but presents some rich experiences in the development of constructed manifestations of structural behaviors. The full potentialities of graphical methods are presented here in Part Five.

Design of trusses today is usually supported by computer-aided methods. Use of the computer has extended possibilities for optimization of the design of simple trusses and for reliable investigations of highly complex systems.

For architects, the truss is a highly visual expression of natural structural form with a fabric that is rich in its rhythms and geometry. Many trusses are used for exposed structures, making them equally desired and subject to design considerations for their engineering efficiency and their viewed forms. The large number of illustrations in this book represents a reflection of the strong interests for the designer in *seeing* the truss, both for development of the logical structural design and for shaping of a viewed form.

UNITS OF MEASUREMENT

The work in this book is presented primarily in English units (feet, pounds, etc.). These are now more appropriately called U.S. units, since England no longer uses them officially. The construction industry in the United States still uses U.S. units, although a slow switch-over to SI units seems to be creeping along. It is hard to estimate how long it will take to get rid of the wood two by four, the 16-in. concrete block, and the 4-ft by 8-ft plywood panel. For some of the computational work here we have chosen to give equivalent metric-based data in brackets (thus) following the U.S. unit data. This may be of some aid for those readers still struggling with the dual-unit situation.

Table 1 lists the standard units of measurement in the U.S. system with the abbreviations used in this work and a description of the type of the use in structural work. In similar form, Table 2 gives the corresponding units in the SI system. The conversion units used in shifting from one system to the other are given in Table 3.

For some of the work in this book, the units of measurement are not significant. What is required in such cases is simply to find a numerical answer. The visualization of the problem, the manipulation of the mathematical processes for the solution, and the quantification of the answer are not related to the specific units—only to their relative values. In such situations we have occasionally chosen not to present the work in dual units, to provide a less confusing illustration for the reader. Although this procedure may be allowed for the learning exercises in this book, the structural designer is generally advised to develop the habit of always indicating the units for any numerical answers in structural computations.

TABLE 1 Units of Measurement: U.S. System

Name of Unit	Abbreviation	Use
Length		
Foot	ft	Large dimensions, building plans, beam spans
Inch	in.	Small dimensions, size of member cross sections
Area		
Square feet	ft^2	Large areas
Square inches	$in.^2$	Small areas, properties of cross sections
Volume		
Cubic feet	ft^3	Large volumes, quantities of materials
Cubic inches	$in.^3$	Small volumes
Force, Mass		
Pound	lb	Specific weight, force, load
Kip	k	1000 pounds
Pounds per foot	lb/ft	Linear load (as on a beam)
Kips per foot	k/ft	Linear load (as on a beam)
Pounds per square foot	lb/ft^2, psf	Distributed load on a surface
Kips per square foot	k/ft^2, ksf	Distributed load on a surface
Pounds per cubic foot	lb/ft^3, pcf	Relative density, weight
Moment		
Foot-pounds	ft-lb	Rotational or bending moment
Inch-pounds	in.-lb	Rotational or bending moment
Kip-feet	k-ft	Rotational or bending moment
Kip-inches	k-in.	Rotational or bending moment
Stress		
Pounds per square foot	lb/ft^2, psf	Soil pressure
Pounds per square inch	$lb/in.^2$, psi	Stresses in structures
Kips per square foot	k/ft^2, ksf	Soil pressure
Kips per square inch	$k/in.^2$, ksi	Stresses in structures
Temperature		
Degree Fahrenheit	°F	Temperature

TABLE 2 Units of Measurement: SI System

Name of Unit	Abbreviation	Use
Length		
Meter	m	Large dimensions, building plans, beam spans
Millimeter	mm	Small dimensions, size of member cross sections
Area		
Square meters	m^2	Large areas
Square millimeters	mm^2	Small areas, properties of cross sections
Volume		
Cubic meters	m^3	Large volumes
Cubic millimeters	mm^3	Small volumes
Mass		
Kilogram	kg	Mass of materials (equivalent to weight in U.S. system)
Kilograms per cubic meter	kg/m^3	Density
Force (Load on Structures)		
Newton	N	Force or load
Kilonewton	kN	1000 newtons
Stress		
Pascal	Pa	Stress or pressure (1 pascal = 1 N/m^2)
Kilopascal	kPa	1000 pascals
Megapascal	MPa	1,000,000 pascals
Gigapascal	GPa	1,000,000,000 pascals
Temperature		
Degree Celsius	°C	Temperature

TABLE 3 Factors for Conversion of Units

To Convert from U.S. Units to SI Units Multiply by	U.S. Unit	SI Unit	To Convert from SI Units to U.S. Units Multiply by
25.4	in.	mm	0.03937
0.3048	ft	m	3.281
645.2	in.^{2}	mm^2	1.550×10^{-3}
16.39×10^3	in.^{3}	mm^3	61.02×10^{-6}
416.2×10^3	in.^{4}	mm^4	2.403×10^{-6}
0.09290	ft^2	m^2	10.76
0.02832	ft^3	m^3	35.31
0.4536	lb (mass)	kg	2.205
4.448	lb (force)	N	0.2248
4.448	kip (force)	kN	0.2248
1.356	ft-lb (moment)	N-m	0.7376
1.356	kip-ft (moment)	kN-m	0.7376
1.488	lb/ft (mass)	kg/m	0.6720
14.59	lb/ft (load)	N/m	0.06853
14.59	kips/ft (load)	kN/m	0.06853
6.895	psi (stress)	kPa	0.1450
6.895	ksi (stress)	MPa	0.1450
0.04788	psf (load or pressure)	kPa	20.93
47.88	ksf (load or pressure)	kPa	0.02093
$0.566 \times (°F - 32)$	°F	°C	$(1.8 \times °C) + 32$

COMPUTATIONS

In professional design firms, structural computations are most commonly done with computers, particularly when the work is complex or repetitive. Anyone aspiring to participation in professional design work is advised to acquire the background and experience necessary to the application of computer-aided techniques. The computational work in this book is simple and can be performed easily with a pocket calculator. The reader who has not already done so is advised to obtain one. The "scientific" type with eight-digit capacity is quite sufficient.

For the most part, structural computations can be rounded off. Accuracy beyond the third place is seldom significant, and this is the level used in this work. In some examples more accuracy is carried in early stages of the computation to ensure the desired degree in the final answer. All the work in this book, however, was performed on an eight-digit pocket calculator.

Symbols

The following "shorthand" symbols are frequently used:

Symbol	Reading
$>$	is greater than
$<$	is less than
$\geq$	equal to or greater than
$\leq$	equal to or less than
$6'$	six feet
$6''$	six inches
Σ	the sum of
ΔL	change in L

Notation

Use of standard notation in the general development of work in mechanics and strength of materials is complicated by the fact that there is some lack of consistency in the notation currently used in the field of structural design. Some of the standards used in the field are developed by individual groups (notably those relating to a single basic material, wood, steel, concrete, masonry, etc.) which each have their own particular notation. Thus the same type of stress (e.g., shear stress in a beam) or the same symbol (f_c) may have various representations in structural computations. To keep some form of consistency in this book, we use the following notation, most of which is in general agreement with that used in structural design work at present.

a	(1) Moment arm; (2) increment of an area
A	Gross (total) area of a surface or a cross section
b	Width of a beam cross section
B	Bending coefficient
c	Distance from neutral axis to edge of a beam cross section
d	Depth of a beam cross section or overall depth (height) of a truss
D	(1) Diameter; (2) deflection
e	(1) Eccentricity (dimension of the mislocation of a load resultant from the neutral axis, centroid, or simple center of the loaded object); (2) elongation
E	Modulus of elasticity (ratio of unit stress to the accompanying unit strain)
f	Computed unit stress
F	(1) Force; (2) allowable unit stress
g	Acceleration due to gravity
G	Shear modulus of elasticity
h	Height
H	Horizontal component of a force
I	Moment of inertia (second moment of an area about an axis in the plane of the area)

J	Torsional (polar) moment of inertia
K	Effective length factor for slenderness (of a column: KL/r)
M	Moment
n	Modular ratio (of the moduli of elasticity of two different materials)
N	Number of
p	(1) Percent; (2) unit pressure
P	Concentrated load (force at a point)
r	Radius of gyration of a cross section
R	Radius (of a circle, etc.)
s	(1) Center-to-center spacing of a set of objects; (2) distance of travel (displacement) of a moving object; (3) strain or unit deformation
t	(1) Thickness; (2) time
T	(1) Temperature; (2) torsional moment
V	(1) Gross (total) shear force; (2) vertical component of a force
w	(1) Width; (2) unit of a uniformly distributed load on a beam
W	(1) Gross (total) value of a uniformly distributed load on a beam; (2) gross (total) weight of an object
Δ (delta)	Change of
Σ (sigma)	Sum of
θ (theta)	Angle
μ (mu)	Coefficient of friction
ϕ (phi)	Angle

(FRONT) Still dominating the Near North Side of Chicago, the John Hancock Building is a celebration of trussed building structures: a sort of Eiffel Tower for Chicago. The form of the monumental size X-bracing is fully displayed on the building surfaces, even though the actual steel structure is encased in finish jackets. Provides a contrast with the tower of the old Water Tower building in the foreground—a survivor of the sensational 1870 fire. Architects: Skidmore, Owings, and Merrill, Chicago; structural engineer: Fazlur Kahn.

PART ONE

HISTORIC DEVELOPMENT

Trussing as a structural means has surely been discovered hundreds of times over in the history of the human species. Invented, stolen, borrowed, or inherited, it has been used and developed by many cultures. This part deals briefly with the past and some of the ideas that have come down to our present times as useful and enduring.

1

EARLY USES OF TRUSSES

The primary uses of trussing are so fundamentally logical that their origins are surely buried in the prehistoric millennia of our emergence as a species. Not to belabor historic accuracy, we consider here how some of the most fundamental principles may have emerged from simple experimentation with building processes by very early people.

1.1 ORIGINS AND BASIC CONCEPTS

Trussing—that is, triangulating of a framework—was probably invented expediently by various people in the process of producing frameworks that were stable. Thus the idea of the truss as a stabilizing device (the *braced frame* in present lateral force design parlance) probably preceded the idea of the truss as a spanning structure.

Early frame structures were produced in many cultures by the lashing together of individual tree branches and trunks. For buildings, natural frame patterns involved the use of vertical members (for walls) and horizontal members (for door and window lintels, roof joists, etc.). This naturally produced rectilinear patterns for the frames, which experience quickly demonstrated to be inherently unstable in response to in-frame planar distortions (see Figure 1.1).

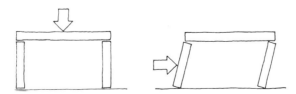

FIGURE 1.1 Stable for gravity loads, but essentially nonresistive to lateral effects, the post and beam needs some form of bracing.

4   EARLY USES OF TRUSSES

All the basic means for bracing the laterally unstable rectilinear frame were available to the early builders, but the simplest one to achieve with the framing materials themselves was the triangulation of the frame pattern (see Figure 1.2). A completely triangulated frame was possible, but for purposes of bracing only, it probably soon became evident that only a few of the rectangles of a multiunit frame needed to be braced to achieve stability for the whole, connected framework.

This use of trussing is evident in the meaning of the word truss as a verb: to "tie up or bind." This relates to the adding of triangulating (diagonal) members to a rectilinear framework, to tie it against lateral distorting effects.

One could write a scenario for the progression from bracing to spanning capabilities for the truss, with some minor speculations. For example, a most dramatic demonstration of the triangulated bracing occurs when it is used to laterally stabilize a tower-like structure, with considerable height in proportion to its width. "Towers" were undoubtedly first achieved with single, large tree trunks. However, where tall trees were less available, enterprising builders soon developed the built-up framework for the tower form—with height limits established only by daring.

An elementary means for erecting a pole from a prone to erect position is to first lift one end (Figure 1.3a) and then pull it up with a rope (Figure 1.3b). Some wise person probably observed, during such an operation for a trussed tower, that the trussed frame retained a stiffened character during this operation and inferred the potential spanning capability for the truss Figure 1.3c).

By this, or other means, the spanning truss was conceived and innovated—whether first in lashed animal bones, tree branches, or metal rods.

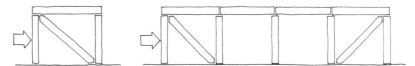

FIGURE 1.2 Use of diagonal members is an obvious means for bracing; developing the fundamental geometry of the nondeformable triangle (can't change the angles of the corners without changing the lengths of the sides).

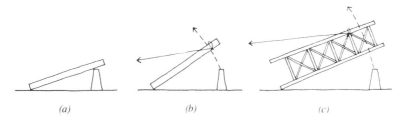

(a)   *(b)*   *(c)*

FIGURE 1.3 Erecting tall structures before the existence of machinery: (*a*) One end could be lifted only as high as a single person; (*b*) With rope, the tall element could be pulled up.; (*c*) The trussed tower could also be assembled in a horizontal position and pulled up.

1.2 APPLICATIONS TO BUILDINGS

For buildings, the principal early applications of trusses were in achieving the basic gable-form roof, with two opposed sloping flat sides meeting at a ridge at the top and forming two draining edges (eaves) at their opposite bottom sides (see Figure 1.4). A horizontal tie across this building at the eave level held the two edges from thrusting outward and thus formed a single triangle truss, with two compression members (the rafters) and one tension member (the horizontal tie).

If no division of the interior space was desired, the ties may have just hung there in the enclosed space. However, they were eventually surely used for hanging things—lights, decorations, game from the hunt (out of range of the dogs, etc.), and sleeping hammocks.

In due time, however, the horizontal members became used as framing for an attic floor or a ceiling surface for the lower space. As the building sizes increased, the sag of the horizontal members required some consideration, and the multiplication of the internalized triangulation of various basic truss forms began.

Figure 1.5 shows progression of multiplication of internal units of the basic gable-form truss. The basic inference in this progression is that the truss span is steadily increasing. However, another consideration was the ability to develop a large structure with a multiple of relatively small and short members. While something other than a single log may be used for the horizontal tie, a limit for spans with most early trusses was the length of a single cut timber that could be obtained for the rafters.

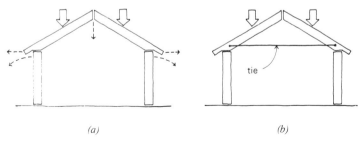

FIGURE 1.4 Making the pitched roof. (*a*) The natural placement of rafters results in the tendency to push the top of the wall outward. (*b*) One way to hold the bottoms of the rafters is to use a tie, resulting in the definition of a simple, single triangle truss.

FIGURE 1.5 Progression of subdivision of the truss, as the span increases and the truss members get longer.

6 EARLY USES OF TRUSSES

1.3 OTHER TRUSSED STRUCTURES

The same forms of spanning structures that were utilized for building roofs were also applied to other spanning situations—primarily bridges for pedestrian or vehicular traffic. Trusses were used for bridges in a parallel development of their use for roofs (see Figure 1.6). Some of the notable technological developments and advances were first achieved by bridge builders, before being applied to buildings. Long spans for trussed bridges built more than a hundred years ago have still never been exceeded for trussed roof structures.

Weather resistance and general durability were of greater concern for bridges, making more critical the rotting of wood and rusting of iron. The development of the covered bridges, with two trusses forming walls to support a roof, was for the purpose of protecting the trusses from the elements, thereby extending the life of the construction.

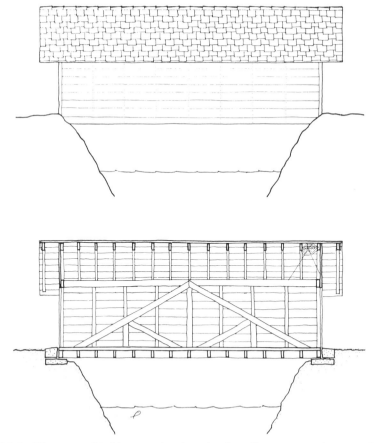

FIGURE 1.6 The covered bridge was developed primarily to protect the spanning bridge structure and the deck from the weather. While these are still major concerns, our efforts now go to making the construction itself weather resistant.

Exposure conditions, including effects of freezing, thermal expansion and contraction, precipitation, wind-blown sand, and salt air, are still major concerns for exposed structures. A major maintenance problem for all exposed steel truss structures is that of the recurring need to renew rust-resistant treatments.

As with most other forms of spanning structures, the longest span trusses have been built for bridges. A form developed to achieve the longest spans is that of the double cantilever with a simple span truss achieving the center span (see Figure 1.7). However, for every gigantic size bridge structure, there have been hundreds of smaller span constructions, down to very modest ones for spanning small creeks or culverts.

A very early use of trussing was for the bracing or otherwise enhancing of other structures. Towers with three or more legs were braced with combinations of horizontal and diagonal members, permitting the use of more slender legs in some cases, but mostly just to achieve lateral stability and stiffness. Truss-braced towers were thus used for elevated water tanks, bridge piers, watch towers, and towers for the support of suspended tension systems (see Figure 1.8).

An early use of trusses was for the support of formwork for the construction of masonry arches. The masonry arch is not stable until the last stone is in place, so the arch form is first described with a support structure, called *centering*. Figure 1.9 shows a timber bowstring arch used for the support of formwork; a method used when the space beneath the arch did not allow for placing of direct supports.

Early iron bridges were mostly of trussed arch forms, producing a novel, open, lacy appearance, in contrast to the historic heavy, solid form of stone construction. With the development of steel, some major spans for both bridges and building structures were achieved with trussed arches.

The development and rapid expansion of the Iron Horse (steam-powered trains) corresponded with the development of various iron and steel structures, resulting in many trestles, bridges, roundhouses, and train sheds for stations of trussed steel construction. The great train sheds of the nineteenth century, with their glass-paneled roof surfaces, produced some truly impressive longspan roofs.

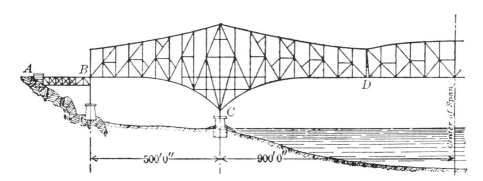

FIGURE 1.7 Profile of a large cantilever bridge. The major part of the span is achieved by the cantilevers that are balanced on piers, projecting in one direction back to the bank, and the other way to support the simple span truss that completes the center of the span. This is the form of the first Quebec bridge, that collapsed during erection in 1906. Ten years later, a second bridge with a modified form was erected, achieving an 1800-ft clear span; a record still unsurpassed for a truss.

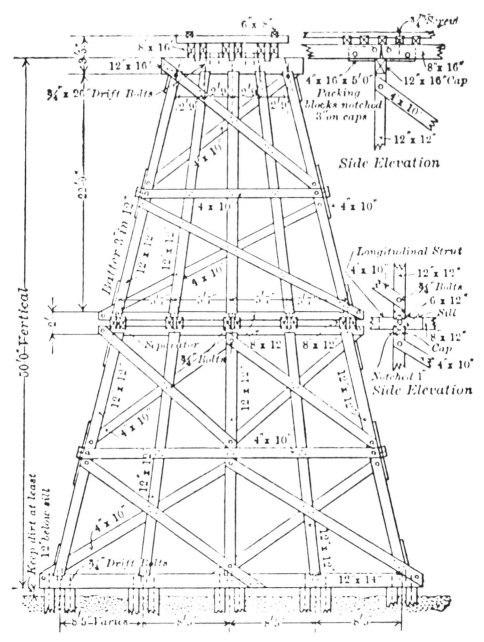

FIGURE 1.8 Drawing for a timber trestle for a railroad bridge. (Reproduced from *American Civil Engineers' Handbook*, 1918 edition, with permission of the publishers, John Wiley & Sons.)

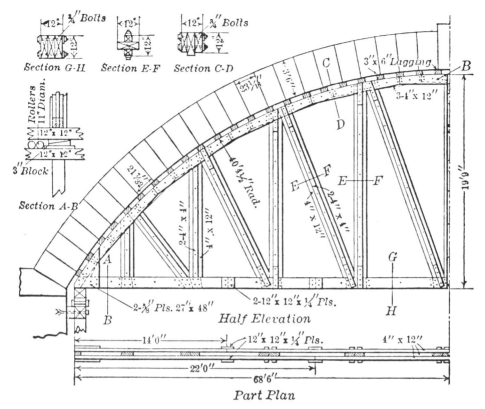

FIGURE 1.9 Drawing for a timber truss with a flat bottom chord and an arched top chord. Now called a bowstring truss when used as a spanning roof structure (see Figure 9.42), but originally developed to provide temporary support (called *centering*) for a stone arch where the spanned space beneath the arch had to be kept open during construction. (Reproduced from *American Civil Engineers' Handbook*, 1918 edition, with permission of the publishers, John Wiley & Sons.)

1.4. EARLY BUILDING EXAMPLES

As discussed previously, trussing (triangulation of framing) was probably first used as a means of bracing; providing both stability and stiffness for wobbly stick frames. Thus single diagonals, or possibly criss-crossed members (X-bracing), were added to otherwise rectangular arrangements of vertical and horizontal framing. This still represents a major means for bracing of frames, simply for basic stability, or for the specific effects of lateral forces from wind or earthquakes.

Early frameworks of heavy timber routinely had some diagonal members, typically placed at the ends of walls and at building corners (see Figure 1.10). When exterior walls consisted of heavy masonry, this was not a concern, but in a completely framed structure with infill walls some diagonal bracing was almost always present.

10 EARLY USES OF TRUSSES

FIGURE 1.10 Half-timber construction, consisting of a truss-braced timber frame with wall infill usually of bricks and plaster.

The spanning truss most likely first emerged from the basic tying of the rafters in a simple gable-form roof (see Figures 1.4 and 1.11). If interior walls did not support the horizontal tie, some interior truss members were used—essentially to keep the tie from sagging under directly applied loads. With roofing mostly of the high-slope type (shingles, thatch, tiles, etc.), the triangular truss profile was the most common form for building trusses.

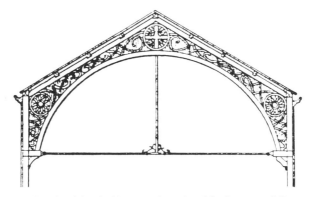

FIGURE 1.11 A tied arch with a built-up, raftered, gable-form roof. Or, maybe a tied truss with a curved lower chord? A lot of structural variations are possible for this general architectural form, with flat, sloping roof surfaces and an arched ceiling. [From *Architects and Builders Handbook*, 1931 (Ref. 14), reproduced with permission of the publishers, John Wiley & Sons.]

Other truss profiles were developed in response to usage considerations. Where an attic was not desired with the gable-form roof, the ties were sometimes objectionable on the interior. This resulted in the development of such special forms as the hammer-beam truss (Figure 1.12), and the scissors truss (Figure 1.13).

Flat profile trusses, with parallel top and bottom chords, were most likely first used for bridges, where the peaked, gable-form had no direct purpose. As waterproofing of flat roofs became more reliable, however, there was a need for the flat profile truss for longspan roofs. A later application was the use of trusses for floors, where the floor above and the ceiling below formed an interstitial space with many purposes for containment of building service elements. The floor trusses allowed for easier passage of wiring, piping, and ducts, as compared to beams with solid webs.

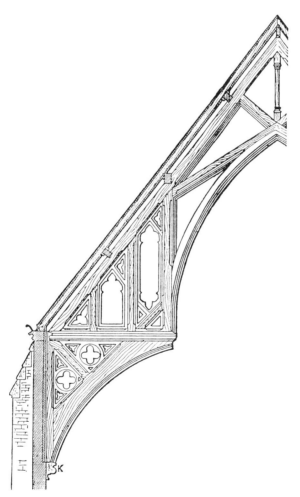

FIGURE 1.12 A composite structure, with a scissor truss at the center span, and arched ceiling rib, and an inwardly cantilevered support designed to counterbalance the outward thrust at the wall; a real courageous balancing act. [From *Architects and Builders Handbook*, 1931 (Ref. 14), reproduced with permission of the publishers, John Wiley & Sons.]

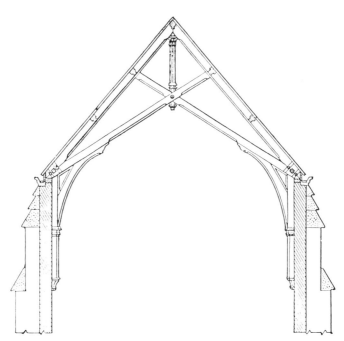

FIGURE 1.13 The scissors truss. A popular form that permits a raised ceiling space; with some infill fillet arches in this case that partly emulate a gothic arch interior form. [From *Architects and Builders Handbook*, 1931 (Ref. 14), reproduced with permission of the publishers, John Wiley & Sons.]

The truss was largely a timber structure in early years, and spans were generally modest in size. The arch, dome, and multipart vault—all executed in masonry—were the choices for long spans. With the development of iron and steel in the eighteenth and nineteenth centuries, however, the truss took new forms and was steadily pushed to longer and longer spans. However, the longest spans were at first ones achieved with trussed arches, rather than with beam-like spanning trusses.

2
THE NINETEENTH CENTURY

Truss construction made enormous advances during the industrial expansion and general technological growth that affected all man-made work during the nineteenth century. The principal thrust of this was the general exploitation of the usage of iron and steel. Slender iron and steel columns replaced heavy timber posts and the even heavier piers and massive walls of masonry. Steel beams replaced timber beams. And thin truss members of steel replaced heavier ones of wood. Diagonal X bracing with slender rods permitted very light structures.

This overall thinning down and lightening of the form of structures affected the styles of design. The open, light character of the Eiffel Tower (1884) contrasted with the solid form of the Washington Monument (completed in 1889); the one marking the emerging of a style and the other the fading of one.

The first really long span structures were steel: the steel cable suspension bridges and the trussed arches of the great train sheds and exhibition halls (see Figure 2.1). Reaching in another direction, the push upward of the tall buildings (skyscrapers) at the end of the century was made possible only because of the development of the rolled shapes of structural steel—stacked ever higher and higher in daring acts of courage by the designers and builders.

2.1 NEW MATERIALS AND FORMS

Steel was the "new" material that most affected development of new truss forms and the steady increase of spans of trussed structures. Wood endured as the economical material of choice for modest size buildings—then as now—and short span trusses of classic forms continued to be used for houses and other small buildings. But the new demands for longer spans and taller buildings were met by steel structures.

Concrete emerged significantly in the last part of the nineteenth century as a challenging material for building structures; being promoted from its formerly lowly status principally for the crude elements of foundations, core filler for cavity

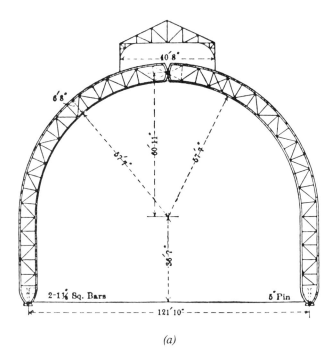

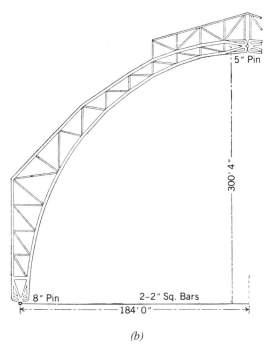

FIGURE 2.1 Three-hinged trussed arches used for exhibition buildings at the Chicago World's Fair of 1893. A stilt-legged, semicircular form for the 121-ft, 10-in. span of the Machinery Hall (*a*), and a straight-edged bent form for the dramatic 368 ft span of the Manufacturers and Liberal Arts Building (*b*). [From *Architects and Builders Handbook*, 1931 (Ref. 14), reproduced with permission of the publishers, John Wiley & Sons.]

masonry walls, pavements, and so forth. Three technical advances were largely responsible for this. The first was the invention of the form of cement known as portland cement; a highly industrialized product in comparison to the crude natural cements of former eras, permitting higher strengths, finer finishes, and a generally much refined material. This is the basic form still used, almost 170 years later.

The other two developments had to do with product forms. Some of the earliest uses for building superstructures were as precast products—primarily concrete masonry units (cast stone, as it was called to imply the resemblance to a more noble material). Very soon, the modular forms of hollow concrete blocks were developed, predicting very closely the forms commonly used today.

Finally, was the development of concrete reinforced with steel rods, which quickly led to the various classic forms of sitecast concrete frames and flat-spanning systems: three-dimensional beam and column frames, slab and beam systems emulating those of wood and steel, and more uniquely, concrete joists, two-way joists (waffles), and the flat slab.

The developments in concrete, however, had very little effect on trussed construction, except that the stronger abutment and pier foundations possible made some of the long-span trusses more feasible. It was steel that produced the tremendous expansion of new truss forms and long-span trussed construction.

Besides the steel elements themselves—mostly hot rolled shapes of the classic forms now still in use (angles, rods and bars, I-shapes, etc.)—there was the development of fastening methods, eventually using mostly rivets in both the fabricating shops and at the erection sites (see Figure 2.2). This strongly affected the logical choices for truss members, relating to the ease with which their forms facilitated the achieving of joints. And with its many joints, trussed construction is generally only feasible when the jointing is carefully and efficiently developed.

2.2 ORIGINS OF MODERN TRUSSED CONSTRUCTION

Trussing has a great deal to do with fundamental geometry and its relations to resolution of forces and the establishment of stability. These basic relationships are enduring and will always be a part of the core of fundamental technology. What time brings is continued experimentation, which results in some fads, some faulted trials, and some really new concepts and useful applications.

For trussed structures, the nineteenth century brought the development of improvements in the use of iron. Iron had been around for many centuries, but had seen little application in building construction before the improvements developed in the eighteenth and nineteenth centuries. With the development of ductile steel, hot rolled structural shapes, and riveting—all in the nineteenth century—the facility was available for extensive use of steel frameworks for building (and other) construction.

Some of the earliest applications were for trussed structures, for the simple reason that the size of individual steel elements that could be produced was at first quite limited. Therefore, any large structures had to be made up from many small parts; a situation that lead to extensive use of trussing. The largest structures were at first trussed towers and trussed arches. However, some magnificent trussed bridges were also produced. Using the experience gained with trussed towers, the first tall

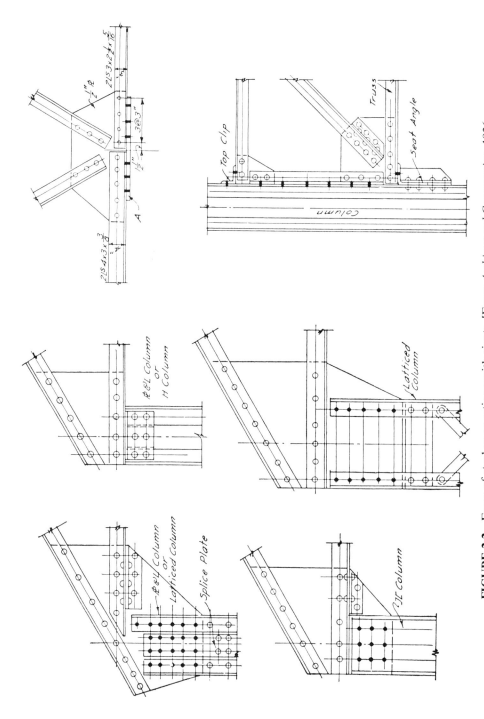

FIGURE 2.2 Form of steel connections with rivets. [From *Architectural Construction*, 1926 (Ref. 19), reproduced with permission of the publishers, John Wiley & Sons.]

buildings were mostly braced by adding diagonal members to the vertical columns and horizontal beams.

Eventually, the ability to roll larger and longer steel shapes reduced the need for some trussing. Large girders and single column members were also produced by riveting together of steel plates and angles. This only meant that the use of trussing retreated somewhat to its enduring realms of practical application. Despite the many advances since the turn of the twentieth century, most of the forms of trussed structures developed in the middle and late nineteenth century still endure.

A somewhat similar situation occurred with the use of wood trussing. The use of trusses was partly limited by the available size of solid-sawn timber pieces, but more significantly by the means of achieving joints. Eventually, the use of steel tension members and steel connecting devices had more effect on the development of wood trusses than any technological advances relating to the wood materials. Use of glued-laminated members in the twentieth century has been the only really significant advance for wood. Even now, the use of combination trusses with wood chords and steel web members is the most frequent application for wood trusses.

3

EARLY TWENTIETH CENTURY CONSTRUCTION

Many of the classic forms of truss construction now widely used were developed by the early part of the twentieth century. This chapter presents a brief discussion of that development.

3.1 GENERIC TRUSS CONSTRUCTION

As a basic class, trussing is most naturally developed to perform two special forms of structure. First is the fundamental use of triangulation as a stabilizing geometric form. This results in the simple use of diagonals, X-braces, knee-braces, and so forth for the lateral bracing of otherwise rectangular frameworks. This was discovered by wood frame builders thousands of years ago, and the basic concept continues in many of the latest steel frames for large towers and high-rise buildings.

The other peculiar aspect of trussing is its capability for highly efficient production of a structural form with minimal use of materials; defining a lot of space for spanning or enclosure with relatively slender, widely spaced elements. This aspect has been exploited most dramatically with the development of steel and its highly sophisticated jointing methods.

Enduring practical forms of truss structures, such as the simple, single triangle, gable truss, the knee-braced frame, and the x-braced tower, are so classic in form that they will undoubtedly proliferate without a break well beyond our time. The basic parallel-chorded flat-spanning truss, in its classic form as a steel open web joist or a combination wood and steel element, is also most likely destined for some kind of immortality—like the paper clip, safety pin, and zipper.

Still, the marvel of trussing is its ubiquitous, all-performing, virtually endless applicability and variety. Just when we assume that all basic uses have been defined, some tinkerer produces a new form or application. Trussing is indeed a marvel; both in its historic continuity and its open-ended potential for exploration.

Who would ever have thought at the end of the nineteenth century that there would be concrete trusses, plastic trusses, space frames, delta trusses, trussed mullions, or interstitial trusses in multistory buildings?

3.2 CONSTRUCTION WITH SPANNING TRUSSES

The efficiency aspect of the truss (a small amount of material concentrated for maximum space definition) leads to the possibility for development of long-spanning roof structures of various forms. This was first exploited in trussed arch structures for the great bridges and then the exhibition halls, train sheds, and other large steel structures in the late nineteenth century. Exposed trusses of increasing span were also used for various forms of roofs for industrial buildings and the sky-lighted spaces of building atriums and covered spaces in urban areas.

Some large trusses were used for buildings, but the most dramatic structures were those for bridges, with spans achieved that still stand as records in some cases. The need for a reasonably horizontal roadway or train track was different than the usual requirement for building roofs, where a significant slope represented a well-drained surface. Thus arched and domed forms were the principal means for achieving long-span trussed roofs.

While long-spanning structures are often dramatic and attract attention, the greatest use of trusses by far was—and is—in the medium-span range, say, from 50 ft to something less than 200 ft. At this range it is mostly beyond a solid beam structure in feasibility for a roof, but not yet suffering from the competition of even more inherently efficient long-span systems, such as steel cable, pneumatic, or arch. Today, the greatest use of trusses for building roofs is in the form of manufactured products, which dominate the market for the medium span range just defined (see Figure 3.1).

However, as the long-held position of solid-sawn wood slowly yields for the short-span range (say 15 ft to 30 ft), and the use of various fabricated elements of wood (fiber, laminated, trussed) proliferates, the lower end of the feasibility range for trusses for building roofs and floors steadily stretches down. For floors, an advantage of significance is the open form of the truss, often useful for installation of wiring and piping.

Not a small attraction for the truss, however, has always been its light-appearing nature and the actual open space it represents for encompassing of various elements often required in overhead structures (see Figure 3.2). Elements for wiring, lighting, roof drains, sprinkler piping, signs, access catwalks, and whatever can thread through and between a trussed structure with ease and often a reasonable lack of intrusive visibility. Not so if they must all be suspended beneath a fabric roof or a concrete dome shell.

3.3 TRUSSED BRACING FOR HIGH-RISE STRUCTURES

For buildings, the natural form of a framed structure is most often one that produces rectangular arrangements of the frame members. Vertical members for natural orientation of supports in response to gravity loads, and horizontal members to

FIGURE 3.1 Combination trusses with wood chords and steel web members, used here for a commercial building to provide a structurally open, interior space that permits a lot of random, easily rearranged, interior space division. Although it appears to support the trusses in the photo, the steel column and beam structure in the foreground are for a separate construction beneath the trusses. The lightness and low cost of these trusses permits them to be used at spacings close enough to permit use of the chords for direct support of roof decking and panel ceiling materials, while still achieving some notable spans.

FIGURE 3.2 Use of a two-way spanning, modular, space grid system for a roof results in the definition of an interstitial space between the planes of the horizontal top and bottom chords. As in other structures, this space can be used to incorporate many service elements. However, the space is more fully useable with a truss system. It is also open to view if no ceiling surfacing is used, and the real ceiling is thus perceived as the underside of the roof, considerably extending the visual spatial experience beneath the structure.

form floors. Horizontal members for roofs are also natural, if roof surface drainage is not a critical factor.

The rectangular pattern with individual frame members is not naturally stable, and will barely stand even under the effects of gravity alone. Add some real lateral loads, and the lack of stability is critical. Adding more individual frame members in a short time leads to the discovery of the effectiveness of having some diagonal members. Every culture that developed frameworks with tree parts soon discovered this simple engineering principle.

As very tall frame structures came into use in the late nineteenth century, trussing became a natural means for achieving lateral stability. Into the early twentieth century, the tall buildings in Chicago, New York City, and other places mostly used this basic bracing system. The classic forms of truss bracing used today were all developed in the very early high-rise structures.

For pure truss action that produces only direct forces of tension or compression in the members, the truss frame members must all meet at shared joints. Thus the single diagonal or the criss-crossed (X-bracing) diagonals were the basic forms for pure truss bracing. Placed in vertical planes with building columns, however, these often intrude on the construction or on the use of the building surface or internal spaces. A less intrusive means of trussing that works in some cases is the knee brace, which only claims a corner of a rectangular space in a frame. On the building interior, this brace may be kept above the heads of the occupants, so that not really useful space is lost in the rectangle of the frame. On the building exterior, the placing of large windows may be possible with the knee brace, whereas more restriction occurs with full diagonals or X-bracing.

While the knee brace may work effectively for a limited degree of lateral bracing, it results in the development of bending and shear in the frame members to which it is attached. Thus these members are no longer *pure* truss members, but are now actually constituted as a modified rigid frame. The knee brace is actually a form of moment-resistive connection. If the connected frame members are not capable of rigid frame action, neither the knee-brace nor any other form of moment-resistive connection will make the lateral bracing system work.

In the lower columns of a high-rise structure (say 20 stories plus) the accumulation of gravity loads will produce quite large column members. Adding a modest amount of shear and bending to these columns may be relatively insignificant. Thus the knee brace at these locations can be used to develop what is really a combined system that has some aspects of a trussed frame and some of a rigid frame working simultaneously. The form of trussing that develops this action is referred as *eccentric bracing*, which points to the fact that the ends of some diagonals do not connect to general framing joints. (See Figure 3.3 and discussion in Section 8.7.)

We are talking for the moment about steel frames, since wood is not likely to be used for high-rise construction (for reasons of fire safety), and concrete frames are not much truss braced. The selected forms of bracing for a building will therefore derive from the practical considerations for steel construction as well as other concerns. Primary considerations in this regard are the form of the frame members (W, tee, angle, channel, pipe, tube, or built up) and the means of fastening (welds or high-strength bolts).

22 EARLY TWENTIETH CENTURY CONSTRUCTION

FIGURE 3.3 Structure with steel columns and beams developed basically for gravity loads, and added diagonal members to produce resistance to horizontal effects. Diagonals may produce various structural actions, from simple trussing to various combinations of truss and rigid frame behavior (see discussion of *eccentric bracing* in Section 8.7).

A limitation in the early high-rise structures was the limited size of individual steel rolled shapes. This often resulted in using multiple pieces to produce a built-up column in the lower stories of frames of more than a few stories. If horizontal spans were significant, the main floor beams that joined the columns to produce the basic frame skeleton might also be built up (plate girders mostly). Adding the bracing to such a framework became simply an extension of an already developed complex assemblage of many steel pieces.

Fastening of large steel frames, up to about the middle of the twentieth century, was mostly achieved with hot-driven rivets. The Eiffel Tower, the Empire State Building, the tower supports for the Golden Gate Bridge, and other notable monumental structures were achieved in this manner. Obviously the choice of members with flat parts that could be mated and lapped for riveting, as well as other considerations for use of this basic system of fastening influenced the design of the frameworks. Accommodation for large numbers of rivets (which were limited to quite small individual size) and use of gusset plates were two major considerations.

While the high-rise buildings from that era remain as impressive constructions, their frames were visible only to the people at the time of their construction. Somewhat more impressive as demonstrations of the builders achievements are the great towers, which still display the open-trussed frameworks and the exposed riveted connections.

3.4 TRUSSES AS INDUSTRIAL PRODUCTS

Trusses are not always the most feasible means for achieving structural tasks. They may be the only good choice when a particular span, building form, or special bracing task must be achieved. However, in ordinary situations a frequent disadvantage for trusses may be the amount of jointing that must be achieved. And, indeed, a considerable amount of designer time is required for the investigation, design development, construction detailing, and specifying of materials.

Turning an entire individual truss into a single manufactured product is one means for reducing fabrication costs. As with doors, windows, and other basic building components, this process is now fully evolved, and its roots go back to the last century.

An obvious advantage to the building designer is the ability to use units of the construction that come fully designed, detailed, specified, and ready to plop into place. A lot of design effort is saved and some level of responsibility is passed to the manufacturer or supplier. This is frankly the practical way to go for trusses, if an appropriate one can be found in somebody's catalog.

Fabricated steel joists for modest spans were produced as soon as reliable welding was developed (see Figure 3.4). In a short time, a major portion of the total truss production became automated, including eventually even the welding of joints. The earliest forms of these products were widely used in the early twentieth century. The basic systems are still extensively used and little changed in form from the early examples.

Wood construction yields in general to relatively simple tasks for handling of the basic materials, modifying of common forms of elements, and effective fastening systems. Thus the putting together of a small, simple wood truss is easily a backyard operation or a business for a small lumberyard. However, the truss with nailable top and bottom chords is an advantage for incorporation into a general wood construction, and thus offers some possibilities of use where steel may be less feasible. The fabricated trusses first developed in steel were more or less imitated in wood in later times; eventually resulting in wide use of the combination truss with wood chords and steel web members. These trusses are now completely produced by factory processes.

Large trusses of steel or wood may also be basically factory produced from predesigned patterns and details. These may even be customized to some degree, within the feasible adjustability of the basic truss type and form. Large companies involved in structural fabricated products can become the suppliers of a wide range of structural elements, possibly including a relatively unique, customized system. A major advantage here is a continuity of experience and expertise that comes with a business that concentrates on a system. Individual designers seldom do this and may encounter a truss design only infrequently. The *design* is thus most likely to be very much a shared, cooperative experience for most truss construction.

Use of trusses as primary structural elements in a whole building structure often means some coordinated development of the whole construction. Suppliers of individual fabricated trusses often also produce other elements required for a full system development. Thus the trusses may stand alone as available products, but often also are only one component in a *system* that is designed to work as a whole.

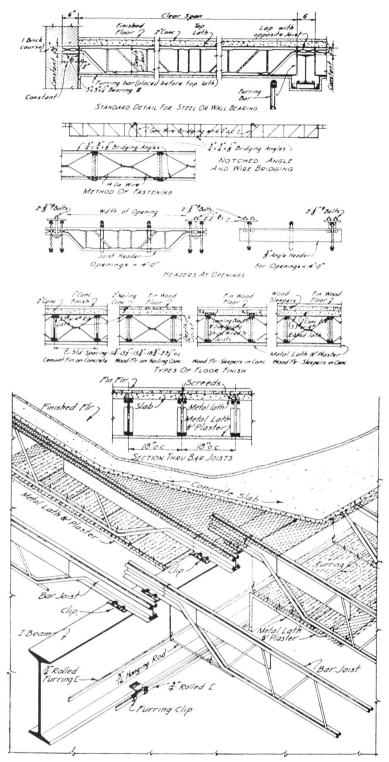

FIGURE 3.4 Details for use of so-called bar joists: light trusses made by welding of straight and bent steel rods. Originally developed for use primarily for scaffolding, their capabilities were eventually employed for actual spanning floor and roof structures, as shown here. [From *Architectural Construction*, 1926 (Ref. 19), reproduced with permission of the publishers, John Wiley & Sons.]

PART TWO

MODERN CONSTRUCTION WITH TRUSSES

Current use of trussing builds on a long history of developments in technology and on the accumulated experience with creating buildings for many purposes. Use of trussing today uses that experience, but must relate directly to current problems, demands for new types of buildings, and the context of competing forms of construction. This part briefly considers the context for use of trussing for buildings today.

4

NEW DESIGN CONCERNS

Trussed structures of many kinds remain in use, but the technology of building continues to advance. So do the desires, perceived needs, and eventually the demands for new forms of structures. These changes bring about the emergence of new materials, new types of construction, and many innovative developments of structural alternatives. This chapter presents some considerations for new design issues and approaches in applications of trussed structures for modern buildings.

4.1 COMPETING LONG-SPAN SYSTEMS

Classic forms of construction must withstand the constant onslaught of competing ideas for new structural systems. Largely because of the development of steel in the mid-nineteenth century, trussing held dominance as a means for achieving long-span roof structures through the late nineteenth century and the early twentieth century. Arches were used for some of the longest spans, but even then mostly in the form of trussed steel arches.

Actually the steel cable, draped in catenary form, set span records for spanning in that period and still holds the records; although the great spans have been achieved for bridges, not buildings. Nevertheless, suspended structures were used increasingly for building roofs and some very long spans have been obtained. In many cases, however, the steel cable systems have been combined with other systems for development of complete roof structures.

Use of air pressure to sustain the stability of a thin fabric or membrane surface has been used in a variety of ways to produce roofs. Some of the largest such roofs have used steel cables to control the roof form, generally developed by opposing the outward pressure of an air-supported membrane surface against a restraining set of cables.

As concrete was steadily improved as a basic material and new ways of using it were explored, various means emerged for producing longspan structures. Concrete arches, vaults, and domes became increasingly thinner and lighter and a whole array of new geometries have been developed in concrete shells and folded plates.

Development of glue lamination for large timber elements has lead to the use of some very large timber beams, well beyond the feasibility for solid-sawn lumber. Lamination with relatively thin lumber pieces also permits curved forms, and many arches and bent frames have been extended to considerable spans that are also beyond the range of other solid timber structures.

This has resulted in some adjustments of the feasibility of trusses for some uses, but hardly the complete obsolescence of the truss as a basic system.

4.2 USE OF INTERSTITIAL SPACE

A major development that has occurred fairly recently is the steady increase of the need for incorporation of building service elements into the construction. Almost every building has electric wiring, a lighting system, piping for water and waste, and some elements for handling ventilation and heating. Increasingly common also are elements for air conditioning (cooling), communication, fire protection, automated control systems, security, and so on. All, or a major part, of this often must be stuffed somewhere into the construction (in the unoccupied space inside the construction, called the *interstitial space*) in order to protect it or to have uncluttered interior spaces.

An inherent characteristic of trussing is its ability to define large structured space without occupying much of the space. Trusses allow for a lot of sharing of the total space they define. Most significantly, they allow for wiring, piping, and ducts to wander through the space (see Figure 3.2). This is a major consideration for helping to keep the overhead clear under a trussed roof structure; lighting, fire sprinkler piping, signs, access catwalks, and so forth, can be taken up into the space defined by the top and bottom chords of the trusses.

Parallel-chorded trusses used for floor structures create the most usable interstitial space between the floor surface above and the ceiling surface below. A special application of this occurs with the use of a deliberately deep truss system for the creation of an interstitial space that permits access for maintenance and alteration on an ongoing basis (see Figure 4.1). Multistory buildings with alternating levels of occupied space and deep trussed interstitial spaces have been used for medical centers, laboratories, and other occupancies that have extensive services and need for continuous modification.

4.3 LONGER SPANS

It is always difficult to establish clearly whether technological development leads design use or follows design demand. An expressed desire for long-span structures surely pushes designers and inventors to produce more and better alternatives for achieving long spans. Then, once the capability is demonstrated (by dare-devil designers and builders), everybody wants one.

Whether basically as expressions of daring and monumental size, or as truly needed, increasing use is steadily made of long-spanning structures, for sports

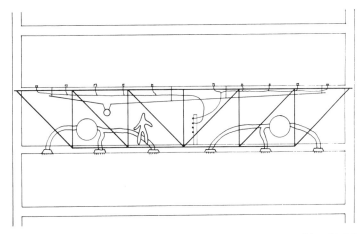

FIGURE 4.1 Use of a long-spanning truss of full story height to provide a highly accessible interstitial space for continual maintenance and modification of service systems that serve highly flexible, structure-free, interior spaces.

FIGURE 4.2 While the steel cable is an impressive structural element, having achieved the longest spans by far, the suspension bridge also requires some impressively sized tower structures. Early bridges, such as the Brooklyn Bridge, used massive masonry towers. Later bridges used steel towers, often of truss-braced form, as shown here for the Oakland Bay Bridge, connecting Oakland and San Francisco across the San Francisco Bay.

arenas, convention center halls, large atrium spaces, covered malls, and so on. This is a frontier of design innovation where many of the most creative designers expend their efforts and where somebody subsidizes a major amount of design time and effort.

As the span gets long (say over 100 ft), spanning efficiency becomes a major issue. While other systems (cable-suspended, pneumatic, etc.) have great potential range, various trussed systems remain in use for a considerable range of spans. Not a small consideration is the adaptability of trussing to many forms. Flat spans, sloped roof surfaces, arches, and domes can all be achieved with trussed systems.

Even when not used for the major spanning system, trussing may be used for its lightness to produce secondary elements of spanning systems. Large columns or towers for suspended structures are often trussed structures. Suspending cables may be used to support a trussed secondary structure; for example, the roadway deck structure of a suspension bridge, or the flat roof structure for a building (see Figure 4.2).

5

NEW TRUSS SYSTEMS AND DETAILS

The pace of technological development has continued strongly throughout the twentieth century. While some forms of construction have become obsolete, or at least much less competitive, new materials, construction methods, and design concepts have emerged and flourished. This chapter considers some of the more recent developments in the applications of trussing to building structures.

5.1 NEW FABRICATION METHODS

Both modest and monumental trussed structures of the nineteenth century and early twentieth century were completed with the available materials and processes of the times. This inventory of available technology was constantly expanded and refined as the basic steel, wood, concrete, and other major industries advanced.

Early steel trusses were assembled almost entirely by riveting, which consisted of placing a red-hot rivet with a head formed on one end like a pin, into matched holes in flat sets of steel members and beating the unformed end of the rivet until it formed a head against the opposite side. As the rivet cooled, it shrunk slightly, squeezing the parts tightly together. This was a process that was relatively easy to achieve in a factory or on the ground at a building site, but became a challenge to perform at great distances above the ground on an open framework. Beating the heads was also noisy, and the streets of cities rung constantly with the machine gun-like sounds of riveting hammers for many years.

Bolting had been around for a long time, but was not capable of producing the tightness of the joints. Holes had to be cut slightly larger than the bolts (then and now) for insertion of the bolts, and some slippage in the joints during loading of the structure was inevitable. Worse yet, under load reversals, the joints could slip back and forth, producing a general looseness of the assembled structure. For the truss, with its many joints, this was highly unacceptable. Bolting was therefore limited to use for temporary connections during erection or for only a few field connections to simplify erection.

The development of high-strength bolts changed all of this. These were bolts designed to be highly tightened with a torque wrench, to a degree that developed yield stress levels of tension in the bolt shaft. The result was a joint so tightly squeezed that slippage would occur only at load levels well above the service load conditions. In a short time, this process entirely replaced the use of rivets for connections made at the building site. The principal advantage was in the ability to achieve the joints with only two men, a torque wrench, and a pocket full of bolts; a lot easier several hundred feet in the air on an open steel framework.

During this same period, fusion welding was also refined and accepted for structural applications. As bolting was replacing riveting for field work, welding replaced the riveting performed in fabrication shops. This now stands as the usual procedure for steel frame structures: shop welding and field bolting to achieve connections. Field welding is also performed, but only when absolutely required and usually at great cost.

For small trusses, almost entirely assembled in the shop, this situation favors the use of members, truss forms, and joint shapes that facilitate welding. It also means the feasible use of steel elements not easily assembled with rivets, such as round rods, pipe, and tubular shapes (see Figure 5.1).

Another factor that has strongly affected truss design in recent times is the development of automated processes that make complex cutting of the ends of members feasible. It is thus possible to bring round steel pipes of several different diameters into a meeting at a joint with many members (such as that in a space frame—possibly eight members or more at a single joint), custom cut the member ends to fit, and weld them all together, all in an automated process. For the several hundred, or even thousand, joints in a large space frame, this is a major consideration.

For wood trusses of various sizes, jointing methods have also been refined. Simple nailing, bolting, or fitting of cut pieces together is no longer necessary. A wide array of hardware items are available for achieving stronger, tighter, less labor-intensive jointing of wood trusses. For manufactured trusses, very sophisticated jointing may be used with the mass production economies and quality controlled conditions possible in a factory situation.

5.2 NEW FORMS AND MATERIALS

As always, technology and architectural design have continued their dance; one leading the other and vice versa. Development of new materials, fabrication methods, and erection processes, combined with new ideas for structural forms and systems, presents opportunities for usage in buildings which invites architects to experiment with their applications to buildings. And, architects spin their dreams as usual, inviting engineers to try to build them.

As an exposed structure, the truss invites sculptural play with its rhythms of repeating modules and intricate geometries (see Figure 5.2). It also can have a visual lightness matching its structural lightness, permitting it to be seen through. When exposed on building interiors, trusses do not intimidate with their massiveness. When exposed on the exterior, they appear more as exterior sculpture than as massive supports. These potentialities have been explored considerably by architects for buildings, both large and small.

FIGURE 5.1 A high-tech interior with an exposed steel structure using welded frames and large trusses of welded steel pipes. United Airlines terminal, O'hare International Airport, Chicago; Helmut Jahn, architect.

FIGURE 5.2 Use of a highly visible structure with modular units invites play with the potentiality of its modifications for geometric variation. These may be derived with some structural logic, as in the case of the pyramidal supports, or for purely architectural purposes, such as opening up the space, creating raised roof areas, or incorporating various elements.

While the very complex bolted joint, with gusset plates and numerous bolts, has some high-tech charm, it usually accompanies a relatively massive truss in general. When truss members are capable of relatively slender form, the riveted or bolted joints may become somewhat cumbersome in appearance. However, direct welding—especially of round pipe or tubes—can produce a very clean, smooth appearance, even when the members are relatively stout.

For wood trusses, in addition to new jointing methods, a major development has been in the use of other than solid-sawn lumber for the truss members. For very large timber trusses it is possible to use glued-laminated timber members; permitting also the possibilities for considerable individual member length or curved forms. A new use for the manufactured truss is a chord member of laminated thin wood pieces, permitting a single chord piece of considerable length. In development are increasingly sophisticated applications of wood fiber products, which will surely eventually reach the quality level necessary for truss members.

A novelty of sorts, but with real applications, is the concrete truss. Small individual trusses may be singly cast in one piece, or components may be cast and assembled with various methods used for precast concrete structures. The development of increasingly high quality of precast concrete and use of prestressing has extended the possibilities for concrete trusses.

New truss forms have also been developed; a major one being the two-way spanning systems achieved with three-dimensional, rather than simple planar, trusses. These are discussed briefly in the next section and in Chapter 12. Other applications of special truss forms and structures are also discussed in Chapter 12.

5.3 SPACE FRAMES

The term *space frame* is badly applied, but has been commonly used to describe a truss system that works other than in simple planar action. The first extension of this is in achieving a two-way spanning structure, versus a one-way spanning action as occurs with individual beams or slabs or simple, planar trusses (see Figure 5.3).

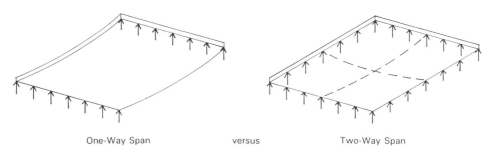

FIGURE 5.3 Depending on its supports, a flat spanning structure may develop either one-way bending (single curvature) or two-way bending (double curvature, domed, dished, etc.). Systems using linear elements, such as wood beams or corrugated steel decks, are essentially limited to one-way actions. Concrete slabs or waffle joist systems have a potentiality for two-way action—if supports truly develop the actions. An intersecting truss system can also achieve a two-way span.

Various geometries for arrangement of the members and joints have been used for the two-way truss, but two primary ones are those using triangulation in general and those using forms with top and bottom chords in rectangular patterns (see Figure 5.4). The latter tend to mesh with architectural planning in a rectangular grid more simply, so they have seen somewhat greater application for buildings.

Some economy accrues in the form of the natural structural efficiency of two-way versus one-way spanning. However, the multiplicity of joints, much greater complexity of structural investigation and design, and sometimes challenging assemblage and erection problems make this an area of design requiring considerable expertise. Very large, expensive buildings may afford this cost. For modest projects, a manufactured system or a generous sponsor is usually required.

The fullest use of the space frame is in the fully three dimensional structure, not merely one with two-way spanning. Here the geometry of the structural system becomes even more dominating for architectural planning, but the trussing may be used in a much more sculptural or playful way. Most trussed towers are three-dimensional space frames, although they may be conceived, arranged, and designed as a set of planar truss components (see Figure 5.5).

It should be noted that the building codes reserve the term *space frame* for the combined, three-dimensional, vertical column and horizontal beam structure of a multistory building with moment-resistive joints in the frame; not a truss at all.

5.4 BRACED FRAMES FOR LATERAL STABILITY

The development of lateral bracing of buildings in general has seen considerable development in recent times. This is a response somewhat to major windstorms and earthquakes, but is also an increase in wider concerns for reliable design standards and control of construction quality.

FIGURE 5.4 Light steel truss system using a pattern called an offset grid. Square form arrangements of top and bottom chords have their intersection points corresponding to the center of the space in the opposite grid. See discussion in Section 12.1.

36 NEW TRUSS SYSTEMS AND DETAILS

FIGURE 5.5 Tower structure produced with multiple faces of planar trusses, constituting a three-dimensional trussed structure. The individual trusses do not imply a spatial system, but their connected assembly defines one.

Use of trussed bracing, both in historic and new forms is a significant option in many situations. A particular advantage of trussing that now is receiving more appreciation is its relative stiffness. A truss is much closer to a shear wall than a rigid frame in stiffness, which is not necessarily so significant to structural safety, but has major implications for nonstructural damage. Translation: Maybe the window glazing, plastered ceilings, piping, and other elements will survive, not just the structure.

Trusses may intrude with their diagonals, but they do have a visual openness not offered by a solid shear wall. Stiffness is achieved without massiveness or total visual blocking.

The general use of trussed bracing for buildings is presented in Section 8.7. This is probably the oldest application of trussing, but it has major current significance and applications.

PART THREE

INVESTIGATION AND DESIGN OF PLANAR TRUSSES

By far, the major use for trusses—both historically and currently—involves a simple spanning task that can be resolved with a single planar truss. In any event, most of what needs to be learned about the design of trusses as structures can be developed for this simple case. This part presents the complete background for the investigation and design of such trusses, including basic information for the two materials used most often for truss construction: *wood* and *steel*. Other succeeding parts of this book present special topics relating to usage and design of trusses, but this part is the takeoff point for those who desire to learn what trusses do in a fundamental way.

6

ANALYSIS FOR STATIC FORCES

6.1 PROPERTIES OF FORCES

Static forces are those that can be dealt with adequately without considering the time-dependent aspects of their action. This limits considerations to those dealing with the following properties:

Magnitude, or the amount, of the force, which is measured in weight units such as pounds or tons.

Direction of the force, which refers to the orientation of its path or line of action. Direction is usually described by the angle that the line of action makes with some reference, such as the horizontal.

Sense of the force, which refers to the manner in which it acts along its line of action (e.g., up or down). Sense is usually expressed algebraically in terms of the sign of the force, either plus or minus.

Forces can be represented graphically in terms of these three properties by the use of an arrow, as shown in Figure 6.1. Drawn to some scale, the length of the arrow represents the magnitude of the force. The angle of inclination of the arrow represents the direction of the force. The location of the arrowhead determines the sense of the force. This form of representation can be more than merely symbolic, since actual mathematical manipulations may be performed using the vector representation that the force arrows constitute. In the work in this book arrows are used in a symbolic way for visual reference when performing algebraic computations, and in a truly representative way when performing graphical analyses.

In addition to the basic properties of magnitude, direction, and sense, some other concerns that may be significant for certain investigations are

The *position of the line of action* of the force with respect to the lines of action of other forces or to some object on which the force operates, as shown in Figure 6.2.

40 ANALYSIS FOR STATIC FORCES

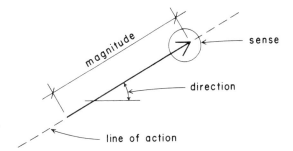

FIGURE 6.1 Basic properties of a single force.

The *point of application* of the force along its line of action may be of concern in analyzing for the specific effect of the force on an object, as shown in Figure 6.3.

When forces are not resisted, they tend to produce motion. An inherent aspect of static forces is that they exist in a state of *static equilibrium*, that is, with no motion occurring. In order for static equilibrium to exist, it is necessary to have a balanced system of forces. An important consideration in the analysis of static forces is the nature of the geometric arrangement of the forces in a given set of forces that constitute a single system. The usual technique for classifying force systems involves consideration of whether the forces in the system are

Coplanar. All acting in a single plane, such as the plane of a vertical wall.
Parallel. All having the same direction.
Concurrent. All having their lines of action intersect at a common point.

Using these three considerations, the possible variations are given in Table 6.1 and illustrated in Figure 6.4. Note that variation 5 in the table is really not possible, since a set of coacting forces that is parallel and concurrent cannot be noncoplanar; in fact, they all fall on a single line of action and are called collinear.

It is necessary to qualify a set of forces in the manner just illustrated before proceeding with any analysis, whether it is to be performed algebraically or graphically.

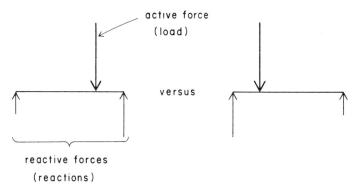

FIGURE 6.2 Effect of the location of the line of action of a force.

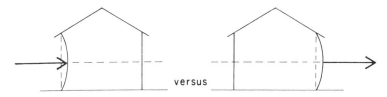

FIGURE 6.3 Effect of the position of a force along its line of action.

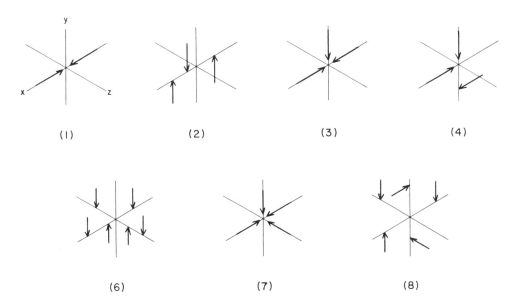

FIGURE 6.4 Classification of force systems—orthogonal reference axes.

TABLE 6.1. Classification of Force Systems

System Variation	Qualifications		
	Coplanar	Parallel	Concurrent
1	yes	yes	yes
2	yes	yes	no
3	yes	no	yes
4	yes	no	no
5	no[a]	yes	yes
6	no	yes	no
7	no	no	yes
8	no	no	no

[a] Not possible if forces are parallel and concurrent.

42 ANALYSIS FOR STATIC FORCES

6.2 COMPONENTS OF FORCES

In structural analysis it is sometimes necessary to perform either addition or subtraction of force vectors. The process of addition is called *composition*, or combining of forces. The process of subtraction is called *resolution*, or the resolving of forces into *components*. A component is any force that represents part, but not all, of the effect of the original force.

In Figure 6.5 a single force is shown, acting upward toward the right. One type of component of such a force is the net horizontal effect, which is shown as F_h at (a) in the figure. The vector for this force can be found by determining the side of the right-angled triangle, as shown in the illustration. The magnitude of this vector may be calculated as $F(\cos \theta)$ or may be measured directly from the graphic construction, if the vector for F is placed at the proper angle and has a length proportionate to its actual magnitude.

If a force is resolved into two or more components, the set of components may be used to replace the original force. A single force may be resolved completely into its horizontal and vertical components, as shown at (b) in Figure 6.5. This is a useful type of resolution for algebraic analysis, as will be demonstrated in the examples that follow. However, for any force, there are an infinite number of potential components into which it can be resolved.

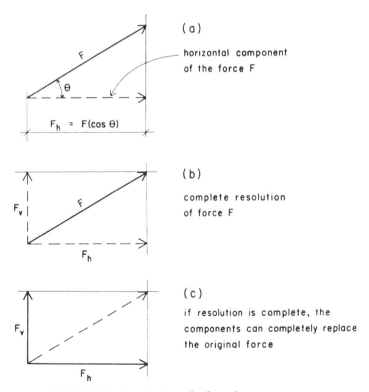

FIGURE 6.5 Resolution of a force into components.

6.3 COMPOSITION AND RESOLUTION OF FORCES

Whether performed algebraically or graphically, the combining of forces is essentially the reverse of the process just shown for resolution of a single force into components. Consider the two forces shown at (*a*) in Figure 6.6. The combined effect of these two forces may be determined graphically by use of the parallelogram shown at (*b*) or the force triangle shown at (*c*). The product of the addition of force vectors is called the *resultant* of the forces. Note that we could have used either of the two triangles formed at (b) to get the resultant. This is simply a matter of the sequence of addition, which does not affect the answer; thus if $A + B = C$, then $B + A = C$ also.

When the addition of force vectors is performed algebraically, the usual procedure is first to resolve the forces into their horizontal and vertical components (or into any mutually perpendicular set of components). The resultant is then expressed as the sum of the two sets of components. If the actual magnitude and direction of the resultant are required, they may be determined as follows. [See Figure 6.6(*d*).]

$$R = \sqrt{(\sum F_v)^2 + (\sum F_h)^2}$$

$$\tan \theta = \frac{\sum F_v}{\sum F_h}$$

When performing algebraic summations, it is necessary to use the sign (or sense) of the forces. Note in Figure 6.6 that the horizontal components of the two forces are both in the same direction and thus have the same sense (or algebraic sign) and that the summation is one of addition of the two components. However, the vertical components of the two forces are opposite in sense and the summation consists of finding the difference between the magnitudes of the two components. Keeping track of the algebraic signs is a major concern in algebraic analyses of forces. It is necessary to establish a sign convention and to use it carefully and consistently throughout the calculations. Whenever possible, the graphic manifestation of the force analysis should be sketched and used for a reference while performing algebraic analyses, since this will usually help in keeping track of the proper sense of the forces.

When more than two forces must be added, the graphic process consists of the construction of a force polygon. This process may be visualized as the successive addition of pairs of forces in a series of force triangles, as shown in Figure 6.7. The first pair of forces is added to produce their resultant; this resultant is then added to the third force; and so on. The process continues until the last of the forces is added

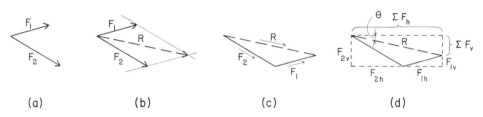

FIGURE 6.6 Composition of forces by vector addition.

44 ANALYSIS FOR STATIC FORCES

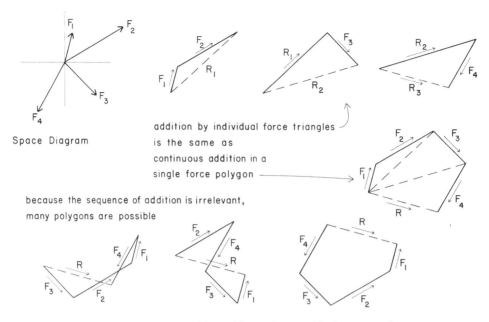

FIGURE 6.7 Composition of forces by graphical construction.

to the last of the intermediate resultants to produce the final resultant for the system. This process is shown in Figure 6.7 merely for illustration, since we do not actually need to find the intermediate resultants, but may simply add the forces in a continuous sequence, producing the single force polygon. It may be observed, however, that this is actually a composite of the individual force triangles.

For an algebraic analysis, we simply add up the two sets of components, as previously demonstrated, regardless of their number. The true magnitude and direction of the resultant can then be found using the equations previously given for R and $\tan \theta$. The equivalent process in the graphic analysis is the closing of the force polygon, the resultant being the vector represented by the line that completes the figure, extending from the tail of the first force to the head of the last force.

As mentioned previously. the sequence of the addition of the forces is actually arbitrary. Thus there is not a single polygon that may be constructed, but rather a whole series of polygons, all producing the same resultant.

6.4 EQUILIBRIUM

The natural state of a static force system is one of equilibrium. This means that the resultant of any complete interactive set of static forces must be zero. For various purposes it is sometimes desirable to find the resultant combined effect of a limited number of forces, which may indeed be a net force. If a condition of equilibrium is then desired, it may be visualized in terms of producing the *equilibrant*, which is the force that will totally cancel the resultant. The equilibrant, therefore, is the force that is equal in magnitude and direction, but opposite in sense, to the resultant.

In structural design the typical force analysis problem begins with the assumption that the net effect is one of static equilibrium. Therefore, if some forces in a system are known (e.g., the loads on a structure), and some are unknown (e.g., the forces generated in members of the structure in resisting the loads), the determination of the unknown forces consists of finding out what is required to keep the whole system in equilibrium. The relationships and procedures that can be utilized for such analyses depend on the geometry or arrangement of the forces, as discussed in Section 6.1.

For a simple concentric, coplanar force system, the conditions necessary for static equilibrium can be stated as follows:

$\sum F_v = 0$ (sum of the vertical components equals zero)

$\sum F_h = 0$ (sum of the horizontal components equals zero)

In other words, if both components of the resultant are zero, the resultant is zero and the system is in equilibrium.

In a graphic solution for the concurrent, coplanar system, the resultant will be zero if the force polygon closes on itself, that is, if the head of the last force vector coincides with the tail of the first force vector.

6.5 ANALYSIS OF CONCURRENT, COPLANAR FORCES

The forces that operate on individual joints in planar trusses constitute sets of concurrent, coplanar forces. The following discussion deals with the analysis of such systems, both algebraically and graphically, and introduces some of the procedures that will be used in the examples of truss analysis in this book.

In the preceding examples, forces have been identified as F_1, F_2, F_3, and so on. However, a different system of notation will be used in the work that follows. This method consists of placing a letter in each space that occurs between the forces or their lines of action, each force then being identified by the two letters that appear in the adjacent spaces. A set of five forces is shown at (*a*) in Figure 6.8. The common intersection point is identified as *BCGFE* and the forces are *BC, CG, GF, FE,* and *EB*. Note particularly that the forces have been identified by reading around the joint in a continuous clockwise manner. This is a convention that will be used throughout

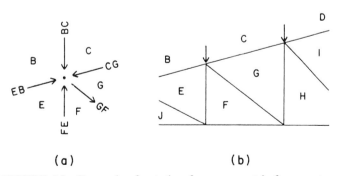

FIGURE 6.8 Example of notation for a concentric force system.

46 ANALYSIS FOR STATIC FORCES

this book, since it has some relevance to the methods of graphic analysis that will be explained later.

At (b) in Figure 6.8, a portion of a truss is shown. Reading around the joint *BCGFE* in a clockwise manner, the upper chord member between the two top joints is read as member *CG*. Reading around the joint *CDIHG*, the same member is read as *GC*. Either designation may be used when referring to the member itself. However, if the effect of the force in the member on a joint is being identified, it is important to use the proper sequence for the two-letter designation.

In Figure 6.9 a weight is shown hanging from two wires that are attached at separate points to the ceiling. The two sloping wires and the vertical wire that supports the weight directly meet at joint *CAB*. The "problem" in this case is to find the tension forces in the three wires. We refer to these forces that exist within the members of a structure as *internal forces*. In this example it is obvious that the force in the vertical wire will be the same as the magnitude of the weight: 50 lb. Thus the solution is reduced to the determination of the tension forces in the two sloping wires. This problem is presented as (b) in the figure, where the force in the vertical wire is iden-

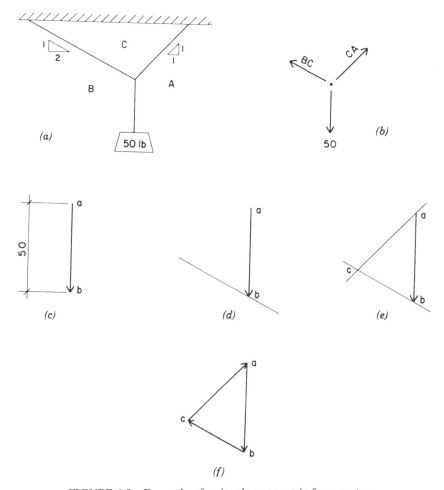

FIGURE 6.9 Example of a simple concentric force system.

tified in terms of both direction and magnitude, while the other two forces are identified only in terms of their directions, which must be parallel to the wires. The senses of the forces in this example are obvious, although this will not always be true in such problems.

A graphic solution of this problem can be performed by using the available information to construct a force polygon consisting of the vectors for the three forces: *BC*, *CA*, and *AB*. The process for this construction is as follows:

1. The vector for *AB* is totally known and can be represented as shown by the vertical arrow with its head down and its length measured in some scale to be 50.
2. The vector for force *BC* is known as to direction and must pass through the point *b* on the force polygon, as shown at (*d*) in the figure.
3. Similarly, we may establish that the vector for force *CA* will lie on the line shown at (*e*), passing through the point *a* on the polygon.
4. Since these are the only vectors in the polygon, the point *c* is located at the intersection of these two lines, and the completed polygon is as shown at (*f*), with the sense established by the continuous flow of the arrows. This "flow" is determined by reading the vectors in continuous clockwise sequence on the space diagram, starting with the vector of known sense. We thus read the direction of the arrows as flowing from *a* to *b* to *c* to *a*.

With the force polygon completed, we can determine the magnitudes for forces *BC* and *CA* by measuring their lengths on the polygon, using the same scale that was used to lay out force *AB*.

For an algebraic solution of the problem illustrated in Figure 6.9, we first resolve the forces into their horizontal and vertical components, as shown in Figure 6.10. This increases the number of unknowns from two to four. However, we have two extra relationships that may be used in addition to the conditions for equilibrium, because the directions of forces *BC* and *CA* are known. As shown in Figure 6.9, force *BC* is at an angle with a slope of 1 vertical to 2 horizontal. Using the rule that the hypotenuse of a right triangle is related to the sides such that the square of the hypotenuse is equal to the sum of the squares of the sides, we can determine that the length of the hypotenuse of the slope triangle is

$$l = \sqrt{(1)^2 + (2)^2} = \sqrt{5} = 2.236$$

We can now use the relationships of this triangle to express the relationships of the force *BC* to its components. Thus, referring to Figure 6.11,

$$\frac{BC_v}{BC} = \frac{1}{2.236}, \quad BC_v = \frac{1}{2.236}(BC) = 0.447\,(BC)$$

$$\frac{BC_h}{BC} = \frac{2}{2.236}, \quad BC_h = \frac{2}{2.236}(BC) = 0.894\,(BC)$$

These relationships are shown in Figure 6.10 by indicating the dimensions of the slope triangle with the hypotenuse having a value of 1. Similar calculations will pro-

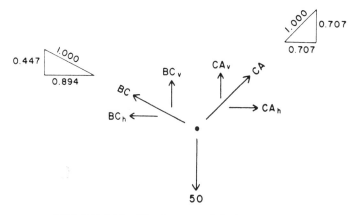

FIGURE 6.10 The forces and their components.

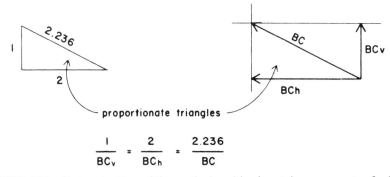

FIGURE 6.11 Determination of the vertical and horizontal components of a force.

duce the values shown for the force *CA*. We can now express the conditions required for equilibrium. (Sense up and right is considered positive.)

$$\sum F_v = 0 = -50 + BC_v + CA_v$$
$$0 = -50 + 0.447(BC) + 0.707(CA) \quad (1)$$

and,

$$\sum F_h = 0 = -BC_h + CA_h$$
$$0 = -0.894(BC) + 0.707(CA) \quad (2)$$

We can eliminate *CA* from these two equations by subtracting equation (2) from equation (1) as follows:

equation (1): $0 = -50 + 0.447(BC) + 0.707(CA)$
equation (2): $0 = + 0.894(BC) + 0.707(CA)$
Combining: $0 = -50 + 1.341(BC)$

ANALYSIS OF CONCURRENT, COPLANAR FORCES

Then: $BC = \dfrac{50}{1.341} = 37.29$ lb

Using equation (2): $0 = -0.894(37.29) + 0.707(CA)$

$$CA = \dfrac{0.894}{0.707}(37.29) = 47.15 \text{ lb}$$

The degree of accuracy of the answer obtained in an algebraic solution depends on the number of digits that are carried throughout the calculation. In this work we will usually round off numerical values to a three- or four-digit number, which is traditionally a level of accuracy sufficient for structural design calculations. Had we carried the numerical values in the preceding calculations to the level of accuracy established by the limits of an eight-digit pocket calculator, we would have obtained a value for the force in member BC of 37.2678 lb. Although this indicates that the fourth digit in our previous answer is slightly off, both answers will round to a value of 37.3, which is sufficient for our purposes. When answers obtained from algebraic solutions are compared to those obtained from graphic solutions, the level of correlation may be even less, unless great care is exercised in the graphic work and a very large scale is used for the constructions. If the scale used for the graphic solution in this example is actually as small as that shown on the printed page in Figure 6.9, it is unreasonable to expect accuracy beyond the second digit.

When the so-called method of joints is used, finding the internal forces in the members of a planar truss consists of solving a series of concurrent force systems.

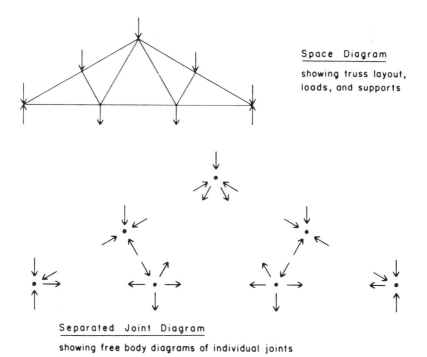

FIGURE 6.12 Examples of diagrams used to represent trusses and their actions.

50 ANALYSIS FOR STATIC FORCES

Figure 6.12 shows a truss with the truss form, the loads, and the reactions displayed in a *space diagram*. Below the space diagram is a figure consisting of the free body diagrams of the individual joints of the truss. These are arranged in the same manner as they are in the truss in order to show their interrelationships. However, each joint constitutes a complete concurrent planar force system that must have its independent equilibrium. "Solving" the problem consists of determining the equilibrium conditions for all of the joints.

6.6 GRAPHICAL ANALYSIS FOR INTERNAL FORCES IN PLANAR TRUSSES

Figure 6.13 shows a single span, planar truss that is subjected to vertical gravity loads. We will use this example to illustrate the procedures for determining the internal forces in the truss, that is, the tension and compression forces in the individual

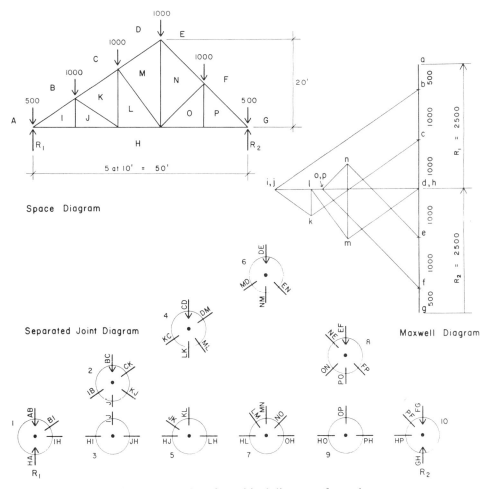

FIGURE 6.13 Examples of graphical diagrams for a planar truss.

members of the truss. The space diagram in the figure shows the truss form, the support conditions, and the loads. The letters on the space diagram identify individual forces at the truss joints, as discussed in Section 6.5. The sequence of placement of the letters is arbitrary, the only necessary consideration being to place a letter in each space between the loads and the individual truss members so that each force at a joint can be identified by a two-letter symbol.

The separated joint diagram in the figure provides a useful means for visualization of the complete force system at each joint as well as the interrelation of the joints through the truss members. The individual forces at each joint are designated by two-letter symbols that are obtained by simply reading around the joint in the space diagram in a clockwise direction. Note that the two-letter symbols are reversed at the opposite ends of each of the truss members. Thus the top chord member at the left end of the truss is designated as *BI* when shown in the joint at the left support (joint 1) and is designated as *IB* when shown in the first interior upper chord joint (joint 2). The purpose of this procedure will be demonstrated in the following explanation of the graphical analysis.

The third diagram in Figure 6.13 is a composite force polygon for the external and internal forces in the truss. It is called a Maxwell diagram after its originator, James Clerk Maxwell, an English engineer. The construction of this diagram constitutes a complete solution for the magnitudes and senses of the internal forces in the truss. The procedure for this construction is as follows.

1. *Construct the force polygon for the external forces.* Before this can be done, the values for the reactions must be found. There are graphic techniques for finding the reactions, but it is usually much simpler and faster to find them with an algebraic solution. In this example, although the truss is not symmetrical, the loading is, and it may simply be observed that the reactions are each equal to one half of the total load on the truss, or $\frac{5000}{2}$ = 2500 lb. Since the external forces in this case are all in a single direction, the force polygon for the external forces is actually a straight line. Using the two-letter symbols for the forces and starting with letter *A* at the left end, we read the force sequence by moving in a clockwise direction around the outside of the truss. The loads are thus read as *AB, BC, CD, DE, EF,* and *FG,* and the two reactions are read as *GH* and *HA.* Beginning at *A* on the Maxwell diagram, the force vector sequence for the external forces is read from *A* to *B, B* to *C, C* to *D,* and so on, ending back at *A*, which shows that the force polygon closes and the external forces are in the necessary state of static equilibrium. Note that we have pulled the vectors for the reactions off to the side in the diagram to indicate them more clearly. Note also that we have used lowercase letters for the vector ends in the Maxwell diagram, whereas uppercase letters are used on the space diagram. The alphabetic correlation is thus retained (*A* to *a*), while any possible confusion between the two diagrams is prevented. The letters on the space diagram designate spaces, while the letters on the Maxwell diagram designate points of intersection of lines.

2. *Construct the force polygons for the individual joints.* The graphic procedure for this consists of locating the points on the Maxwell diagram that correspond to the remaining letters, *I* through *P,* on the space diagram. When all the lettered points on the diagram are located, the complete force polygon for each joint

may be read on the diagram. In order to locate these points, we use two relationships. The first is that the truss members can resist only forces that are parallel to the members' positioned directions. Thus we know the directions of all the internal forces. The second relationship is a simple one from plane geometry: A point may be located as the intersection of two lines. Consider the forces at joint 1, as shown in the separated joint diagram in Figure 6.13. Note that there are four forces and that two of them are known (the load and the reaction) and two are unknown (the internal forces in the truss members). The force polygon for this joint, as shown on the Maxwell diagram, is read as *ABIHA*. *AB* represents the load; *BI* the force in the upper chord member; *IH* the force in the lower chord member; and *HA* the reaction. Thus the location of point *I* on the Maxwell diagram is determined by noting that *I* must be in a horizontal direction from *H* (corresponding to the horizontal position of the lower chord) and in a direction from *B* that is parallel to the position of the upper chord.

The remaining points on the Maxwell diagram are found by the same process, using two known points on the diagram to project lines of known direction whose intersection will determine the location of another point. Once all the points are located, the diagram is complete and can be used to find the magnitude and sense of each internal force.

The process for construction of the Maxwell diagram typically consists of moving from joint to joint along the truss. Once one of the letters for an internal space is determined on the Maxwell diagram, it may be used as a known point for finding the letter for an adjacent space on the space diagram. The only limitation of the process is that it is not possible to find more than one unknown point on the Maxwell diagram for any single joint. Consider joint 7 on the separated joint diagram in Figure 6.13. If we attempt to solve this joint first, knowing only the locations of letters *A* through *H* on the Maxwell diagram, we must locate four unknown points: *L*, *M*, *N*, and *O*. This is three more unknowns than we can determine in a single step, so we must first solve for three of the unknowns by using other joints.

Solving for a single unknown point on the Maxwell diagram corresponds to finding two unknown forces at a joint, since each letter on the space diagram is used twice in the force identifications for the internal forces. Thus for joint 1 in the previous example, the letter *I* is part of the identity for forces *BI* and *IH*, as shown on the separated joint diagram. The graphic determination of single points on the Maxwell diagram, therefore, is analogous to finding two unknown quantities in an algebraic solution. As discussed previously, two unknowns are the maximum that can be solved for in the equilibrium of a coplanar, concurrent force system, which is the condition of the individual joints in the truss.

When the Maxwell diagram is completed, the internal forces can be read from the diagram as follows:

1. The magnitude is determined by measuring the length of the line in the diagram, using the scale that was used to plot the vectors for the external forces.
2. The sense of individual forces is determined by reading the forces in clockwise sequence around a single joint in the space diagram and tracing the same letter sequences on the Maxwell diagram.

Figure 6.14 shows the force system at joint 1 and the force polygon for these forces as taken from the Maxwell diagram. The forces shown initially are shown as solid lines on the force polygon, and the unknown forces are shown as dashed lines. Starting with letter A on the force system, we read the forces in a clockwise sequence as AB, BI, IH, and HA. On the Maxwell diagram we note that moving from a to b is moving in the order of the sense of the force, that is from tail to head of the force vector that represents the external load on the joint. If we continue in this sequence on the Maxwell diagram, this force sense flow will be a continuous one. Thus reading from b to i on the Maxwell diagram is reading from tail to head of the force vector, which tells us that force BI has its head at the left end. Transferring this sense indication from the Maxwell diagram to the joint diagram indicates that force BI is in compression; that is, it is pushing, rather than pulling, on the joint. Reading from i to h on the Maxwell diagram shows that the arrowhead for this vector is on the right, which translates to a tension effect on the joint diagram.

Having solved for the forces at joint 1 as described, we may use the fact that we now know the forces in truss members BI and IH when we proceed to consider the adjacent joints, 2 and 3. However, we must be careful to note that the sense reverses at the opposite ends of the members in the joint diagrams. Referring to the separated joint diagram in Figure 6.13, if the upper chord member shown as force BI in joint 1 is in compression, its arrowhead is at the lower left end in the diagram from joint 1, as shown in Figure 6.14. However, when the same force is shown as IB at joint 2, its pushing effect on the joint will be indicated by having the arrowhead at the upper right end in the diagram for joint 2. Similarly, the tension effect of the lower chord is shown in joint 1 by placing the arrowhead on the right end of the force IH, but the same tension force will be indicated in joint 3 by placing the arrowhead on the left end of the vector for force HI.

If we choose the solution sequence of solving joint 1 and then joint 2, we can transfer the known force in the upper chord to joint 2. Thus the solution for the five forces at joint 2 is reduced to finding three unknowns, since the load BC and the chord force IB are now known. However, we still cannot solve joint 2, since there are two unknown points on the Maxwell diagram (k and j) corresponding to the three unknown forces. An option, therefore, is to proceed from joint 1 to joint 3, at which there are presently only two unknown forces. On the Maxwell diagram we can find

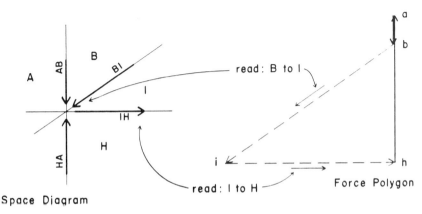

FIGURE 6.14 Graphical solution for Joint 1.

54 ANALYSIS FOR STATIC FORCES

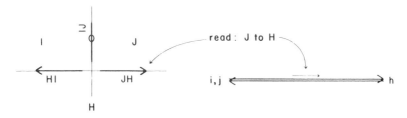

FIGURE 6.15 Graphical solution for Joint 3.

the single unknown point *j* by projecting vector *IJ* vertically from *i* and projecting vector *JH* horizontally from point *h*. Since point *i* is also located horizontally from point *h*, we thus find that the vector *IJ* has zero magnitude, since both *i* and *j* must be on a horizontal line from *h* in the Maxwell diagram. This indicates that there is actually no stress in this truss member for this loading condition and that points *i* and *j* are coincident on the Maxwell diagram. The joint force diagram and the force polygon for joint 3 are as shown in Figure 6.15. In the joint force diagram we place a zero, rather than an arrowhead, on the vector line for *IJ* to indicate the zero stress condition. In the force polygon in Figure 6.15, we have slightly separated the two force vectors for clarity, although they are actually coincident on the same line.

Having solved for the forces at joint 3, we can next proceed to joint 2, since there now remain only two unknown forces at this joint. The forces at the joint and the force polygon for joint 2 are shown in Figure 6.16. As explained for joint 1, we read the force polygon in a sequence determined by reading in a clockwise direction around the joint: *BCKJIB*. Following the continuous direction of the force arrows on the force polygon in this sequence, we can establish the sense for the two forces *CK* and *KJ*.

It is possible to proceed from one end and to work continuously across the truss from joint to joint to construct the Maxwell diagram in this example. The sequence in terms of locating points on the Maxwell diagram would be *i-j-k-l-m-n-o-p*, which would be accomplished by solving the joints in the following sequence: 1, 3, 2, 5, 4, 6,

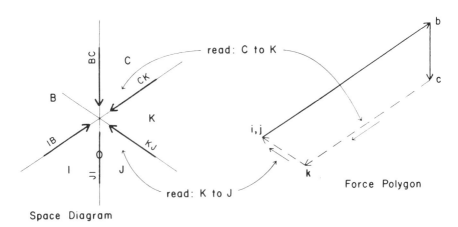

FIGURE 6.16 Graphical solution for Joint 2.

7, 9, 8. However, it is advisable to minimize the error in graphic construction by working from both ends of the truss. Thus a better procedure would be to find points *i-j-k-l-m*, working from the left end of the truss, and then to find points *p-o-n-m*, working from the right end. This would result in finding two locations for the point *m*, whose separation constitutes the error in drafting accuracy.

6.7 ALGEBRAIC ANALYSIS FOR INTERNAL FORCES IN PLANAR TRUSSES

Graphic solution for the internal forces in a truss using the Maxwell diagram corresponds essentially to an algebraic solution by the so-called *method of joints*. This method consists of solving the concentric force systems at the individual joints using simple force equilibrium equations. We will illustrate the method and the corresponding graphic solution using the previous example.

As with the graphic solution, we first determine the external forces, consisting of the loads and the reactions. We then proceed to consider the equilibrium of the individual joints, following a sequence as in the graphic solution. The limitation of this sequence, corresponding to the limit of finding only one unknown point in the Maxwell diagram, is that we cannot find more than two unknown forces at any single joint. Referring to Figure 6.17, the solution for joint 1 is as follows.

The force system for the joint is drawn with the sense and magnitude of the known forces shown, but with the unknown internal forces represented by lines without arrowheads, since their senses and magnitudes initially are unknown. For forces that are not vertical or horizontal, we replace the forces with their horizontal and vertical components. We then consider the two conditions necessary for the equilibrium of the system: The sum of the vertical forces is zero and the sum of the horizontal forces is zero.

If the algebraic solution is performed carefully, the sense of the forces will be determined automatically. However, we recommend that whenever possible the sense be predetermined by simple observation of the joint conditions, as will be illustrated in the solutions.

The problem to be solved at joint 1 is as shown at (*a*) in Figure 6.17. At (*b*) the system is shown with all forces expressed as vertical and horizontal components. Note that although this now increases the number of unknowns to three (IH, BI_v, and BI_h), there is a numeric relationship between the two components of BI. When this condition is added to the two algebraic conditions for equilibrium, the number of usable relationships totals three, so that we have the necessary conditions to solve for the three unknowns.

The condition for vertical equilibrium is shown at (*c*) in Figure 6.17. Since the horizontal forces do not affect the vertical equilibrium, the balance is between the load, the reaction, and the vertical component of the force in the upper chord. Simple observation of the forces and the known magnitudes makes it obvious that force BI_v must act downward, indicating that BI is a compression force. Thus the sense of BI is established by simple visual inspection of the joint, and the algebraic equation for vertical equilibrium (with upward force considered positive) is

$$\Sigma F_v - 0 = +2500 \ -500 \ -BI_v$$

56 ANALYSIS FOR STATIC FORCES

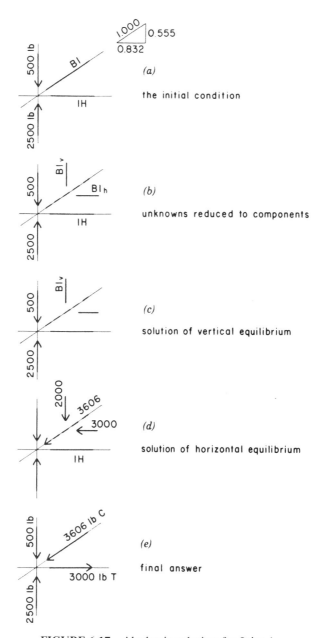

FIGURE 6.17 Algebraic solution for Joint 1.

From this equation we determine BI_v to have a magnitude of 2000 lb. Using the known relationships between BI, BI_v, and BI_h, we can determine the values of these three quantities if any one of them is known. Thus

$$\frac{BI}{1,000} = \frac{BI_v}{0.555} = \frac{BI_h}{0.832}$$

ALGEBRAIC ANALYSIS FOR INTERNAL FORCES IN PLANAR TRUSSES 57

$$BI_h = \frac{0.832}{0.555}(2000) = 3000 \text{ lb}$$

$$BI = \frac{1.000}{0.555}(2000) = 3606 \text{ lb}$$

The results of the analysis to this point are shown at (d) in Figure 6.17, from which we can observe the conditions for equilibrium of the horizontal forces. Stated algebraically (with force sense toward the right considered positive) the condition is

$$\sum F_h = 0 = IH - 3000$$

from which we establish that the force in IH is 3000 lb.

The final solution for the joint is then as shown at (e) in the figure. On this diagram the internal forces are identified as to sense by using C to indicate compression and T to indicate tension.

As with the graphic solution, we can proceed to consider the forces at joint 3. The initial condition at this joint is as shown at (a) in Figure 6.18, with the single known force in member HI and the two unknown forces in IJ and JH. Since the forces at this joint are all vertical and horizontal, there is no need to use components. Consideration of vertical equilibrium makes it obvious that it is not possible to have a force in member IJ. Stated algebraically, the condition for vertical equilibrium is

$$\sum F_v = 0 = IJ \quad \text{(since } IJ \text{ is the only vertical force)}$$

It is equally obvious that the force in IJ must be equal and opposite to that in HI, since they are the only two horizontal forces. That is, stated algebraically,

$$\sum F_h = 0 = JH - 3000$$

The final answer for the forces at joint 3 is as shown at (b) in Figure 6.18. Note the convention for indicating a truss member with no internal force.

If we now proceed to consider joint 2, the initial condition is as shown at (a) in Figure 6.19. Of the five forces at the joint only two remain unknown. Following the procedure for joint 1, we first resolve the forces into their vertical and horizontal components, as shown at (b) in Figure 6.19.

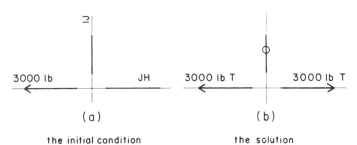

FIGURE 6.18 Algebraic solution for Joint 3.

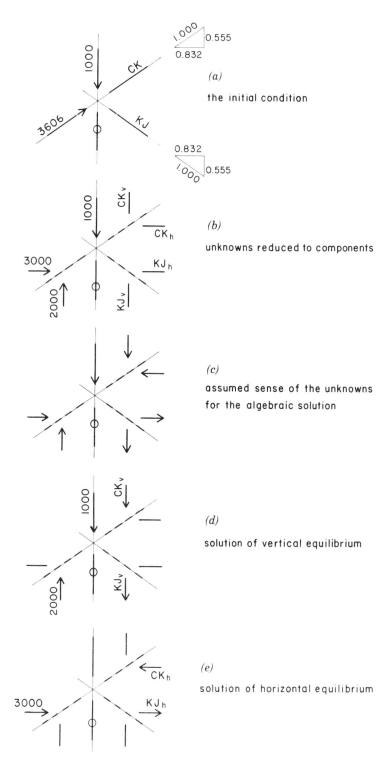

FIGURE 6.19 Algebraic solution for Joint 2.

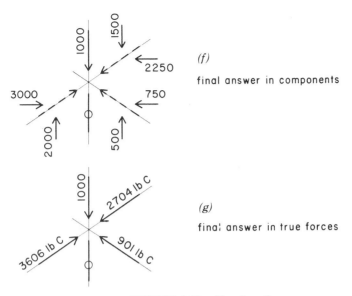

FIGURE 6.19 (*Continued*)

Since we do not know the sense of forces *CK* and *KJ*, we may use the procedure of considering them to be positive until proven otherwise. That is, if we enter them into the algebraic equations with an assumed sense, and the solution produces a negative answer, then our assumption was wrong. However, we must be careful to be consistent with the sense of the force vectors, as the following solution will illustrate.

Let us arbritarily assume that force *CK* is in compression and force *KJ* is in tension. If this is so, the forces and their components will be as shown at (*c*) in Figure 6.19. If we then consider the conditions for vertical equilibrium, the forces involved will be those shown at (*d*) in Figure 6.19, and the equation for vertical equilibrium will be

$$\sum F_v = 0 = -1000 + 2000 - CK_v - KJ_v$$

or

$$0 = +1000 - 0.555\ CK - 0.555\ KJ \qquad (1)$$

If we consider the conditions for horizontal equilibrium, the forces will be as shown at (*e*) in Figure 6.19, and the equation will be

$$\sum F_h = 0 = +3000 - CK_h + KJ_h$$

or

$$0 = +3000 - 0.832\ CK + 0.832\ KJ \qquad (2)$$

Note the consistency of the algebraic signs and the sense of the force vectors, with positive forces considered as upward and toward the right. We may solve these two equations simultaneously for the two unknown forces as follows.

1. Multiply equation (1) by $\dfrac{0.832}{0.555}$.

Thus: $0 = \dfrac{0.832}{0.555}(+1000) + \dfrac{0.832}{0.555}(-0.555\ CK)$

$\qquad + \dfrac{0.832}{0.555}(-0.555\ KJ)$

$0 = +1500 - 0.832\ CK - 0.832\ KJ$

2. Add this equation to equation (2) and solve for CK.

Thus: $\quad 0 = +1500 - 0.832\ CK - 0.832\ KJ$

$\qquad 0 = +3000 - 0.832\ CK + 0.832\ KJ$

Adding: $\quad 0 = +4500 - 1.664\ CK$

Therefore: $CK = \dfrac{4500}{1.664} = 2704\ \text{lb}$

Note that the assumed sense of compression in CK is correct, since the algebraic solution produces a positive answer. Substituting this value for CK in equation (1),

$$0 = +1000 - 0.555(2704) - 0.555(KJ)$$
$$= +1000 - 1500 - 0.555(KJ)$$

Then

$$KJ = -\dfrac{500}{0.555} = -901\ \text{lb}$$

Since the algebraic solution produces a negative quantity for KJ, the assumed sense for KJ is wrong and the member is actually in compression.

The final answers for the forces at joint 2 are thus as shown at (g) in Figure 6.19. In order to verify that equilibrium exists, however, the forces are shown in the form of their vertical and horizontal components at (f) in the illustration.

When all of the internal forces have been determined for the truss, the results may be recorded or displayed in a number of ways. The most direct way is to display them on a scaled diagram of the truss, as shown in the upper part of Figure 6.20. The force magnitudes are recorded next to each member with the sense shown as T for tension or C for compression. Zero stress members are indicated by the conventional symbol consisting of a zero placed directly on the member.

When solving by the algebraic method of joints, the results may be recorded on a separated joint diagram as shown in the lower portion of Figure 6.20. If the values for the vertical and horizontal components of force in sloping members are shown, it is a simple matter to verify the equilibrium of the individual joints.

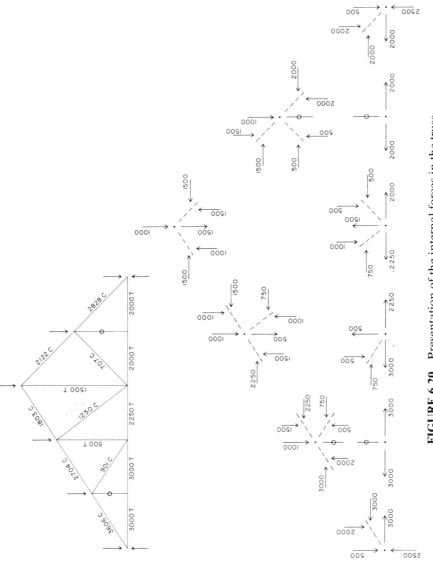

FIGURE 6.20 Presentation of the internal forces in the truss.

62 ANALYSIS FOR STATIC FORCES

6.8 VISUALIZATION OF THE SENSE OF INTERNAL FORCES

It is often possible to determine the sense of the internal forces in a truss with little or no quantified calculations. Where this is so, it is useful to do so as a first step in the truss analysis. If a graphic analysis is performed, the sense determined by the preliminary inspection serves as a cross check on the sense determined from the Maxwell diagram. If an algebraic analysis is performed, the preliminary inspection will aid greatly in keeping track of minus signs in the equilibrium equations. In addition to these practical uses, the preliminary analysis for the sense of internal forces is a good exercise in the visualization of truss behavior and of equilibrium conditions in general. The following examples will illustrate procedures that can be used for such an analysis.

Consider the truss shown in Figure 6.21a. For a consideration of the sense of internal forces, we may proceed in a manner similar to that for the graphic analysis or the algebraic analysis by the method of joints. We thus consider the joints as follows:

Joint 1. (see Figure 6.21b). For vertical equilibrium the vertical component in *BH* must act downward. Thus member *BH* is in compression and its horizontal component acts toward the left. For horizontal equilibrium the only other horizontal force, that in member *HG*, must act toward the right, indicating that *HG* is in tension.

Joint 2. (see Figure 6.21c). Member *HB* is the same as *BH* as shown at joint 1. The known compression force in the member, therefore, is transferred to joint 2, as shown. If we use a rotated set of reference axes (*x* and *y*) it may be observed that the force in member *IH* is alone in opposing the load effect. Therefore, *IH* must be in compression. (IH_y must oppose the *y* component of the load.)

The sense of the force in member *CI* is difficult to establish without some quantified analysis. One approach is to consider the action of the whole truss and to attempt to visualize what the result would be if these members were removed. As shown in Figure 6.21d, it is fairly easy to visualize that the force in these members must be one of compression to hold the structure in place.

Joint 3. (see Figure 6.21e) We first transfer the known conditions for members *GH* and *HI* from the previous analysis of joints 1 and 2. Considering vertical equilibrium, we observe that the vertical component in *IJ* must oppose that in *HI*. Therefore, member *IJ* is in tension. For member *JG* we may use the technique illustrated for member *CI* at joint 2. As shown in Figure 6.21f it may be observed that this member must act in tension for the structure to function.

Because of the symmetry of the truss in this example, the sense of all internal forces is established with the consideration of only these three joints. The results of the analysis are displayed on the truss figure shown in Figure 6.21g using *T* and *C* to indicate internal forces of tension and compression, respectively.

For a second example of this type of analysis we consider the truss shown in Figure 6.22a. As before, we begin at one support and move across the truss from joint to joint.

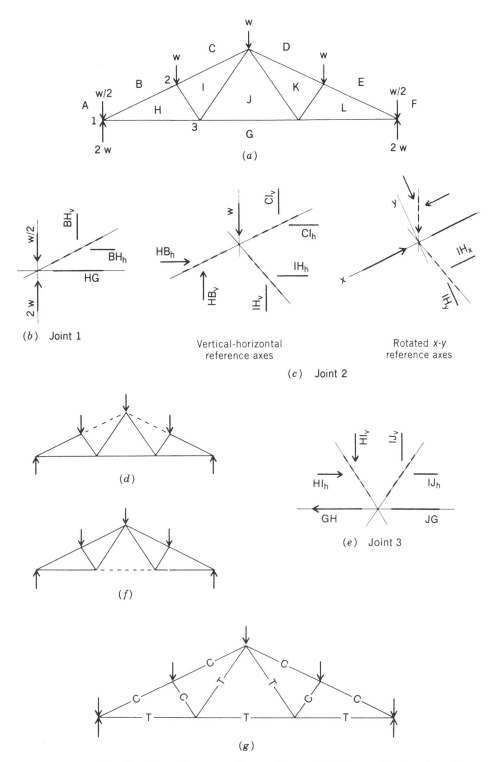

FIGURE 6.21 Visualization of the sense of internal forces. (*a*) The truss loads and reactions. (*b*) Force resolution at Joint 1. (*c*) Force actions at Joint 2. (*d*) Visualization of actions of members *CI* and *DK*. (*e*) Force actions at Joint 3. (*f*) Visualization of action of member *JG*. (*g*) Answers for the sense of internal forces.

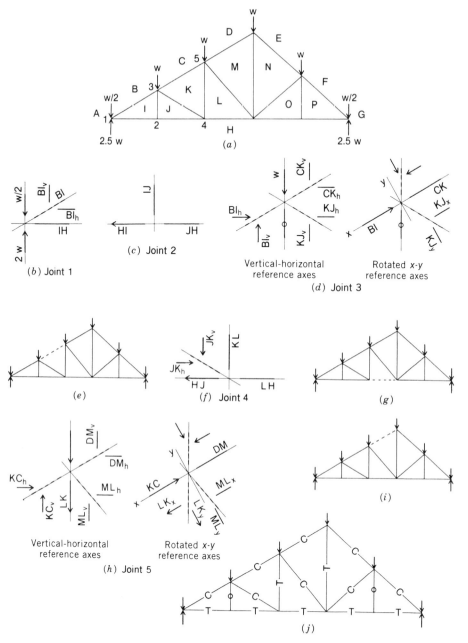

FIGURE 6.22 Visualization of the sense of internal forces. (*a*) The truss, loads, and reactions. (*b*) Force actions at Joint 1. (*c*) Force actions at Joint 2. (*d*) Force actions at Joint 3. (*e*) Visualization of action of member *CK*. (*f*) Force actions at Joint 4. (*g*) Visualization of action of member *LH*. (*h*) Force actions at Joint 5. (*i*) Visualization of action of member *DM*. (*j*) Answers for the sense of the internal forces.

Joint 1. (see Figure 6.22b). As in the previous example, consideration of vertical equilibrium establishes the compression in member *BI*, after which consideration of horizontal equilibrium establishes the tension in member *IH*.

Joint 2. (see Figure 6.22c). Since *IJ* is the only potential vertical force, it must be zero. Since *JH* alone opposes the tension in *HI*, it must be in tension also.

Joint 3. (see Figure 6.22d). With member *IJ* essentially nonexistent, due to its zero stress condition, this joint is similar to joint 2 in the previous example. Thus the use of rotated reference axes may be a means of establishing the condition of compression in member *KJ*. Also, as before, we may observe the necessity for compression in member *CK* by considering the result of removing it from the truss, as illustrated in Figure 6.22e.

Joint 4. (see Figure 6.22f). Because of the known compression force in member *JK*, it may be observed that *KL* must be in tension. Since the two known horizontal forces are opposite in sense, it is not possible to visualize the required sense of member *LH*. As before, the device of removing the member, shown in Figure 6.22g, may be used to establish the required tension in this member.

Joint 5. (see Figure 6.22h). Using the rotated *x-y* references axes, it is possible to establish the sense of the force in member *ML*. Since the other two *y*-direction forces have the same sense, the *y* component of *ML* must oppose them, establishing the requirement for a compression force in the member. Neither set of reference axes can be used to establish the sense of the force in member *DM*, however. As before, we can use the device of removing the member to establish its required compression action, as shown in Figure 6.22i.

Working from the other end of the truss, it is now possible to establish the sense of the force in the remaining members. The final answers for the senses in all members are as shown on the truss figure in Figure 6.22j.

6.9 ANALYSIS OF NONCONCURRENT FORCES

In the analysis of concurrent forces, it is sufficient to consider only the basic vector properties of the forces: magnitude, direction, and sense. However, when forces are not concurrent, it is necessary to include the consideration of another type of force action called the *moment*, or rotational effect.

Consider the two interacting vertical forces shown at (*a*) in Figure 6.23. Since the forces are concurrent, the condition of equilibrium is fully established by satisfying the single algebraic equation: $\sum F_v = 0$. However, if the same two forces are not concurrent, as shown at (*b*) in Figure 6.23, the single force summation is not sufficient to establish equilibrium. In this case the force summation establishes the same fact as before: There is no net tendency for vertical motion. However, because of their separation, the forces tend to cause a counterclockwise motion in the form of a rotational effect, called the moment. The moment has three basic properties:

1. It exists in a particular plane—in this case the plane defined by the two force vectors.
2. It has a magnitude, expressed as the product of the force magnitude times the

66 ANALYSIS FOR STATIC FORCES

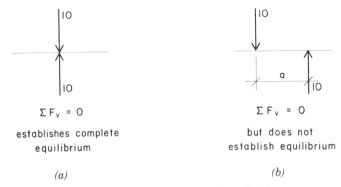

FIGURE 6.23 Equilibrium of parallel forces.

distance between the two vectors. In the example shown at (b) in Figure 6.23, the magnitude of the moment is (10)(a). The unit for this quantity becomes a compound of the force unit and the distance unit: lb-in. kip-ft, and so on.

3. It has a sense of rotational direction. In the example the sense is counterclockwise.

Because of potential moment effects, the consideration of equilibrium for noncurrent forces must include another summation: $\sum M = 0$. Rotational equilibrium can be established in various ways. One way is shown in Figure 6.24. In this example a second set of forces whose rotational effect counteracts that of the first set has been added. The complete equilibrium of this general coplanar force system (nonconcurrent and nonparallel) now requires the satisfaction of three summation equations:

$$\sum F_v = 0 = +10 - 10$$
$$\sum F_h = 0 = +4 - 4$$
$$\sum M = 0 = +(10)(a) - (4)(2.5\,a)$$
$$= +(10\,a) - (10\,a)$$

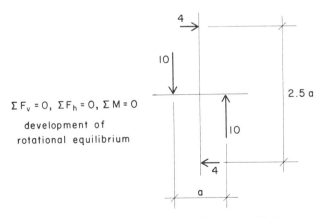

FIGURE 6.24 Establishing rotational equilibrium.

ANALYSIS OF NONCONCURRENT FORCES   67

Since all of these summations total zero, the system is indeed in equilibrium.

Solution of equilibrium problems with nonconcurrent forces involves the application of the available algebraic summation equations. The following example illustrates the procedure for the solution of a simple parallel force system.

Figure 6.25 shows a 20 kips force applied to a beam at a point between the beam's supports. The supports must generate the two vertical forces, R_1 and R_2, in order to oppose this load. (In this case we will ignore the weight of the beam itself, which will also add load to the supports, and we will consider only the effect of the distribution of the load added to the beam.) Since there are no horizontal forces, the complete equilibrium of this force system can be established by the satisfaction of two summation equations:

$$\Sigma F_v = 0$$
$$\Sigma M = 0$$

Considering the force summation first,

$$\Sigma F_v = 0 = -20 + (R_1 + R_2) \qquad \text{[sense up considered positive]}$$

Thus

$$R_1 + R_2 = 20$$

This yields one equation involving the two unknown quantities. If we proceed to write a moment summation involving the same two quantities, we then have two equations that can be solved simultaneously to find the two unknowns. We can simplify the algebraic task somewhat if we use the technique of making the moment summation in a way that eliminates one of the unknowns. This is done simply by using a moment reference point that lies on the line of action of one of the unknown forces. If we choose a point on the action line of R_2, as shown at (b) in Figure 6.25, the summation will be as follows:

$$\Sigma M = 0 = -(20)(7) + (R_1)(10) + (R_2)(0) \qquad \text{(clockwise moment plus)}$$

Then

$$(R_1)(10) = 140$$
$$R_1 = 14 \text{ k}$$

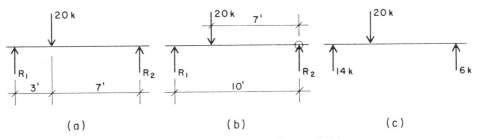

FIGURE 6.25 Investigation of a simple parallel force system.

68 ANALYSIS FOR STATIC FORCES

Using the relationship established from the previous force summation:

$$R_1 + R_2 = 20$$
$$14 + R_2 = 20$$
$$R_2 = 6 \text{ k}$$

The solution is then as shown at (c) in Figure 6.25.

In the structure shown in Figure 6.26a, the forces consist of a vertical load, a horizontal load, and some unknown reactions at the supports. Since the forces are not all parallel, we may use all three equilibrium conditions in the determination of the unknown reactions. Although it is not strictly necessary, we will use the technique of finding the reactions separately for the two loads and then adding the two results to find the true reactions for the combined load.

The vertical load and its reactions are shown in Figure 6.26b. In this case, with the symmetrically placed load, each reaction is simply one-half the total load.

For the horizontal load, the reactions will have the components shown at (1) in Figure 6.26c, with the vertical reaction components developing resistance to the moment effect of the load, and the horizontal reaction components combining to resist the actual horizontal force effect. The solution for the reaction forces may be accomplished by finding these four components; then, if desired, the actual reaction forces and their directions may be found from the components, as explained in Section 6.3.

Let us first consider a moment summation, choosing the location of R_2 as the point of rotation. Since the action lines of V_2, H_1, and H_2 all pass through this point, their moments will be zero and the summation is reduced to dealing with the forces shown at (2) in Figure 6.26c. Thus

$$\sum M = 0 = +(6)(12) - (V_1)(10) \quad \text{(clockwise moment considered positive)}$$
$$V_1 = \frac{(6)(12)}{10} = 7.2 \text{ k}$$

We next consider the summation of vertical forces, which involves only V_1 and V_2, as shown at (3) in Figure 6.26c. Thus

$$\sum F_v = 0 = -V_1 + V_2 \quad \text{(sense up considered positive)}$$
$$0 = -7.2 + V_2$$
$$V_2 = +7.2 \text{ k}$$

For the summation of horizontal forces the forces involved are those shown at (4) in Figure 6.26c. Thus

$$\sum F_h = 0 = +6 - H_1 - H_2 \quad \text{(force toward right considered positive)}$$
$$H_1 + H_2 = 6 \text{ k}$$

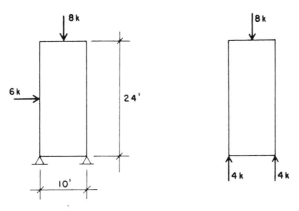

(a) Space diagram (b) Resolution of the vertical load

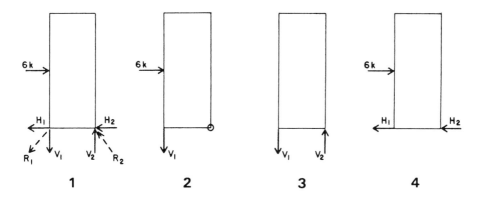

(c) Resolution of the horizontal loading

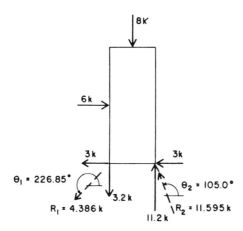

(d) Reactions for the combined loading

FIGURE 6.26 Structure subjected to a general planar force system. (*a*) Space diagram. (*b*) Resolution of the vertical forces. (*c*) Resolution of the horizontal forces. (*d*) Reactions for the combined loading; shown both in component form (solid arrows) and resultant form (dashed arrows).

70 ANALYSIS FOR STATIC FORCES

This presents an essentially indeterminate situation that cannot be solved unless some additional relationships can be established. Some possible relationships are the following.

1. R_1 offers resistance to horizontal force, but R_2 does not. This may be the result of the relative mass or stiffness of the supporting structure or the type of connection between the supports and the structure above. If a sliding, rocking, or rolling connection is used, some minor frictional resistance may be developed, but the support is essentially without significant capability for the development of horizontal resistance. In this case $H_1 = 6$ k and $H_2 = 0$.
2. The reverse of the preceding; R_2 offers resistance, but R_1 does not. $H_1 = 0$ and $H_2 = 6$ k.
3. Details of the construction indicate an essentially symmetrical condition for the two supports. In this case it may be reasonable to assume that the two reactions are equal. Thus, $H_1 = H_2 = 3$ k.

For this example we will assume the symmetrical condition for the supports with the horizontal force being shared equally by the two supports. Adding the results of the separate analyses we obtain the results for the combined reactions as shown in Figure 6.26d. The reactions are shown both in terms of their components and in their resultant form as single forces. The magnitudes of the single force resultants are obtained as follows:

$$R_1 = \sqrt{(3)^2 + (3.2)^2} = \sqrt{19.24} = 4.386 \text{ k}$$

$$R_2 = \sqrt{(11.2)^2 + (3)^2} = \sqrt{134.44} = 11.595 \text{ k}$$

The directions of these forces are obtained as follows:

$$\theta_1 = \arctan \frac{3.2}{3} = \arctan 1.0667 = 46.85°$$

$$\theta_2 = \arctan \frac{11.2}{3} = \arctan 3.7333 = 75.0°$$

Note that the angles for the reactions as shown on the illustration in Figure 6.26d are measured as counterclockwise rotations from a right-side horizontal reference. Thus, as illustrated, the angles are actually

$$\theta_1 = 180 + 46.85 = 226.85°$$

and

$$\theta_2 = 180 - 75.0 = 105.0°$$

If this standard reference system is used, it is possible to indicate both the direction and sense of a force vector with the single value of the rotational angle. The technique is illustrated in Figure 6.27. At (a) four forces are shown, all of which are

ANALYSIS OF NONCONCURRENT FORCES 71

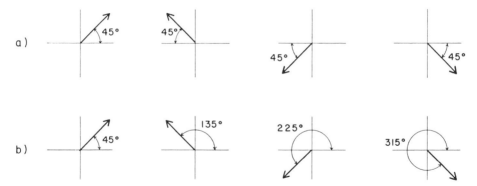

FIGURE 6.27 Reference notation for angular direction: (*a*) with horizontal reference axis; (*b*) with polar reference axis.

rotated 45° from the horizontal. If we simply make the statement, "the force is at an angle of 45° from the horizontal," the situation may be any one of the four shown. If we use the reference system just described, however, we would describe the four situations as shown at (*b*) in the figure, and they would be identified unequivocally.

Analysis of a Trussed Tower

The structure shown in Figure 6.28*a* consists of a vertical planar truss utilized as a cantilevered tower. This single truss may be used as one of a series of bracing elements in a building or as one side of a freestanding tower. The loading shown is due to a combination of gravity effects and wind force directed from left to right (referred to as wind load left, as shown in the illustration).

The truss form in this case is that of a typical x-braced rectangular frame, consisting of vertical and horizontal elements that are braced against the lateral load (the horizontal forces) by the addition of the diagonal elements. The trussed bracing could actually be accomplished by a single diagonal in each of the three bays, or stories, of the frame. If this is done, however, the single diagonal must have a dual functioning capability in order to brace the structure for wind from both directions. This dual functioning is demonstrated in Figure 6.28*b*, in which the sense of the reactions and internal forces is shown for the two wind loadings. Note the reversal of sense of the reactions and internal forces that occurs with the change in the direction of the wind.

Although the use of the single diagonals is possible, it results in a design situation that is often questionable, that is, the necessity to provide a very long compression element. Because of its great length and the requirements for minimum stiffness against buckling, this element is likely to be quite heavy in proportion to the actual load it must resist. This is a major reason for the use of x-bracing. Theoretically the addition of the extra diagonals creates a situation that makes the structure statically indeterminate, since the extra diagonals are redundant for simple static equilibrium of the structure. However, the design of such a structure is usually done on the basis of the behavior illustrated in Figure 6.28*c*. The individual diagonals are designed as tension members, which normally results in their being excessively slender for compression utilization. Thus while the wind from a single direction will

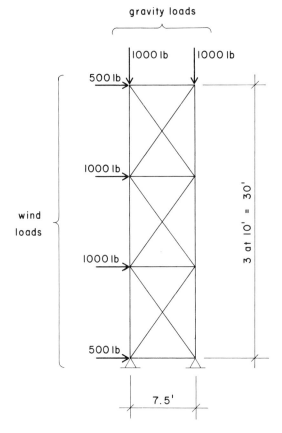

(a) The Trussed Tower: Form and Loads

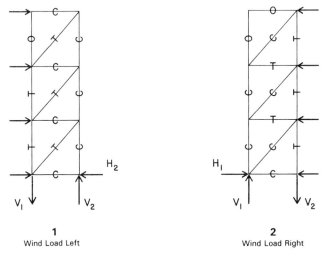

(b) Action of Single Diagonal Bracing

FIGURE 6.28 Investigation of a planar trussed tower frame. (*a*) Space diagram. (*b*) Action of single diagonal bracing under reversible horizontal (wind) loads. (*c*) Assumed behavior of the slender X-bracing under the reversible loading.

ANALYSIS OF NONCONCURRENT FORCES 73

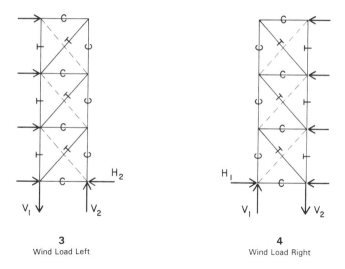

(c) Assumed Behavior of Slender X-Bracing

FIGURE 6.28. (*Continued*)

actually tend to induce tension in one of the diagonals of a bay and compression in the opposing diagonal, the compression diagonal is assumed to buckle under the compression force, leaving the tension diagonal to do the entire bracing job. With this assumption the wind load from the left, as shown at (*a*) in Figure 6.28c, is assumed to be taken by one diagonal in each bay, while the opposing diagonal rests (as indicated by the dashed lines in the illustration). With wind load from the right, the roles of the diagonals reverse.

The complete analysis of the structure for gravity load plus wind from both directions is summarized in Figure 6.29. Separate analyses are made for each loading, as was illustrated in the preceding section. Since the structure and the wind loadings are both assumed to be symmetrical, the analysis for wind load has been done for wind from one direction only. The results obtained from this analysis may then be simply reversed on the structure for wind from the opposite direction.

The analysis for the reactions is shown at (*b*) in Figure 6.29. Note that we have made the simplifying assumption that the entire horizontal force is resisted by the support on the side opposite the wind. With the reversal of the wind, this relationship is also reversed. This results in some redundancy in the design loads for the supports and the bottom horizontal element. If any conditions exist that allow for more specific qualification of the supports, they should be considered in the analysis.

Analysis for the sense of the internal forces is shown at (*c*) in Figure 6.29. Note that the gravity loads produce internal forces only in the vertical elements. When the wind direction is reversed, the sense of internal force will reverse in all the elements except the bottom horizontal element.

The space diagram and Maxwell diagram for wind load left are shown at (*d*) and (*e*) in Figure 6.29. The gravity loading in this example produces a simple condition of a constant 1000 lb compression in all the vertical elements. The reactions and internal forces caused by the combined gravity and wind load left loadings are

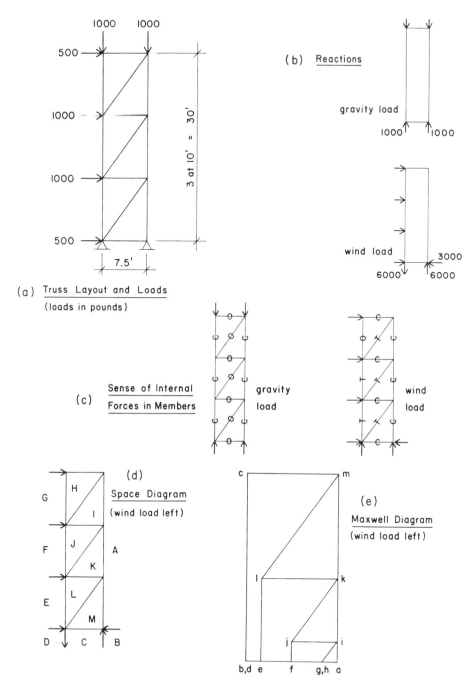

FIGURE 6.29 Investigation of the trussed tower. (*a*) Space diagram. (*b*) Development of the external support reactions. (*c*) Assumed sense of the internal forces in the members. (*d*) Space diagram for the horizontal wind load only. (*e*) Graphical solution for the internal forces due to the horizontal loads only. (*f*) Separated joint diagram for the combined loading. (*g*) Table for determination of the critical load conditions for the members.

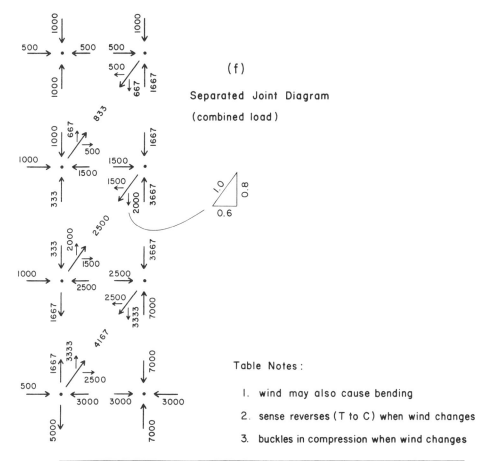

(f) Separated Joint Diagram (combined load)

Table Notes:

1. wind may also cause bending
2. sense reverses (T to C) when wind changes
3. buckles in compression when wind changes

(g)

member	gravity load	wind left	wind right	combinations minimum	maximum	notes
A H	0	500 C	500 C	0	500 C	
I J	0	1500 C	1500 C	0	1500 C	
K L	0	2500 C	2500 C	0	2500 C	
M C	0	3000 C	3000 C	0	3000 C	
GH, IA	1000 C	0	667 C	1000 C	1667 C	1
FJ, KA	1000 C	667 T	2667 C	333 C	3667 C	1
EL, MA	1000 C	2667 T	6000 C	1667 T	7000 C	1,2
H I	0	833 T	—	0	833 T	3
J K	0	2500 T	—	0	2500 T	3
L M	0	4167 T	—	0	4167 T	3

Design Values for Members — internal forces in pounds

FIGURE 6.29. (*Continued*)

shown on the separated joint diagram at (*f*) in Figure 6.29. The values for the horizontal and diagonal elements may be taken directly from the Maxwell diagram for the wind load. For the vertical elements, a constant 1000 lb compression is added to the values taken from the Maxwell diagram. The forces in the diagonal elements are shown in both component and net resultant form in the illustration.

When a structure is subjected to a combination of gravity and wind loads, it is usually necessary to consider the effects on the structure of four different load combinations:

1. Gravity dead (permanent) load only. This loading is used for consideration of any long-term load effects on the structure or its supports.
2. Gravity dead load plus gravity live (transient) load. This load must be designed for using the maximum limiting stresses for the structure. In some cases this loading may be critical for certain parts of the structure, even though the gravity plus wind load combinations may produce higher values for internal forces. This is due to the fact that most codes permit an increase in allowable stresses when the loading combination includes wind effects.
3. Gravity load plus wind load left. Depending on the code used, this may include only the dead load, or it may have to include some portion of the live load as well.
4. Gravity load plus wind load right. This is necessary only if the structure is not symmetrical.

Consideration of these various combinations produces the critical design forces for the supports and elements of the structure. In most cases it is necessary to consider two different net results for each element of the structure. The first result is the maximum total force; the second result is the so-called minimum design force. Of course each element must be designed for the maximum force utilizing the maximum allowable stresses for the element. However, when another loading combination produces a force of opposite sense, it may also be critical. This design problem is explained more fully in the examples in Chapters 9 and 10. In the table shown at (*g*) in Figure 6.29, the results of the individual loadings are given for the truss elements. These results are then combined to produce the minimum and maximum combinations. Since the gravity load was not distinguished as live or dead, it has been treated as a single condition in the table. For this example, if the combinations shown are the true design values, the minimum design force is not likely to be critical for any of the truss elements. The only really possible situation is that of the bottom vertical member (*EL*, *MA*) for which a stress sense reversal occurs as the wind reverses. However, if this element is designed as a compression member for the 7000 lb load, it is not likely to be critical as a tension member for only 1667 lb. The more critical concern would be for the support, which must sustain a vertical uplift of 5000 lb as well as the compression of 7000 lb.

6.10 BEAMS

This section presents the behavior of beams and some of the aspects of beam behavior that are related to trussed structures.

Aspects of Beam Behavior

A beam is a linear element subjected to loading that is lateral (or perpendicular) to its major axis. The primary internal forces that are developed are bending and shear, and the external forces, consisting of the loads and reactions, normally constitute a parallel, coplanar force system. A beam is ordinarily supported at isolated points; the distance between supports is called the beam's span. A special case is the cantilever beam, which is supported at only one end of its span and requires a rotational (moment) resistive support.

In structural systems, beams are often subjected to combinations of forces that produce other actions in addition to the beam actions just described. The top chord of a truss may be loaded in a manner that makes it function as a beam between truss joints while it also develops tension or compression due to the action of the truss. The design of such an element must consider these actions as occurring simultaneously. In such cases the use of the term *beam action* refers only to the efforts involved in spanning and resisting loads perpendicular to the span.

Beam is the general name for a spanning element. However, several other names are used also:

Girder. Usually a large beam or one that supports a series of smaller beams.

Joist. Usually a small beam (or sometimes a light truss) that is used in a closely spaced set, as in the ordinary light wood floor system.

Rafter. A roof beam or joist.

Purlin. Used in describing elements of a roof framing system in which girders or trusses support the purlins, the purlins support the joists or rafters, and the joists or rafters support the deck.

Girt. Usually used to describe a light beam used in a wall framing system to span between columns or bents in order to resist the wind loads on the wall.

Beams are ordinarily loaded by one of two types of load: distributed or concentrated. Distributed loads include the dead weight of the beam itself and any loads that are applied through a deck that is supported continuously along the beam length. When a series of smaller beams is supported by a larger beam, the end reactions of the smaller beams become concentrated loads on the larger beam. Other concentrated loads may result from columns on top of a beam or various objects hung from a beam.

Figure 6.30 shows a single span beam that sustains a uniformly distributed loading. The term *simple beam* is often used to describe such a beam, referring to the facts that the beam is of single span and the supports resist only forces that are perpendicular to the beam span. The primary concerns for such a structural element are

The Reactions. These are the forces developed at the supports. In this case each reaction is simply equal to one half of the total load on the beam. With the load quantified in terms of a constant unit w per unit of length of the beam span, as shown in the illustration, the reactions thus become

78 ANALYSIS FOR STATIC FORCES

$$R_1 = R_2 = \tfrac{1}{2}(w)(L) = \frac{wL}{2}$$

The Variation of Internal Shear Along the Beam. This may be determined by a summation of the external forces along the beam length. At an intermediate point in the span, x distance from the left support, the value of the shear thus is

$$V_x = R_1 - (w)(x) = \frac{wL}{2} - wx$$

or

$$V_x = w\left(\frac{L}{2} - x\right)$$

The Variation of Internal Moment Along the Beam. This may be determined at any point by a summation of the moments due to the loads and the reactions on either side of the point. Again, referring to the point at x distance from the left support and using the forces to the left of this point,

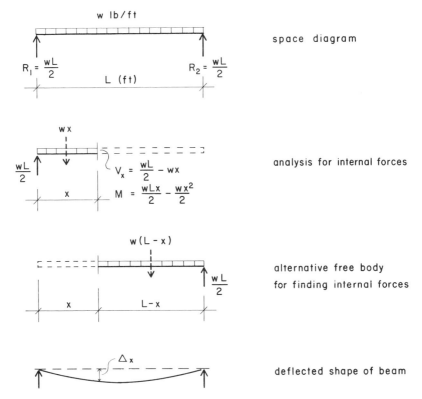

FIGURE 6.30 Considerations for the development of internal forces in a simple beam under uniformly distributed loading.

$$M_x = \left(\frac{wL}{2}\right)(x) - (wx)\left(\frac{x}{2}\right) = w\left(\frac{Lx}{2} - \frac{x^2}{2}\right)$$

If we use the forces to the right of the point,

$$M_x = \left(\frac{wL}{2}\right)(L - x) - (w)(L - x)\left(\frac{L - x}{2}\right)$$

$$= \frac{wL^2}{2} - \frac{wLx}{2} - (w)\left(\frac{L^2 - 2Lx + x^2}{2}\right)$$

$$= \frac{wL^2}{2} - \frac{wLx}{2} - \frac{wL^2}{2} + wLx - \frac{wx^2}{2}$$

$$= w\left(\frac{Lx}{2} - \frac{x^2}{2}\right)$$

The Deflected Shape of the Beam. Assuming the beam initially is straight, the bending action will cause it to take a curved form. This curve is visualized in terms of the lateral movement (called deflection) of points along the beam length away from their original positions. As shown in Figure 6.30, the uniformly loaded simple beam will assume a single, symmetrically curved form, with the maximum deflection at the center of the span.

6.11 GRAPHIC REPRESENTATION OF BEAM BEHAVIOR

The structural action of a beam often is described by using a set of graphic representations, the major elements of which are the diagrams shown in Figure 6.31. The diagrams in the figure describe the action of the simple span uniformly loaded beam whose behavior was discussed in the preceding section.

The space diagram shows the loading, the span, and the support conditions for the beam. In this example the values for the magnitudes of the reactions are also shown on the space diagram.

The shear diagram is a graph of the external loads on the beam. Its use, however, is in determining the conditions of internal shear along the beam length. A review of the method used for finding internal shear (described in the preceding section) will reveal that the graph of external forces is also the graph of the internal shear force in the beam. Thus the diagram is called the shear diagram, although the technique used for its determination is simply that of graphing the external forces.

The convention ordinarily used for construction of the shear diagram is that of starting at the left end of the beam. As shown in Figure 6.31, the graph of external forces thus begins with a plot of the upward force due to the left reaction. As we proceed along the beam length, the graph then descends uniformly as the uniformly distributed loading is encountered. This line continues as a straight sloping line until the reaction at the right end is encountered, at which point it jumps upward, representing the upward force of the reaction. Since the external forces are in equilibrium, the diagram is closed; that is, it starts at zero and ends at zero. At some

80 ANALYSIS FOR STATIC FORCES

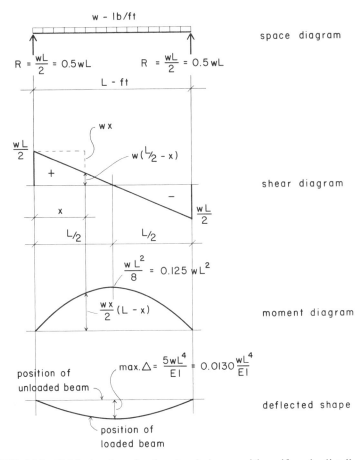

FIGURE 6.31 Critical values for the simple beam with uniformly distributed load.

arbitrary point between the left end and the right end, the value of the internal shear on the graph may be seen to be

$$\frac{wL}{2} - wx$$

which is the same value that was determined algebraically in the preceding section.

It may be seen that the shear diagram has both positive and negative values, relating to the net sense of the internal shear force. The sign of shear in the diagram in Figure 6.31 relates to the convention used for the construction of the diagram, that of beginning from the left end. If we started from the right end, the resulting diagram would have a similar form, but would be the mirror image of the diagram shown, with the sense of the shear values reversed. Either diagram would be correct, as long as the convention used for its construction is known.

The moment diagram is a graph of the internal bending moment in the beam as it varies along the beam length. At any point along the beam, the value of the moment can be found by a summation of the moments on either side of the point, as was described in the preceding section.

It may be seen that the equation for the moment at some point along the beam contains a value of x to the second power. Thus the characteristic form for the moment diagram for a beam with a uniformly distributed load will be a series of segments of a second degree curve, that is, a parabola.

If the beam is discontinuous at its ends, that is, if it is not attached rigidly to its supports, the moment diagram will begin and end with zero values. This indicates that there is no external moment.

For practical purposes it is sometimes useful to use a relationship that exists between the shear and moment diagrams. This relationship may be stated as follows: The change in the internal moment between any two points along the beam length is equal to the area of the shear diagram between the two points. Some of the uses of this relationship are the following:

1. *To find actual values for the moment graph.* Referring to Figure 6.31, it may be observed that the maximum value of the moment at the center of the span will be equal to the area of the triangular portion of the shear diagram between the left end and the middle of the span. It should be noted that the relationship as stated refers to changes in the moment values. In this example the moment at one point is zero, thus the change is also the true value. However, in general the shear areas must be dealt with as algebraic quantities having both magnitude and sense (sign), and they must be added algebraically to the specific moment values that exist at the ends of the beam segment being considered.

2. *To observe the direction of change of the moment diagram.* Where the shear values are positive, the positive values of shear areas will produce positive changes (upward due to our convention for the moment diagram), and where the shear values are negative the changes will be negative. Use of this relationship requires that the convention of beginning from the left end must be used for both the shear and moment graphs.

3. *To establish the peaks on the moment diagram.* Since the shear values indicate moment change, when the sign of the shear changes, the direction of moment change will reverse. Thus the moment graph will arrive at its high values at points that correspond to shifts of sign of the shear.

The deflection diagram is an exaggeration of the actual loaded profile of the beam. The original (unloaded) position of the beam is used as a reference. Using the moment diagram and the loading and support conditions, deflections can be calculated by various means.

6.12 ANALYSIS OF STATICALLY DETERMINATE BEAMS

The two preceding sections have described the behavior of the most ordinary beam—the simple span, uniformly loaded beam. In this section we will discuss and illustrate some additional beams that fall in the category of statically determinate, that is, analyzable with the use of simple static equilibrium considerations alone.

82 ANALYSIS FOR STATIC FORCES

If the end of a single span beam is projected over one of the supports to form an overhang, or a cantilevered end, the beam will be bent into an S shape, as opposed to the single-curved form of the simple span beam. A uniformly loaded beam of this type is shown in Figure 6.32, with the cantilevered distance expressed as a percent a of the beam span L. The cantilever produces a reversal of the sign of the bending, which results in the S shape of the deflected beam. The point at which the curvature of the beam reverses is called the inflection point. This point corresponds to the point on the moment diagram at which the sign of the moment changes. Thus on

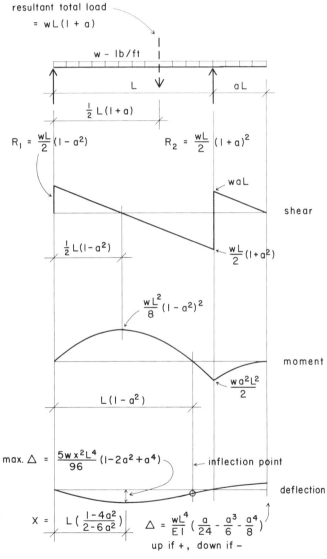

FIGURE 6.32 Critical values for a uniformly loaded beam with one overhanging (cantilevered) end.

one side of the inflection point the beam curves in one direction, and on the other side it curves in the opposite direction.

Critical values for the reactions, shears, moments and deflections are shown on the diagrams in Figure 6.32. The effects of the cantilevering of the beam end are as follows:

1. There is a shift in the symmetry of the shear diagram, with a slight decrease of the reaction and end shear value at the support opposite the cantilevered end. The point at which the shear diagram changes sign moves off center, away from the cantilevered end.
2. The value of the maximum positive moment in the beam span decreases by a factor of $(1 - a^2)$, as shown in the figure.
3. The maximum deflection in the span decreases and its location moves slightly off center of the span, away from the cantilevered end.
4. There is a double effect on the deflection at the cantilevered end. The cantilever effect itself tends to produce a downward deflection. However, the load on the beam span tends to rotate the beam at the support, causing the cantilevered end to deflect upwards. The net deflection is the sum of these too opposed effects.

The diagrams in Figure 6.33 show the behavior of a beam with a uniform load and a single cantilevered end. Critical values are given in Table 6.2 for four different cantilever distances. As the cantilever increases, the following observations can be made.

1. The value of the reaction opposite the cantilever slowly reduces.
2. The value of the maximum positive moment reduces at a rate somewhat more rapid than that for the reaction opposite the cantilever.
3. The value of the maximum negative moment at the cantilevered end increases quite rapidly, eventually exceeding that for the positive moment.
4. The maximum deflection in the beam span reduces quite rapidly, becoming less than one half of that for a simple span for the largest cantilever shown. (See Figure 6.31).
5. The deflection at the cantilevered end is upward until the cantilever distance becomes quite large. The largest value for the cantilever used in the illustration ($a = 0.433L$) corresponds to the condition at which the two opposed effects on the deflection exactly balance each other, producing a net deflection of zero. Further increase of the cantilever, therefore, will produce a net downward deflection.

Figure 6.34 shows a beam with both ends cantilevered over the supports. If both cantilever distances are equal, as shown in the illustration, the beam will behave symmetrically, which somewhat simplifies the analysis. Note that the shear diagram in the center span portion is the same as that for the simple beam, as shown in Figure 6.31. The moment diagram in the center span is also of the same form as that for the simple span, with the reference axis for actual moment values simply being lowered by the value of the cantilever moment in this case.

84 ANALYSIS FOR STATIC FORCES

TABLE 6.2. Beam Behavior Values for Beams with One Cantilevered End[a]

Components of Beam Action[a]	Factors for Values of a Equal to					Multiply Factor by
	0	0.1	0.25	0.333	0.433	
R_1	0.50	0.495	0.469	0.445	0.406	wL
R_2	0.50	0.605	0.781	0.888	1.027	wL
S_1	0.50	0.495	0.469	0.445	0.406	wL
S_2	0.50	0.505	0.531	0.555	0.594	wL
S_3	0	0.1	0.25	0.333	0.433	wL
M_1	0.125	0.123	0.110	0.0990	0.0824	wL^2
M_2	0	0.005	0.031	0.055	0.0937	wL^2
x_1	0.50	0.495	0.469	0.445	0.406	L
x_3	0.50	0.495	0.462	0.417	0.286	L
Δ_1	0.0130	0.0127	0.0101	0.00694	0.00251	$\dfrac{wL^4}{EI}$ [b]
Δ_2	0	0.00399	0.00733	0.00620	0	$\dfrac{wL^4}{EI}$ [b]

[a] See Figure 6.33 for reference.
[b] E and I are normally given in inch units. If w and L are used in ft units, multiply by 1728 to get deflection in inches.

Table 6.3 gives critical values for four symmetrical, uniformly loaded beams with doubly cantilevered ends, as shown in Figure 6.35. The largest cantilever distance shown corresponds to the condition that results in a net deflection of zero at the ends of the cantilevers.

Figure 6.36 shows a beam with a loading condition that occurs quite commonly when beams are used to carry the ends of other beams: The end reactions of the car-

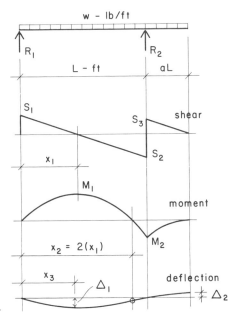

FIGURE 6.33 Reference figure for Table 6.1.

ANALYSIS OF STATICALLY DETERMINATE BEAMS

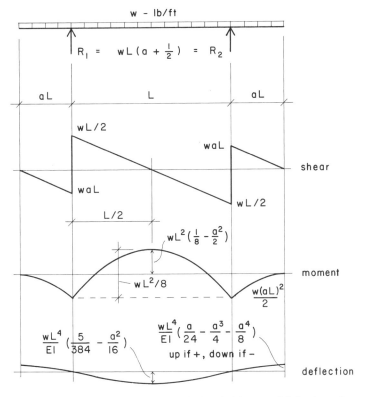

FIGURE 6.34 Critical values for a uniformly loaded beam with both ends cantilevered.

TABLE 6.3. Beam Behavior Values for Doubly Cantilevered Beams[a]

Components of Beam Action[a]	Factors for Values of a Equal to					Multiply Factor by
	0	0.1	0.25	0.333	0.375	
R	0.5	0.6	0.75	0.833	0.875	wL
S_1	0	0.1	0.25	0.333	1.375	wL
S_2	0.5	0.5	0.5	0.5	0.5	wL
M_1	0	0.005	0.031	0.055	0.0703	wL^2
M_2	0.125	0.120	0.094	0.070	0.0547	wL^2
Δ_1	0.0130	0.0124	0.00910	0.00607	0.00421	$\dfrac{wL^4}{EI}$ [b]
Δ_2	0	0.00390	0.00602	0.00311	0	$\dfrac{wL^4}{EI}$ [b]
x	0	0.01	0.066	0.126	0.169	L

[a] See Figure 6.35 for reference.
[b] E and I are normally given in inch units. If w and L are used in ft units, multiply by 1728 to get deflection in inches.

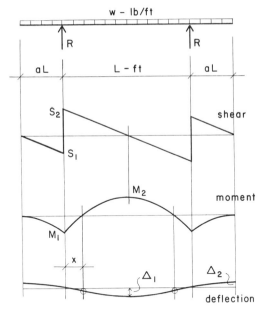

FIGURE 6.35 Reference figure for Table 6.3.

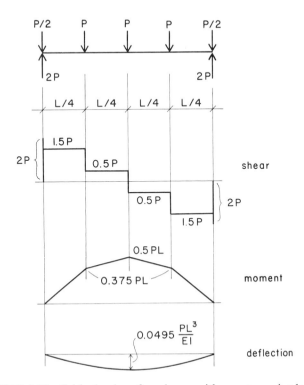

FIGURE 6.36 Critical values for a beam with quarter-point loading.

ried beams become concentrated loads on the carrying beam. If the loads occur at the quarter points of the beam span, the resulting shear, moment, and deflection effects will be as shown in the illustration. Note the characteristic shape of the shear diagram, which becomes a series of rectangular units since the shear value is unchanged between the loads. The moment diagram, instead of being a smooth curve, as with uniform load, becomes a segmented polygon shape.

Figures 6.37 and 6.38 show the conditions for a beam with the ends cantilevered and a loading similar to that for the simple span beam in Figure 3.36. Comparisons may be made of the effects on shifting of the values for reactions and shears and of the reductions of positive moment and deflection in the beam span.

The greater the spacing of concentrated loads on a beam, as a proportion of the beam span, the less similar the behavior of the beam will be to that for a uniform load. As the spacing decreases, the loading eventually becomes essentially uniformly distributed for practical purposes. The illustrations in Figure 6.39 show the effect on the value of the maximum positive moment in a simple span beam with a total load of W that is uniformly distributed or divided into various even units at equal spacings. As may be noted, when the load spacing falls below one tenth of the beam span, the loading may be treated as a uniformly distributed one for all practical purposes.

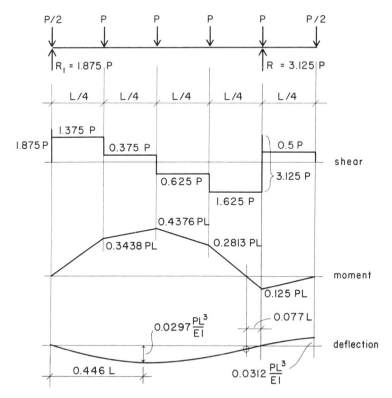

FIGURE 6.37 Critical values for a beam with one cantilevered end and quarter-point loading.

88 ANALYSIS FOR STATIC FORCES

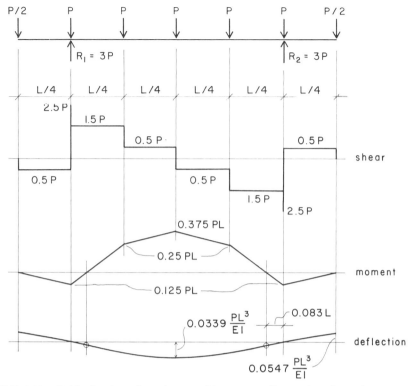

FIGURE 6.38 Critical values for a beam with two cantilevered ends and quarter-point loading.

Tabulated Values for Beam Loadings

Values for reactions, shear, bending moment, and deflection for common beam forms and loadings may be obtained from various handbooks. An extensive set of such tabulations is given in the *AISC Manual* (Ref. 2). Some values for loadings for simple beams are given in this book in Appendix C. For beams of wood and steel, and trusses of simple span, these values may be readily used to avoid extensive computations.

6.13 BEHAVIOR OF INDETERMINATE BEAMS

When beam ends are fixed at their supports or when beams are built as continuous elements through multiple spans, analysis for static equilibrium alone is not sufficient for the determination of the beam behavior. There are various techniques that can be used for the analysis of such beams, but a general discussion of the analysis of indeterminate beams is beyond the scope of this book.

Figure 6.40 shows the conditions that occur when a uniformly loaded beam is continuous through two equal spans. Note that the individual beam spans are

BEHAVIOR OF INDETERMINATE BEAMS   89

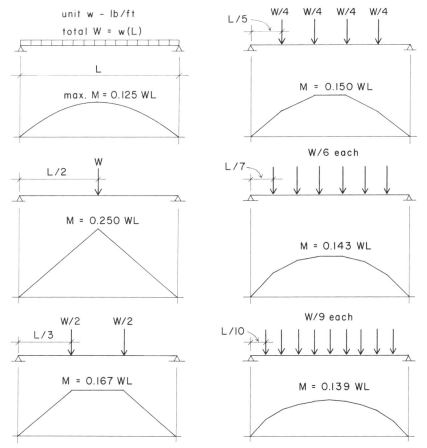

FIGURE 6.39 Effect of load distribution on the value of maximum moment in a simple beam.

similar in their action to a beam with a single cantilevered end, as shown in Figure 6.32. Note that the critical maximum value for moment, $wL^2/8$, is the same as that for a simple span beam, as shown in Figure 6.31, although the moment occurs here as a negative moment at the support rather than as a positive moment at midspan.

When a beam is continuous through more than two spans, there are two different span conditions. The end spans behave in a manner similar to that for the two span beam, or for the beam with a single cantilevered end. Interior spans, however, develop some negative moment at both ends of the span, behaving in a manner similar to that for a beam with two cantilevered ends, as shown in Figure 6.34. The conditions that occur when a uniformly loaded beam is continuous through three and four equal spans are shown in Figures 6.41 and 6.42, respectively. Note that there is some reduction in the maximum value of moment for these beams as compared to that obtained when all spans are simple and there is no continuity of the beams over the supports.

Although continuous beam action may have some advantage in reducing mo-

90 ANALYSIS FOR STATIC FORCES

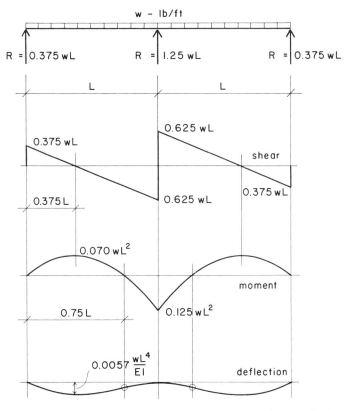

FIGURE 6.40 Critical values for a two-span beam with uniformly distributed loading.

ments, there is often a more significant advantage in the reduction of deflections. The amount of this reduction may be observed by comparing the values for deflection shown in Figures 6.40 through 6.42 with that obtained for the simple beam in Figure 6.31.

Beams With Internal Pins

In the discussion in Section 6.1 it was noted that inflection points in the curved form of the loaded beam correspond to the location of zero moment values on the moment diagram. Because of this phenomenon, it is literally possible to construct a beam with a series of internal joints without moment transfer capability (called pinned joints, or simply pin-joints) and yet have the beam behave as one with full continuity. Figure 6.43 shows a typical three span, continuous beam with its corresponding moment diagram. Shown below the continuous beam is a possibility for construction of the beam with pinned joints in the end spans. The center span element is built as a doubly cantilevered beam, and the end spans each become simple beams, spanning between the end supports and the cantilevered ends of the center span element. Since the pins are located at the points of zero moment in the end spans, the internal conditions of shear, moment, and deflection, and the values of the reactions will be identical to those for the fully continuous beam.

BEHAVIOR OF INDETERMINATE BEAMS

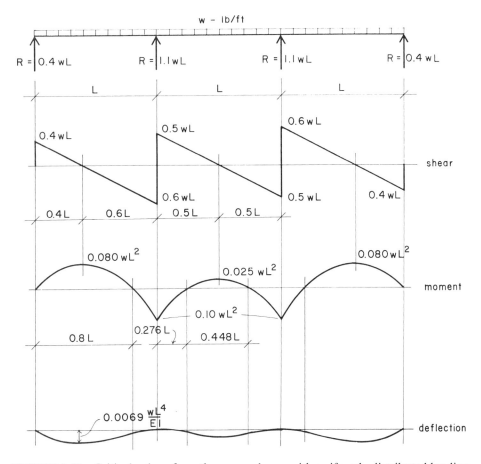

FIGURE 6.41 Critical values for a three-span beam with uniformly distributed loading.

The lower illustration in Figure 6.43 shows a construction for the beam, with the pins located in the center span. Since the pins are located at the points of zero moment in the center span, the effect remains analogous to that of the fully continuous beam.

The use of internal beam joints is often necessary in long rows of beams in both wood and steel structures. Since a pinned joint generally is much easier to construct than a moment transfer joint, the techniques just described are quite useful in such situations. Thus the long beams may be built of relatively short pieces, using simple jointing methods, but obtaining the advantages of reduced moment and deflections possible with continuous beams.

This use of internal pinned joints may also be applied when trusses are made continuous through multiple spans. Figure 6.44 shows a continuous truss with the use of internal pinned joints to reduce the constructed length of elements for the truss. The technique may be used just for construction purposes, or it may be used to make the joints effective as zero moment connections in the completed structure.

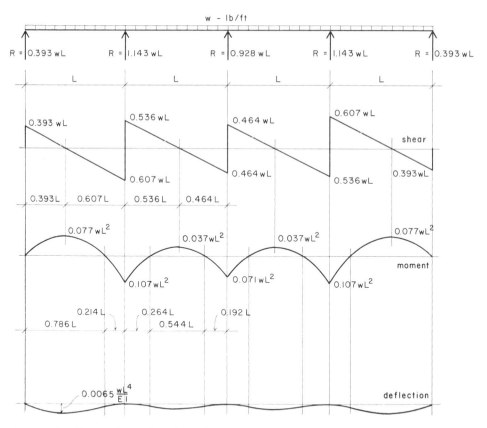

FIGURE 6.42 Critical values for a four-span beam with uniformly distributed loading.

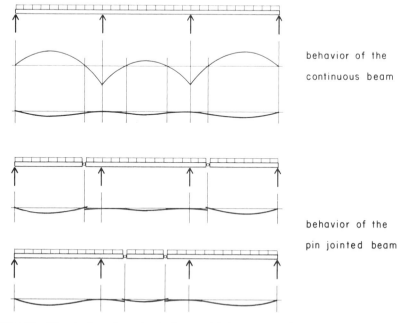

FIGURE 6.43 Simulation of continuity in a multiple-span continuous beam by use of internal pins off the supports.

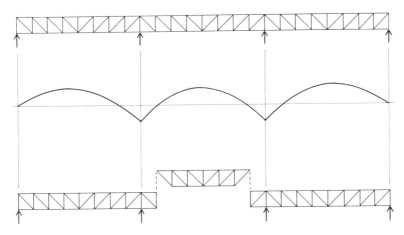

FIGURE 6.44 Continuous truss with internal pins.

6.14 ANALYSIS OF TRUSSES BY THE METHOD OF SECTIONS

The analysis for internal forces in the members of a truss by the method of sections consists of dealing with the truss in a manner similar to that used for beam analysis. The cut section utilized in the explanation of internal forces in a beam, as illustrated in Section 6.1, is utilized here to externalize the forces in the cut members of the truss. The following example will illustrate the technique.

Figure 6.45 shows a simple span, flat chorded truss with a vertical loading on the top chord joints. The Maxwell diagram for this loading and the answers for the internal forces are also shown in the figure. This solution is provided as a reference for comparison with the results that will be obtained by the method of sections.

In Figure 6.46 the truss is shown with a cut plane passing vertically through the third panel. The free body diagram of the portion of the truss to the left of this cut plane is shown at (*a*) in the figure. The internal forces in the three cut members become external forces on this free body diagram, and their values may be found using the following analysis of the static equilibrium of the free body.

At (*b*) we observe the condition for vertical equilibrium. Since *ON* is the only cut member with a vertical force component, it must be used to balance the forces, resulting in the value for ON_v of 500 lb acting downward. We may then establish the value for the horizontal component and the actual force in *ON*.

We next consider a moment equilibrium condition, picking a point for moment reference that will eliminate all but one of the unknown forces. If we select the top chord joint as shown at (*c*) in the figure, both the force in the top chord and the force in the diagonal member *ON* will be eliminated. The only remaining unknown force is then that in the bottom chord and the summation is as follows.

$$\begin{aligned} M_1 = 0 = & + (3000)(24) = + 72{,}000 \\ & - (500)(24) = - 12{,}000 \\ & - (1000)(12) = - 12{,}000 \\ & - (NI)(10) = - 10(NI) \end{aligned}$$

94 ANALYSIS FOR STATIC FORCES

Thus

$$10(NI) = +72{,}000 - 12{,}000 - 12{,}000 = +48{,}000$$

$$NI = \frac{48{,}000}{10} = 4800 \text{ lb}$$

Note that the sense of the force in NI was assumed to be tension, and the sign used for NI in the moment summation was based on this assumption.

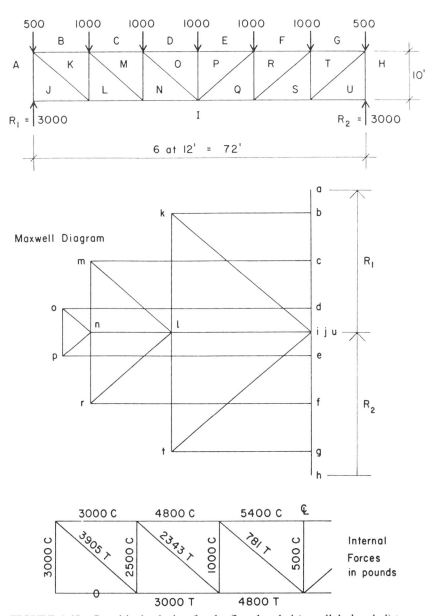

FIGURE 6.45 Graphical solution for the flat-chorded (parallel-chorded) truss.

ANALYSIS OF TRUSSES BY THE METHOD OF SECTIONS   95

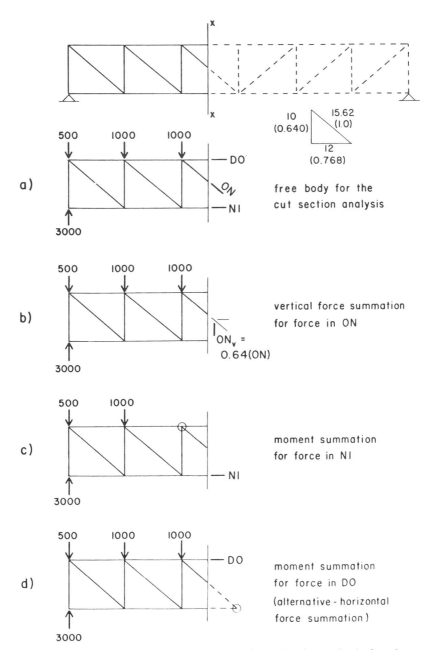

FIGURE 6.46 Investigation of the flat-chorded truss by the method of sections.

One way to find the force in the top chord would be to do a summation of horizontal forces, since the horizontal component of *ON* and the force in *NI* are now known. An alternative method would be to use another moment summation, this time selecting the bottom chord joint shown at (*d*) in order to eliminate *IN* and *ON* from the summation.

$$M_2 = 0 = +(3000)(36) = +108{,}000$$
$$-(500)(36) = -18{,}000$$
$$-(1000)(24) = -24{,}000$$
$$-(1000)(12) = -12{,}000$$
$$-(DO)(10) = -10(DO)$$

Thus

$$10(DO) = +54{,}000$$
$$DO = \frac{54{,}000}{10} = 5400 \text{ lb}$$

The forces in all of the diagonal and horizontal members in the truss may be found by cutting the truss with a series of vertical planes and doing static equilibrium analyses similar to that just illustrated. In order to find the forces in the vertical members of the truss, it is possible to cut the truss with an angled plane, as shown in Figure 6.47. A summation of vertical forces on this free body will yield the internal force of 1500 lb in compression in member *MN*.

The method of sections is sometimes useful when it is desired to find the forces in some members of the truss without performing a complete analysis for the forces in all members. By the method of joints it is necessary to work from one end of a truss all the way to the desired location, while the single cut plane may be used anywhere on the truss.

6.15 ANALYSIS OF TRUSSES BY THE BEAM ANALOGY METHOD

The method of sections may be used on any truss, but it is particularly effective in the investigation of flat, parallel chorded trusses. A special version of the method of sections is the beam analogy method. Figure 6.48 shows the truss that was analyzed in the previous section. Below the truss are shown the shear and moment diagrams as they would be constructed for a beam with the same loading. From this illustration, we make the following observations.

1. Values on the shear diagram indicate the net internal force required in the truss at points between the joints. Since the diagonal members are the only ones with vertical force components, the shear values indicate the vertical

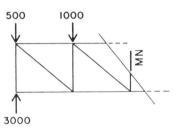

FIGURE 6.47 Cut-section investigation for the internal forces in the vertical web members of the flat-chorded truss.

ANALYSIS OF TRUSSES BY THE BEAM ANALOGY METHOD 97

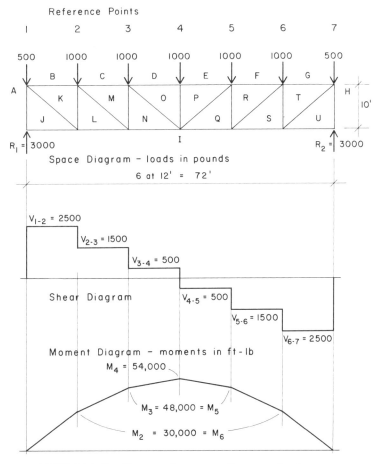

FIGURE 6.48 Shear and moment diagrams for the truss.

force components in the diagonals in the corresponding panels of the truss. The vertical force summation illustrated in (a) of Figure 6.46 may be seen to be the same as the result observed from the shear diagram.

2. Shear diagram values may also be used to obtain the forces in the vertical members of the truss. Referring to the free body obtained by cutting the angled plane in Figure 6.47, we may observe that the force in member MN is the same as the shear value in the adjacent panel to the left of the member.

3. Forces in the chords may be found by dividing the values on the moment diagram by the height of the truss. Referring to the free body shown at (c) in Figure 6.46, we note that the net moment of 48,000 lb-ft found from the summation is the same as the value shown on the moment diagram at line 3 in Figure 6.48. Similarly, the value of 54,000 lb-ft obtained from the free body in (d) of Figure 6.46 is the same as that shown at line 4 on the moment diagram.

Using the relationships just observed, we may do a complete analysis of the forces in the members of the truss using the values from the shear and moment diagrams for the analogous beam. The procedure for this analysis is summarized in Table 6.4.

TABLE 6.4. Analysis of a Truss by the Beam Analogy[a]

Truss Members	Reference Value	Operation and Factor	Force in Member (lb)
Verticals			
AJ	$R_1 = 3,000$ lb	$\times\ 1$	$3,000C$
KL	$V_{1-2} = 2,500$ lb	$\times\ 1$	$2,500C$
MN	$V_{2-3} = 1,500$ lb	$\times\ 1$	$1,500C$
OP	load $= 1,000$ lb	$\times\ 1$	$1,000C$
Diagonals			
JK	$V_{1-2} = 2,500$ lb	$\times\ 1.562$	$3,905T$
LM	$V_{2-3} = 1,500$ lb	$\times\ 1.562$	$2,343T$
NO	$V_{3-4} = 500$ lb	$\times\ 1.562$	$781T$
Top Chords			
BK	$M_2 = 30,000$ lb-ft	$\div\ 10$	$3,000C$
CM	$M_3 = 48,000$ lb-ft	$\div\ 10$	$4,800C$
DO	$M_4 = 54,000$ lb-ft	$\div\ 10$	$5,400C$
Bottom Chords			
JI	$M_1 = 0$		0
LI	$M_2 = 30,000$ lb-ft	$\div\ 10$	$3,000T$
NI	$M_3 = 48,000$ lb-ft	$\div\ 10$	$4,800T$

[a] See Figure 6.48.

7

ANALYSIS OF PLANAR TRUSSES: GENERAL CONCERNS

Considerations for the application of the principles of mechanics to the investigation of truss behaviors are presented in Chapter 6. This chapter deals with the general concerns for the complete process of investigation (analysis) of trusses of planar form as used for buildings. The uses of the data obtained from investigations for applications to the general process of design of trussed construction is discussed in Chapters 9 and 10.

7.1 LOADS ON TRUSSES

The principal sources and types of loads on trusses are the following:

1. *Gravity Dead Loads.* These are the permanent loads on the truss, caused by the weight of the building construction including that of the truss itself.
2. *Gravity Live Loads.* Although live load technically refers to any load other than gravity dead load, the term ordinarily is used to refer only to gravity loads specified as superimposed design loads by building codes. Thus roofs are required to sustain an added load to simulate the effects of snow, ice, and construction activities, and floors are required to sustain a load to simulate the effects of movable furniture, pedestrians, vehicular traffic, and so on.
3. *Wind Loads.* Wind and earthquake (seismic) effects are the principal sources of what are described as lateral loads on a building. Lateral—meaning sideways—refers to effects having a direction at right angles to that of gravity; thus they are characterized as producing unstabilizing effects, tending to topple the building sideways. Wind also exerts a direct pressure on the exterior surfaces of a building, both as an inward and an outward (suction) effect.
4. *Seismic Loads.* Forces developed by ground motions are actually a result of the gravity weight of the building, which becomes a mass in motion resisting

changes in its state of motion. This is similar to the jolting effect on a car and its occupants caused by rapid acceleration or braking.
5. *Temperature Change.* Most materials tend to expand when heated and to contract when cooled. If long continuous elements of a building's construction (such as the chords of a truss) are subjected to a considerable range of temperature, they must be allowed to change length freely; otherwise potentially damaging stresses may be developed in either the restrained structural elements or whatever is restraining them. With regard to this behavior, a major concern is the joint between a truss and its support.

7.2 TRUSS REACTIONS

Once the loading conditions for a truss have been determined, and the specific load amounts and their disposition on the truss established, the next step in the analysis procedure is usually that of finding the forces generated by the truss supports. These forces are called the reactions for the truss. The combination of the loads and the reactions constitutes the entire set of external effects on the structure.

The reactions must develop the necessary vertical and horizontal forces to maintain the external equilibrium of the truss. When a spanning truss is subjected only to gravity loads, the reactions will be limited to vertical forces. When a truss must also resist wind or seismic effects, the reactions must develop horizontal forces as well as vertical forces.

A special problem for the truss supports is that of the necessity of allowing for some length change in the truss. One type of length change is that caused by temperature fluctuation, as described in the preceding section. Another type of length change is that caused by the development of stresses in the top and bottom chords as the truss is loaded. The development of tension stress in the bottom chord results in a stretching of the chord, which requires that the supports permit some outward movement if the truss is supported at its ends. With flat, parallel chord trusses, the support sometimes occurs at the ends of the top chord, in which case the direction of the effect is reversed because of the compression in the top chord. These effects must be considered in the development of the supporting structures and the details of the connections between the truss and its supports.

7.3 STABILITY AND DETERMINACY

A large percentage of the trusses used in buildings are of ordinary form and are used in a limited number of ordinary situations. The basic device of trussing—that is, triangulating a framework—may be used, however, to produce a great range of possible structures. When truss forms are complex or unusual, a basic determination that must be made early in the design is that of the condition of the particular truss configuration with regard to its stability and determinacy. The entire feasibility of the design may hinge on this determination.

Stability is an inherent quality generally having to do with the nature of arrangement of members and joints or with the support conditions. An essential feature of a truss is the need for complete triangulation of the framework. In the truss shown at (*a*) in Figure 7.1, there is a lack of triangulation in the center panel. It is relatively

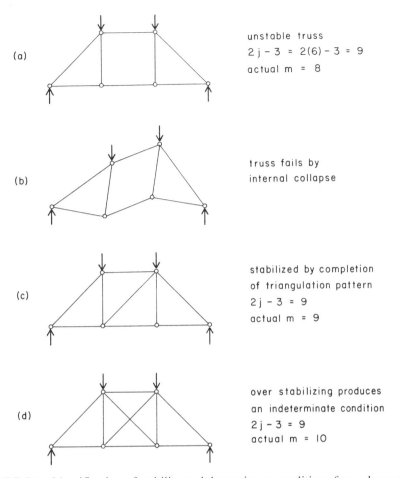

FIGURE 7.1 Identification of stability and determinacy conditions for a planar truss.

easy to visualize that this structure is inherently unstable and that load applied to the structure will cause the center rectangular panel to distort, one form of failure being that shown at (b) in the illustration. Addition of the single diagonal member, as shown at (c) in Figure 7.1, will complete the internal triangulation of the truss and develop a stable structure. Since one diagonal in the center panel is sufficient, if two are used, as shown at (d) in Figure 7.1, the structure will be overstabilized.

Usually it is quite easy to determine the potential stability condition for a truss by simple visual inspection of the truss pattern. If all units of the truss consist of triangles (no rectangles or other polygons), the truss is usually stable. If X forms exist in the truss, it is usually statically indeterminate due to an excess of members.

Where some doubt exists about the stability condition for a particular truss, a simple formula that can be used to analyze the condition is

$$m = 2j - 3$$

in which m is the number of truss members and j is the number of truss joints.

If the actual number of members is less than $2j - 3$, the truss is unstable; if the

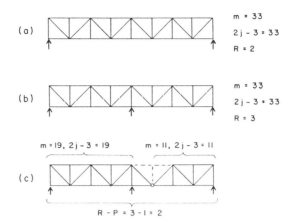

FIGURE 7.2 Effects of external supports and internal pins on the stability and determinacy of a truss.

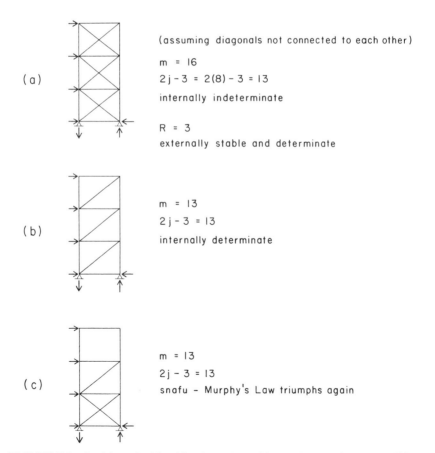

FIGURE 7.3 Problems in identification of stability and determinacy conditions.

same it is stable; if more, it is most likely stable but indeterminate. These conditions are illustrated in Figure 7.1.

The support conditions and the existence of internal pins (as discussed in Section 2.5) also affect the stability conditions for a truss. When the loads and support reactions constitute a parallel force system, as shown in Figure 7.2, the truss will be stable and determinate when there are just two support reactions. Less and the truss will be unstable; more and it will be indeterminate. If internal pins are used, the key number remains 2, but is determined as

$$R - P = 2$$

in which R is the number of support reactions and P the number of internal pins.

Note that when internal pins are used, the internal stability condition is analyzed separately for the truss units between the pins and the end supports.

When the loads and support reactions on a truss constitute a general planar force system, as shown for the tower in Figure 7.3, the external stability formula becomes

$$R - P = 3$$

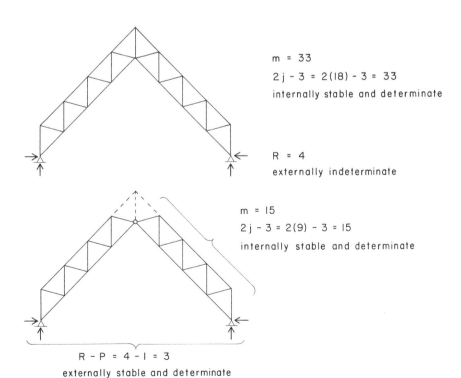

FIGURE 7.4 Effect of an internal pin on a single-span bent. Upper: the two-hinged (two-pinned) bent. Lower: the three-hinged bent. (See also discussion in Section 12.6.)

It is possible for a truss to be internally indeterminate but externally determinate, or vice versa. In Figure 7.3 the x-braced tower is internally indeterminate, with three redundant members. On the other hand, the external support reactions are determinate. If three of the diagonals are removed, as shown at (b), the truss is made determinate. However, if the remaining members are disposed as shown at (c), the member count is correct by the formula, but the truss is obviously unstable in the top panel and indeterminate in the bottom panel, which simply demonstrates that the formula is not a magic one. Thus, whether the formula is used or not, one should look for lack of triangulation and X forms in the pattern.

Another example of the general planar force system is shown in Figure 7.4. In the upper figure the structure is shown to be indeterminate due to an excess of support reaction components. Use of the internal pin in the lower figure is a means for reduction of the indeterminacy, although the number of reaction components remains the same. Note that, as with the flat spanning truss in Figure 7.2, the internal stability is determined independently for units of the truss between the ends and the internal pins.

7.4 ANALYSIS FOR GRAVITY LOADS

Gravity loads on trusses consist of the dead weight of the truss plus the weight of other elements of the building construction that are supported directly by the truss. In addition to this real gravity load, the building design requires some additional gravity load to simulate the conditions of usage. In the structure shown in Figure 7.5, the trusses support directly a series of joists that are placed at the panel joints along the upper chord. These joists in turn support a series of deck elements that constitute the actual roof surface. Determination of the gravity loading for this structure proceeds as follows:

1. The deck is loaded with a uniformly distributed surface loading consisting of the weight of the roof construction and the applied, superimposed design loading required for the roof. The deck itself is typically designed as a series of strips with the strip load applied as a linear load on the deck.
2. The deck delivers a uniformly distributed linear load along the length of the joist. Added to this is the weight of the joist, which is similarly distributed.
3. The joists deliver a series of concentrated loads to the top of the truss. Added to this is the weight of the truss, which is more or less uniformly distributed along the truss span.

There are a number of factors that influence the weight of a truss. The major ones are

1. The material of the truss members.
2. The types of device used for connection. When gusset plates and bolts are used for steel trusses, the plates, bolts, nuts, and washers may constitute as much as 20% of the total truss weight. Welding of joints virtually eliminates this weight.
3. The magnitude of the loads. When the construction is heavy or required live

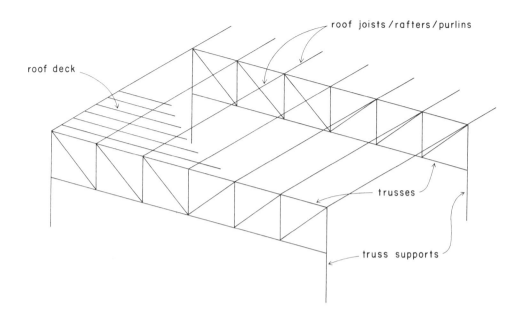

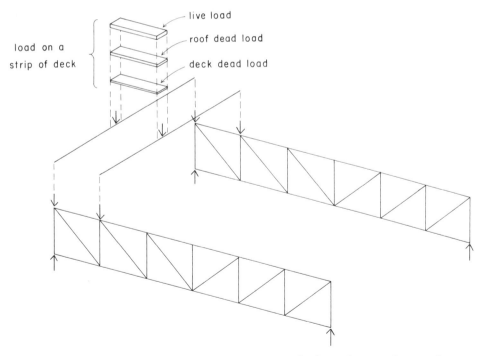

FIGURE 7.5 Gravity load effects on the components of a framed system for a roof structure.

load is high, the truss must work harder and will require more material to reduce stresses.
4. The spacing of trusses. The load accumulated on each truss will be determined by the truss spacing; double the spacing between trusses and the applied loads will double.
5. The span of the truss.

Estimation of truss weight is best done on the basis of observation of actual trusses previously designed for similar conditions. Where a raw first guess must be made, the following formula may be used.

$$w = \text{spacing}\,(0.078\,\sqrt{L}\,)$$

in which

w = the average weight of the truss per ft² of area supported (spacing times span)
spacing is in inches
L is the truss span in feet

This formula provides very approximate values due to the many variables. These include the need for intermediate framing (purlins) when the truss spacing exceeds that which is feasible for available decking. The formula is limited to trusses of light construction (mostly welded steel or light wood systems) generally used for roof structures of spans in a medium range (30–150 ft, or so).

Data for approximate truss weights, derived by use of the preceding formula, is presented in Table 7.1. The truss weight values in the table are given in units of pounds per foot of the truss span, and include the weights of light purlins that may be used for wider truss spacings. The combined loading of 40 psf (DL + LL) is an average for a low slope roof of light construction with no heavy snow pack and no suspended ceiling construction. Weights should be increased accordingly for heavier snow loads, heavier roof construction, or where some suspended loads (ceilings or equipment) are supported by the trusses.

Although the truss weight is actually distributed along the truss span, the usual practice is to consider units of weight as collected at the truss joints, since this is the only type of loading for which the truss can be analyzed directly. For short to medium span trusses (under 200 ft or so in span), the truss weight ordinarily will not be a major part of the total design load. Thus a minor error in the assumption of the truss weight ordinarily will have virtually no effect on design results.

Figure 7.6 illustrates the basis for determination of the loading for the basic components of the structure shown in Figure 7.5. The actual load determination begins with the load on the smallest component (the deck) and proceeds with the transfer of loading to other supporting elements until the forces are finally delivered to the end of the line. In most cases the end of the line is the ground, so that the final load determination is that of the load on the foundations, although our illustration ends with the truss supports in this example.

Although the truss must sustain the total gravity load—both dead and live—the

TABLE 7.1. Average Weight for Short to Medium Span Roof Trusses (Pounds per Feet of Span)[a]

Span (ft) \ Spacing (ft)	2	3	4	6	8	10	15
30	10	15	20	30	(Spacing too wide for span)		
35	11	17	22	34	44		
40	12	18	24	36	48	60	
50	13	20	26	40	53	66	100
60		22	29	44	58	73	109
80			34	50	68	84	126
100			38	56	76	94	141
120	(Spacing too close for span)			62	82	103	154
150					92	115	172

[a] Superimposed load of 40 psf (total live load plus dead load) on roof, including all dead load except the trusses.

two loadings usually are tabulated separately since their individual effects often must be used in the truss design. The following example will illustrate the procedure for determination of the gravity loads on a typical roof truss.

Figure 7.7 shows the truss structure previously illustrated and provides data from which the load tabulation can be found. Referring to Figures 7.6 and 7.7, we proceed as follows.

The Deck

Using the 1 ft wide design "strip," the loading on the deck becomes the per ft^2 load of the roof construction. Thus

$$LL = 20 \text{ lb/ft}$$
$$DL = 9 \text{ lb/ft}$$

This load is applied as a linear distributed load on the 8 ft span of the deck.

108 ANALYSIS OF PLANAR TRUSSES: GENERAL CONCERNS

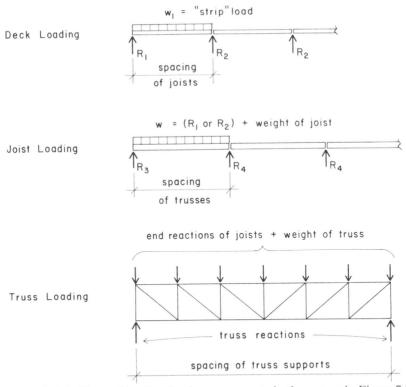

FIGURE 7.6 Form of loading for the components in the system in Figure 7.5.

The Joist

The end support reactions for the deck are applied loads on the joists. There are two different joist loads—that for the end joists on the edge of the structure, and that for the interior joists that support deck from both sides. To this applied load must be added the weight of the joist, which is an estimate until the joist itself is selected. We will assume a weight of 20 lb/ft for the joist. The joist loads are thus as follows:

At the end joists,

$$LL = 4(20) = 80 \text{ lb/ft}$$
$$DL = 4(9) + 20 = 36 + 20 = 56 \text{ lb/ft}$$

At the interior joists,

$$LL = 8(20) = 160 \text{ lb/ft}$$
$$DL = 8(9) + 20 = 72 + 20 = 92 \text{ lb/ft}$$

The Truss

The support reactions for the joists become concentrated loads at the upper chord joints of the truss. If the trusses occur at the ends of the building, there will be two dif-

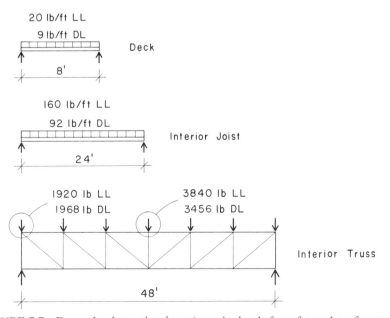

FIGURE 7.7 Example: determination of gravity loads for a framed roof system.

ferent truss loadings, as discussed for the joists. We can determine the loads for an interior truss as follows.

At the end joints,

$$LL = (24)(80) = 1920 \text{ lb}$$
$$DL = (24)(56) = 1344 \text{ lb}$$

At interior joints,

$$LL = (24)(160) = 3840 \text{ lb}$$
$$DL = (24)(92) = 2208 \text{ lb}$$

This is the *applied*, or *superimposed*, load on the truss. The weight of the truss must be added to the dead load. Using the formula previously given, we estimate the weight to be as follows.

$$w = (\text{spacing})(0.078 \sqrt{L})$$
$$= (24 \times 12)(0.078 \sqrt{48})$$
$$= 156 \text{ lb/ft}$$

This is an average weight for the truss and its use as a uniformly distributed load is only an approximation. For the loading to be used for the truss analysis the truss weight is determined as a top chord joint loading on a peripheral basis by multiplying the average weight by the appropriate length along the truss span as follows.

At the end joints.

$$(4)(156) = 624 \text{ lb}$$

At interior joints,

$$(8)(156) = 1248 \text{ lb}$$

Thus the total design loads for the truss, as shown in Figure 7.7, are as follows.

At the end joints,

$$LL = 1920 \text{ lb}$$
$$DL = 1344 + 624 = 1968 \text{ lb}$$

At interior joints,

$$LL = 3840 \text{ lb}$$
$$DL = 2208 + 1248 = 3456 \text{ lb}$$

Required design live loads for roofs are specified by local building codes. Where snow is a potential problem, the load usually is based on anticipated snow accumulation. Otherwise the specified load essentially is intended to provide some capacity for sustaining of loads experienced during construction and maintenance of the roof. The basic required load usually can be modified when the roof slope is of some significant angle and on the basis of the total roof surface area that is supported by the structure. Table 14.2 gives the minimum roof live loads specified by the *Uniform Building Code*, 1991 ed. (Ref. 1), which are based on the situation where snow load is not the critical concern.

7.5 ANALYSIS FOR WIND LOADS

Wind forces produce a number of possible loading conditions on building structures. Specific effects depend on the wind velocity, wind direction, and the building's exterior form and dimensions. As shown in Figure 7.8, the wind ordinarily induces a direct, inward pressure on surfaces of the building facing the wind, and an outward suction pressure on surfaces on sides opposite to the wind. Changing direction of the wind may reverse these effects, requiring most surfaces to be investigated for more than one effect.

Pressures on roof surfaces depend partly on the slope of the surface from the horizontal, as shown in Figure 7.9. Flat and near-flat surfaces tend to have upward, suction pressures, which may actually lift the roof structure or rip off surfacing materials. As the slope increases, the roof surfaces facing the wind develop inward pressures, approaching those on vertical surfaces.

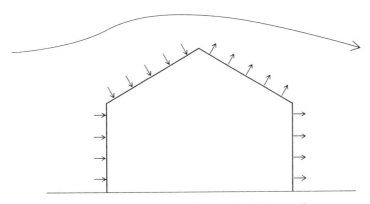

FIGURE 7.8 Wind pressure effects on building surfaces.

112 ANALYSIS OF PLANAR TRUSSES: GENERAL CONCERNS

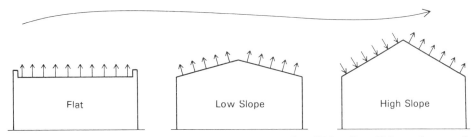

FIGURE 7.9 Wind effects on building roofs. (*a*) Flat roof; uplift loading. (*b*) Low-slope roof; all outward (uplift) loading. (*c*) High-slope roof; inward loading on the windward side slope and outward loading on the leeward side slope.

Application of current wind design criteria is illustrated in the design examples in Chapters 9 and 10. For an explanation of the general problems of designing building structures for wind effects, the reader is referred to *Simplified Building Design for Wind and Earthquake Forces* (Ref. 11).

Because of the increased allowable stresses (or reduced load factors) generally used when wind effects are considered, design of roof structures are often little effected by wind, except in areas of critical windstorm conditions. One concern that frequently occurs, however, is that for the potential of complete uplift (blowing off the roof), due to the light construction dead loads often associated with trussed roof structures. The latter generally requires careful consideration of the anchorage of the roof structure at its supports.

Trussed towers are often strongly influenced by lateral force effects, involving both the internal forces in the members and the overturning effect (moment at the base). For very slender towers, the lateral forces may be the predominant design concerns for the vertical edge members and the tower foundations.

An explanation of current building code requirements for design for wind loads is given in Section 14.4.

7.6 DESIGN FORCES FOR TRUSS MEMBERS

The primary concern in analysis of trusses is the determination of the critical forces for which each member of the truss must be designed. The first step in this process is the decision about which combinations of loading must be considered. In some cases the potential combinations may be quite numerous. Where both wind and seismic actions are potentially critical, and more than one type of live loading occurs (e.g., roof loads plus hanging loads), the theoretically possible combinations of loadings can be overwhelming. However, designers are usually able to exercise judgment in reducing the sensible combinations to a reasonable number. For example, it is statistically improbable that a violent windstorm will occur simultaneously with a major earthquake shock.

Once the required design loading conditions are established, the usual procedure is to perform separate analyses for each of the loadings. The values obtained

DESIGN FORCES FOR TRUSS MEMBERS   113

can then be combined at will for each individual member to ascertain the particular combination that establishes the critical result for the member. This means that in some cases certain members will be designed for one combination and others for different combinations.

In most cases design codes permit an increase in allowable stress for design of members when the critical loading includes forces due to wind or seismic loads. For wood trusses the codes also permit an increase in allowable stress for roof live loads. On the other hand, when the load is permanent (all dead load) the codes require a *decrease* of 10% for wood structures. These factors must be taken into account when considering load combinations. One procedure is to use adjusted values for the various combinations, as illustrated in the following example.

Table 7.2 illustrates a process for summarizing the results of load analysis on a truss. In this case the three basic design loads are dead load, live load, and wind load. Four loadings are produced since a separate analysis must be done for the wind from the two opposite directions, referred to as wind left and wind right. The results of the load analyses for these four conditions are shown for three members of the truss, with force magnitudes given in pounds and sense indicated using plus for tension and minus for compression.

The four load combinations considered in this example are dead load alone, dead load plus live load, dead load plus wind left, and dead load plus wind right.

TABLE 7.2. Examples of Combination of Load Analyses for Critical Design Values

	Truss Members		
	AB	BC	CD
Loadings			
Dead load only	+2478	−1862	+3847
Live load	+3684	−2768	+5719
Wind left	−2862	+894	−2643
Wind right	+2074	+1046	+3427
Combinations			
DL only at $\frac{1}{0.9} = 1.11$	+2751	−2067	+4270
DL + LL at $\frac{1}{1.25} = 0.80$	+4930	−3704	+7653
DL + WL$_L$ at $\frac{L}{1.33} = 0.75$	−288	−726	+903
DL + WL$_R$ at $\frac{R}{1.33} = 0.75$	+3414	−2181	+5456
Design Forces			
Maximum force	+4930	−3704	+7653
Reversal force	−288	—	—

114 ANALYSIS OF PLANAR TRUSSES: GENERAL CONCERNS

These are typical design combinations, but may not be the only ones required in all cases. Some codes require that some or all of the live load be included with the wind load. In this example we assume the truss to be of wood and the structure to be a roof. Thus the following adjustments must be made for all of the combinations.

> *For Dead Load Only.* Allowable stresses must be reduced by 10%. We thus adjust the load by a factor of $\frac{1}{0.9} = 1.11$. With this adjusted load the member can be designed for the full allowable stress, since the reduction has already been performed.
>
> *For Dead Load Plus Live Load.* Allowable stresses may be increased by 25%, assuming the roof live load to be not more than 7 days in duration. (See Ref. 9.) We thus adjust the load by a factor of $\frac{1}{1.25} = 0.8$ to obtain the adjusted load that may be used with the full allowable stresses.
>
> *For Dead Load Plus Wind Load.* Allowable stresses may be increased by 33% and the adjustment factor becomes $\frac{1}{1.33} = 0.75$.

With the four combinations determined, together with their adjustments, we next scan the list of combinations for the critical design values to be used for the actual design of the members. Of first concern is simply the largest number, which is the maximum force in the member. However, of possible equal concern, or in some situations even greater concern, is the case of a reversal of sign in some combination. In Table 7.2 it may be observed that the maximum force in member AB is 4930 lb in tension. However, the combination of dead load plus wind left produces a compression force, albeit of small magnitude. If the member is long, it is possible that the slenderness limitations for compression members may prove to be more critical in the selection of the member, even though the tension force is much larger.

7.7 COMBINED STRESS IN TRUSS MEMBERS

When analyzing trusses the usual procedure is to assume that the loads will be applied to the truss joints. This results in the members themselves being loaded only through the joints and thus having only direct tension or compression forces. In some cases, however, truss members may be directly loaded, as when the top chord of a truss supports a roof deck without benefit of joists. Thus the chord member is directly loaded with a linear uniform load and functions as a beam between its end joints.

The usual procedure in these situations is to accumulate the loads at the truss joints and analyze the truss as a whole for the typical joint loading arrangement. The truss members that sustain the direct loading are then designed for the combined effects of the axial force caused by the truss action and the bending caused by the direct loading.

Figure 7.10 shows a typical roof truss in which the actual loading consists of the roof load distributed continuously along the top chords and a ceiling loading distributed continuously along the bottom chords. The top chords are thus designed for a combination of axial compression and bending and the bottom chords for a

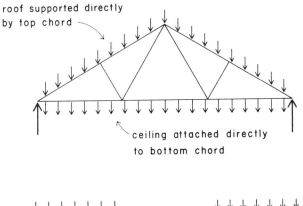

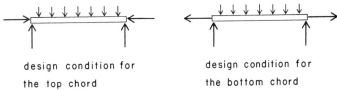

FIGURE 7.10 Effects of loads applied directly to truss chords.

combination of axial tension plus bending. This will of course result in somewhat larger members being required for both chords, and any estimate of the truss weight should account for this anticipated additional requirement.

8
GENERAL CONSIDERATIONS FOR TRUSSED CONSTRUCTION

This chapter discusses the basic uses of trussing for building structures and some general considerations for design of trussed construction.

8.1 PLANAR SPANNING TRUSSES

A common use of trussing is for achieving horizontal spans. This section considers the typical situation of the single, planar truss as a horizontally spanning element.

Elements of Spanning Trusses

Figure 8.1 shows a typical short-span roof truss, used for the dual tasks of providing for gable-form, pitched roof surfaces and the direct support of a flat, horizontal ceiling surface. The common elements of such a truss, as shown in Figure 8.1, are the following:

Chord Members. These are the top and bottom boundary members of the truss, analogous to the top and bottom flanges of a steel beam. For trusses of modest size these members are often made of a single element that is continuous through several joints, with a total length limited only by the maximum ordinarily obtainable for the element selected.

Web Members. The interior members of the truss are called web members. Unless there are interior joints, these members are of a single piece between joints.

Panels. Most trusses have a pattern that consists of some repetitive, modular unit. This unit ordinarily is referred to as the panel of the truss, and joints sometimes are referred to as panel points.

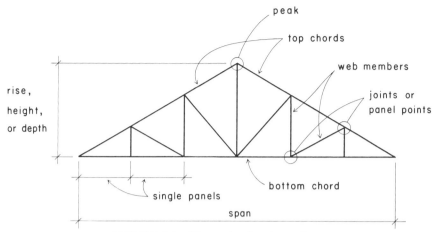

FIGURE 8.1 Elements of a planar truss.

A critical dimension of a truss is its overall height, which is sometimes referred to as its rise or its depth. For the truss illustrated, this dimension relates to the establishment of the roof pitch and also determines the length of the web members. A critical concern with regard to the efficiency of the truss as a spanning structure is the ratio of the span of the truss to its height. Although beams and joists may be functional with span/height ratios as high as 20 to 30, trusses generally require much lower ratios.

Structural Systems with Trusses

Trusses may be utilized in a number of ways as part of the total structural system for a building. Figure 8.2 shows a series of single-span, planar trusses, of the form shown in Figure 8.1, together with the other elements of the building structure that develop the roof system and provide support for the trusses. In this example the trusses are spaced a considerable distance apart. In this situation it is common to use purlins to span between the trusses, supported at the top chord joints of the trusses in order to avoid bending in the chords. The purlins in turn support a series of closely spaced rafters that are parallel to the trusses. The roof deck is then attached to the rafters so that the roof surface actually floats above the level of the top of the trusses.

Figure 8.3 shows a similar structural system that utilizes trusses with parallel chords. This system may be used for a floor or a flat roof.

When the trusses are slightly closer together, it may be more practical to eliminate the purlins and to slightly increase the size of the top chords to accommodate the additional bending due to the rafters. Extending this idea, if the trusses are really close, it may be possible to eliminate the rafters as well, and to place the deck directly on the top chords of the trusses.

For various situations additional elements may be required for the complete structural system. If a ceiling is required, another framing system will be used at the level of the bottom chords or suspended some distance below it. If the roof and ceiling framing do not provide it adequately, it may be necessary to use some bracing system perpendicular to the trusses, as is discussed in Section 8.4.

118 GENERAL CONSIDERATIONS FOR TRUSSED CONSTRUCTION

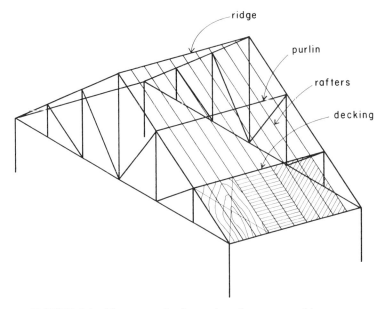

FIGURE 8.2 Elements of a framed roof structure with trusses.

8.2 TRUSS FORMS

The two primary concerns for form of trusses are for the general profile of the truss—as defined by its top and bottom chords—and for the pattern of arrangement of the truss members. Figure 8.4 shows a number of common truss profiles for spanning trusses. These profiles derive in many cases from the need for particular building forms, but also may relate to some basic functions of the trusses themselves.

Truss patterns are derived from a number of considerations, starting with the basic profile of the truss. For various reasons a number of classic truss patterns have evolved and have become standard parts of our structural vocabulary. Some of

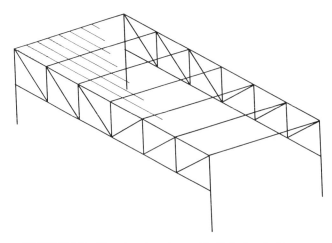

FIGURE 8.3 Structural system with flat-chorded trusses.

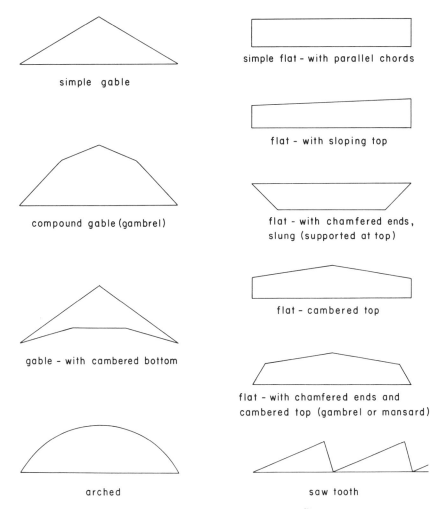

FIGURE 8.4 Common truss profiles.

these carry the names of the designers who first developed them. Several of these common truss forms are shown in Figure 8.5.

Trussing—that is, triangulated arrangement of framing—may be utilized in a number of ways, producing trussed towers, trussed arches, trussed rigid bents, and so on. A few of these special trussed structures are shown in Figure 8.6.

Although many of the classic, regular truss patterns are used widely, trusses are actually highly adaptable to form variations, and are especially advantageous in this regard. Figure 8.7a illustrates several ways in which a simple spanning truss can be extended beyond its supports to provide cantilevers. Although all of these variations utilize the same basic structure, the exterior architectural forms produced are dramatically different.

In a similar manner, Figure 8.7b illustrates how exterior variations can be achieved with a gable-form truss.

Derivation of logical truss patterns and the establishment of the module for panels of the truss may relate to various concerns. In the trusses shown in Figure 8.2, the

120 GENERAL CONSIDERATIONS FOR TRUSSED CONSTRUCTION

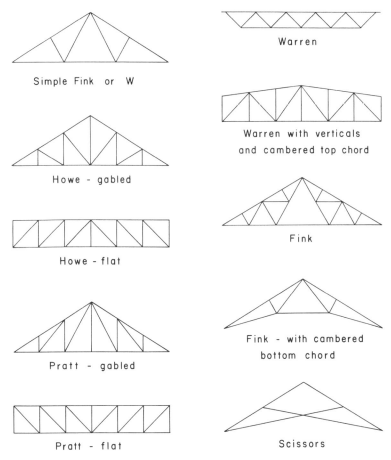

FIGURE 8.5 Truss patterns.

panel length of the trusses relates to the span of the rafters and to the load generated on the purlins. These relationships must be coordinated for optimal efficiency of the system.

Figure 8.8a shows the use of the bottom of a truss for support of a ceiling. If the ceiling is attached directly to the bottom chord, a critical concern for the panel size is the length of span that is created for the bottom chord. On the other hand, if the ceiling is suspended from the bottom joints, the concern shifts to that of the logical spacing of elements of the framing for the suspended ceiling.

Because of their need for significant height, trusses tend to take up considerable space inside the building. Therefore, it is often necessary to be able to utilize some of this space for the incorporation of various elements of the building services, such as ducts, piping, electrical power conduits, and lighting fixtures. As shown in Figure 8.8b, these concerns may affect the choice of the panel size or even the pattern of members.

There is probably no more critical dimension for a truss than its overall height. In terms of the structural efficiency of the truss, this dimension will determine the effectiveness of the chords in developing moment resistance and will establish the

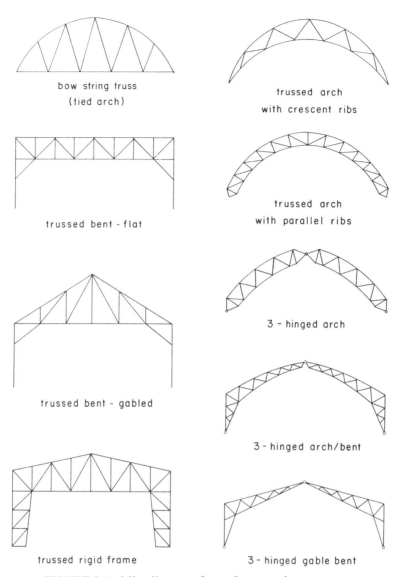

FIGURE 8.6 Miscellaneous forms for trussed structures.

length of the web members. The latter is of major concern for the web members that sustain compression forces. From an architectural point of view, the truss height establishes the amount of space occupied by the structure, which generally has little potential for use architecturally.

Figure 8.9a illustrates the typical conditions of concern with regard to the forces developed in the chords. If the load on the truss is distributed relatively uniformly over the span, the variation of moment essentially will be of the parabolic form shown. If the truss chords are parallel, the critical concern will be for the chord force at midspan. However, for the gable-form truss, even though the height at the peak is critical for the web members and for architectural concerns, the major forces in the

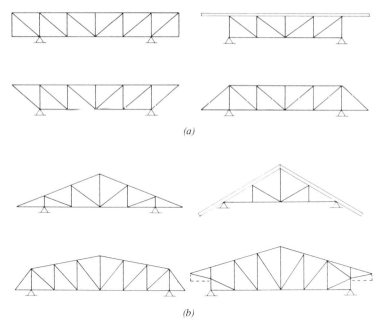

FIGURE 8.7 Achieving cantilevered ends with trusses. (*a*) Trusses with parallel chords (flat chorded). (*b*) Trusses with sloped top chords.

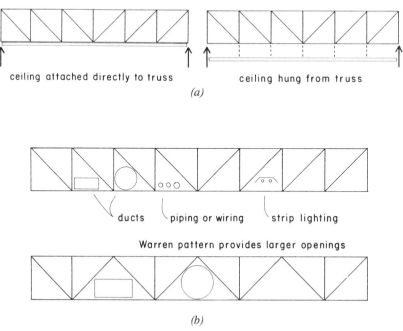

FIGURE 8.8 Facilitation of construction elements and service elements. (*a*) Supported ceilings, directly attached and hung (suspended). (*b*) Incorporation of service elements in the truss interstitial space.

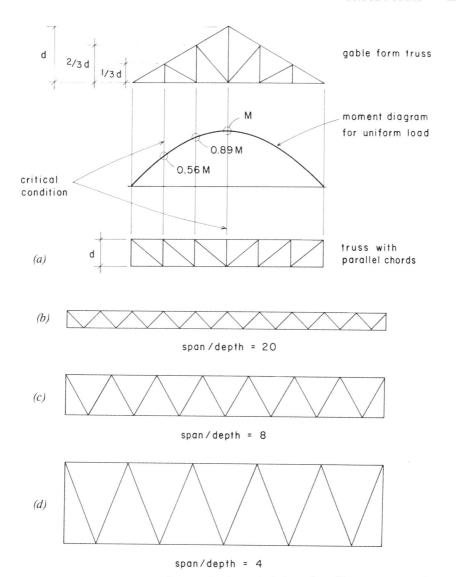

FIGURE 8.9 Effects of truss profile and depth. (*a*) Relation of profile to moment resistance in a simple span truss with uniformly distributed loading. (*b-d*) Relation of the span/depth ratio on the pattern of internal members in a flat-chorded truss.

chords may actually be near the supports, rather than at midspan. Thus it is not possible to generalize about desirable span/height ratios without also discussing truss profile and pattern.

Figure 8.9*b* to *d* illustrates some of the concerns that must be dealt with in deriving logical dimensions for truss height and panel length for a parallel chord truss. At (*b*) the span/height ratio is so high that it must be expected that the chord forces will be considerable. At this ratio the truss is scarcely feasible, since it cannot compare favorably with a solid beam. In addition, there is a profusion of joints and web members that often cannot be achieved feasibly in the production process.

124 GENERAL CONSIDERATIONS FOR TRUSSED CONSTRUCTION

At (*d*) the ratio has been pushed to the other extreme. While chord forces will be extremely low, the web members will be very large, due simply to their length, regardless of the actual forces they must develop.

The truss shown at (*c*) has several desirable features. The chord spacing in relation to the span is likely to make the truss reasonably efficient. The panel length is better related to the height than at (*c*). The number of joints and web members is

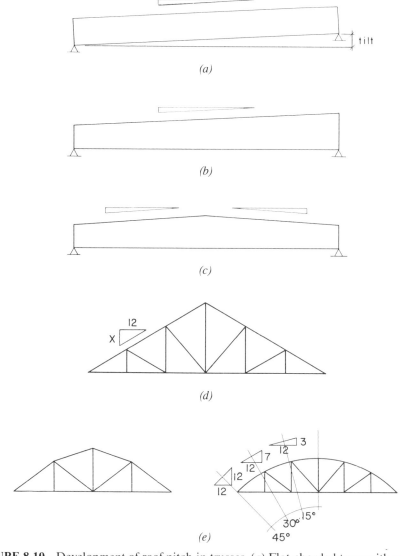

FIGURE 8.10 Development of roof pitch in trusses. (*a*) Flat-chorded truss with constant truss depth; pitched by elevation of one end support. (*b*) Flat-chorded truss with single direction pitch achieved by variable depth. (*c*) Flat-chorded truss with two-way pitch achieved by variable depth. (*d*) Symmetrical gable-form truss with single pitch. (*e*) Multiple-pitched surfaces with flat profile and multiple-pitched surfaces in a continuously developed, arched surface.

reasonable. Although there may be special circumstances that could make any of these trusses acceptable, this range of span/height ratio is generally more feasible.

For gable-form trusses the span/height ratios must be somewhat lower than for the parallel chord truss, primarily for the reason illustrated in Figure 8.9a. Given the same loading and span, therefore, it should be expected that a gable-form truss will be considerably higher than one with parallel chords.

Roof Slope

Roof surfaces must have some slope in order to allow for the drainage of water. The so-called flat roof is usually sloped a minimum of $\frac{1}{4}$ in./ft. If the roof structure is flat, the slope must be achieved by building up the surface on top of the structure. Although this is sometimes done, it is also possible to make the top of the structure itself with the required slope.

Figure 8.10a to c shows three ways to achieve a slope for the top of a truss that is essentially constructed with parallel chords. At (a) the chords are indeed parallel and the whole truss is simply tilted the required amount. If there is no ceiling, or the ceiling is suspended below the trusses, this may be acceptable. However, if a flat ceiling is to be attached to the bottom chords of the trusses, the trusses may be built as shown at (b) or (c). If the slope is as low as the minimum $\frac{1}{4}$ in./ft, the variation of truss height will be minor.

Roof slope relates to the type of water-resistive surfacing that is used. If the slope is as much as 3 in./ft (3:12), various types of shingles may be used. If it is lower, it is usually necessary to use some type of sealed membrane roofing, which is generally both heavier and more expensive than shingles. The gable-form (triangle profile) truss, such as that shown in Figure 8.10d, is one that is commonly used with roof slopes of this magnitude. When the slope becomes as much as 12:12 (45°), the web members of the truss become quite long, and the structure may be better designed simply as a set of rafters with a horizontal tie. In this case the web members would be used merely to suspend the horizontal tie member and to reduce its span if it is used to support a ceiling.

When the roof surface is made with compound slope or with a continuous curve as an arch, as shown in Figure 8.10e, there may be several different problems of slope. Shingles may be used for the steeper slopes, but a more watertight surfacing may be required for lower slopes, especially in the area at the top of the arch.

8.3 MATERIALS FOR TRUSSES

The two materials most used for trusses—and the only ones that are discussed in this book—are wood and steel. Detailed discussions of the design of trusses with these materials are given in Chapters 9 and 10.

Wood trusses of small to medium size are usually built by one of the methods shown in Figure 8.11a. For the light truss with single elements typically of 2 or 3 in. nominal thickness, gusset plates may be plywood or thin sheet steel. When multiple elements are used, the members are bolted together, although special shear developers often are inserted in the faces between the members to strengthen the connections. Trusses with heavy single elements of solid timber or glue laminated construction usually are assembled with heavy steel gusset plates.

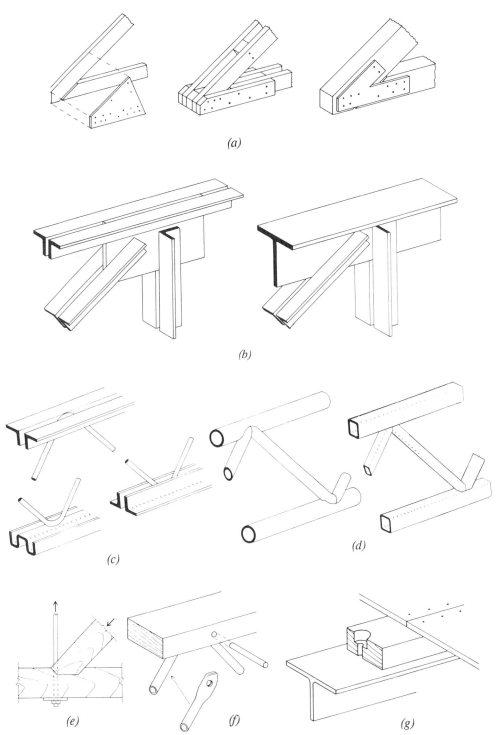

FIGURE 8.11 Typical details for light trusses: (*a*) Wood trusses; (*b*) Steel trusses with light rolled shapes; (*c*) Common forms for steel open-web joists; (*d*) welded trusses with steel pipe or tube members; (*e*) Traditional combination truss with steel rod and wood timber web members; (*f*) Combination truss with wood chord and steel tube web members; (*g*) Development of top of steel truss to accommodate nailing of plywood deck.

MATERIALS FOR TRUSSES

The two most common forms of steel trusses of small to medium size are those shown in Figure 8.11b. In both cases the members may be connected by rivets, bolts, or welds. The most common practice is to use welding for connections that are assembled in the fabricating shop and high strength bolts (torque tensioned) for field connections.

Figure 8.11c shows two popular forms of construction for the light steel truss that is produced as a prefabricated product by various manufacturers—called the open web steel joist. These are made in a wide range of sizes, with the larger sizes usually being of the form shown in the illustration on the right in Figure 8.11b.

When trusses are exposed to view, a popular form is that of the truss with members of tubular steel elements that are directly welded to each other, as in a bicycle frame. As shown in Figure 8.11d, the elements may consist of round pipe or flat-sided, rectangular tubing. Although these trusses are very neat and trim in appearance, they are usually considerably more expensive than those produced by more conventional means, so that their use is mostly limited to situations where appearance is highly valued.

In some cases wood and steel elements are mixed in the same truss; this is called composite construction. In heavy timber trusses, it used to be common to use steel rods for the tension members and wood elements for the compression members in the web of the truss, as shown in Fig. 8.11e. A primary reason for this was the relative difficulty of achieving effective tension connections for the wood members. Use of shear developers (equally functional in tension and compression) or of welded steel gusset plates has reduced the need for this, although sometimes the steel tension member is still used for certain situations.

The most common form of composite wood-steel truss is that shown at (*f*) in Figure 8.11. These trusses are produced as manufactured products with the specific details patented by individual companies. Most of them employ solid wood members for the chords and steel tubular members for the web. A recent innovation is the use of chords made of glue laminated elements, which permits the fabrication of very long, one-piece members for the chords.

The detail shown at (*g*) in Figure 8.11 illustrates a simple solution to the problem of attaching a wood deck to the top of a steel truss. This is not exactly composite construction, since the wood member is not really a functioning element of the truss.

The following are some general considerations that may affect the decision about what materials to use for a particular truss design.

1. *Cost.* If other functional requirements are not influential, the choice is likely to be on the basis of the most economical solution.
2. *Other Structural Elements.* The materials used for the other parts of the building structure—roof deck, columns, walls, and so—may have some influence in terms of logical mixing of the components of the building construction.
3. *Fire Requirements.* The need for a fire-rated structure, or simply for use of noncombustible materials, may be a concern.
4. *Local Availability.* Local competition of manufacturers or contractors, or the availability of specific materials or types of construction work may be factors in the choice of materials or types of truss construction.

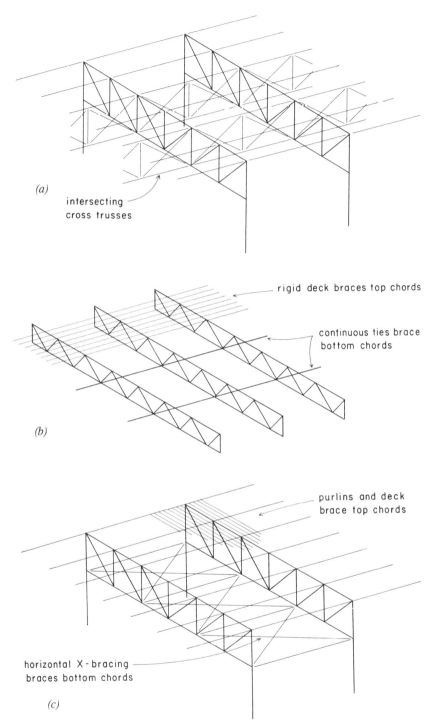

FIGURE 8.12 Lateral bracing for planar trusses: (*a*) Use of perpendicular trusses (called cross bracing); (*b*) Use of attached deck plus continuous bottom struts to brace joist trusses; (*c*) Use of supported framing to brace top chords and horizontal planes of trussing to brace bottom chords.

8.4 BRACING FOR TRUSSES

Single planar trusses are very thin structures that require some form of lateral bracing. The compression chord of the truss must be designed for its laterally unbraced length. In the plane of the truss, the chord is braced by other truss members at each joint. However, if there is no lateral bracing, the unbraced length of the chord in a direction perpendicular to the plane of the truss becomes the full length of the truss. Obviously it is not feasible to design a slender compression member for this unbraced length.

In most buildings other elements of the construction ordinarily provide some or all of the necessary bracing for the trusses. In the structural system shown in Figure 8.12a, the top chord of the truss is braced at each truss joint by the purlins. If the roof deck is a reasonably rigid planar structural element and is adequately attached to the purlins, this constitutes a very adequate bracing of the compression chord—which is the main problem for the truss. However, it is also necessary to brace the truss generally for out-of-plane movement throughout its height. In the example this is done by providing a vertical plane of x-bracing at every other panel point of the truss. The purlin does an additional service by serving as part of this vertical plane of trussed bracing. One panel of this bracing is actually capable of bracing a pair of trusses, so that it would be possible to place it only in alternate bays between the trusses. However, the bracing may be part of the general bracing system for the building, as well as providing for the bracing of the individual trusses. In the latter case, it would probably be continuous.

Light truss joists that directly support a deck, as shown in Figure 8.12b, usually are adequately braced by the deck. This constitutes continuous bracing, so that the unbraced length of the chord in this case is actually zero. Additional bracing in this situation often is limited to a series of continuous steel rods or single small angles that are attached to the bottom chords as shown in the illustration.

Another form of bracing that is used is that shown in Figure 8.12c. In this case a horizontal plane of x-bracing is placed between two trusses at the level of the bottom chords. This single braced bay may be used to brace several other bays of trusses by connecting them to the x-braced trusses with horizontal struts. As in the previous example, with vertical planes of bracing, the top chord is braced by the roof construction. It is likely that bracing of this form is also part of the general lateral bracing system for the building so that its use, location, and details are not developed strictly for the bracing of the trusses.

When bracing between trusses is not desired, it is sometimes possible to use a structure that is self-bracng. One form of self-bracing truss is the so-called delta truss, which is discussed in Section 12.3.

8.5 TRUSS JOINTS

The means used to achieve the connection of truss members at the truss joints depends on a number of considerations, the major ones being

The materials of the members.
The form of the members.

130 GENERAL CONSIDERATIONS FOR TRUSSED CONSTRUCTION

The size of the members.
The magnitude of forces in the members.

Design of ordinary connections for wood and steel trusses is discussed in detail in Chapters 9 and 10. The following discussion deals with a number of general concerns relating to the jointing of the truss.

For small to medium trusses, it is common to use chord members that are continuous through two or more panels. For small truss joists the chords may be a single, unspliced piece for the entire length of the truss. This reduces the number of individual pieces that need to be fabricated, but its chief advantage is in the elimination of a large amount of connecting. Figure 8.13 shows three common arrangements used for small trusses.

One consideration in the choice of a pattern for the truss may be the relative complexity of the joints that the pattern produces. At the lower left in Figure 8.14 the truss patterns result in there being only two web members at each joint, which is generally an easier detailing task than that shown in the upper left figure, where three web members meet at every other joint. An even busier intersection is that shown in the illustration on the right in Figure 8.14, where nine members meet in a three-dimensional traffic jam. This type of joint situation should be avoided.

In most cases it is desirable to apply the loads on a truss at the truss joints. This is especially so when the loads are concentrated, rather than distributed uniformly, as when a deck is directly supported. As shown in Figure 8.15a, the application of concentrated loads to the chords between joints will induce considerable bending, which will reduce the relative efficiency of the truss.

The truss supports also represent concentrated forces on the truss. As shown in Figure 8.15b, these should also be located at the truss joints whenever possible.

When trusses are large or are designed to be continuous through several spans, it is usually necessary that they be fabricated in units in the shop to simplify their transportation to the building site. Thus the truss pattern and the arrangement of members must be developed to facilitate the necessary division of the truss into units and the assemblage of the units at the site. The upper illustration in Figure 8.16

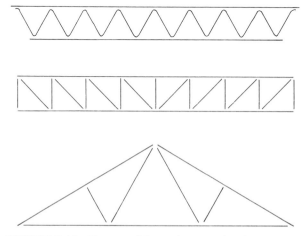

FIGURE 8.13 Trusses with continuous chord members.

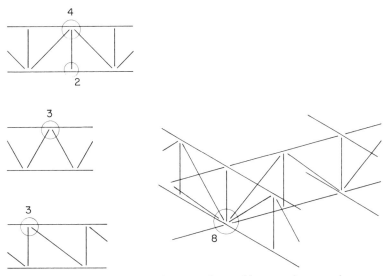

FIGURE 8.14 Truss joints with various numbers of intersecting members, as noted.

shows the splicing for a truss that consists of four field connections in the chords with two of the web members inserted at the site. If this truss cannot be shipped in larger units, there are not many alternatives to this design for the truss. However, the connections will add some cost to the structure.

In the lower illustration in Figure 8.16, a splicing arrangement is shown for the continuous truss that relates to the location of low, or even zero, moment in the con-

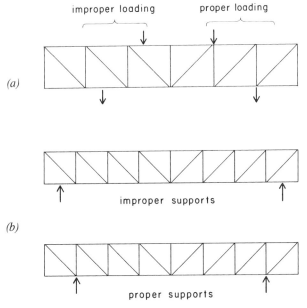

FIGURE 8.15 Truss load and support conditions to ensure pure truss action: (*a*) Placement of concentrated loads; (*b*) Locations for supports.

132 GENERAL CONSIDERATIONS FOR TRUSSED CONSTRUCTION

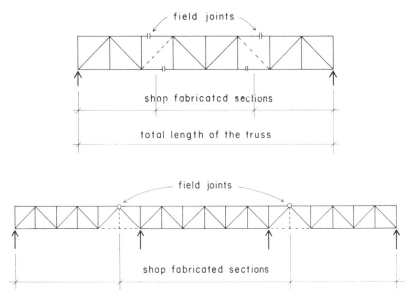

FIGURE 8.16 Considerations for field joints in large trusses.

tinuous spanning structure. (See discussion in Section 6.13.) This makes it possible for the field joints to be designed for relatively low forces in comparison to those in the single span truss.

Truss joints may sometimes be designed for special functions. The structure shown in Figure 8.17 consists of a single span truss that is supported by two columns. Most likely detailing of the structure must allow for field connection of the truss and the columns. If it is not desired that the truss transfer bending to the columns, the two end chord pieces must be attached with loose joints—that is, with joints that are free to slip or slide in the direction of the members. Various means may be used to achieve this performance, depending on the material and form of the members.

If the bent in the previous illustration is required to take lateral forces on the building—as a rigid frame—the connection of the end chords must be rigidly made. However, to reduce the unavoidable bending due to the gravity loads that will result from this connection, it may be possible to delay the making of this connection—or

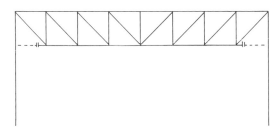

FIGURE 8.17 Delayed field jointing for controlled behavior in a trussed bent. Not required for simple gravity loads, the end members of the bottom chords are left unconnected until construction is almost complete; thus essentially eliminating bending due to dead loads in the support columns.

at least its final tightening—until after the roof construction is in place. Thus the rigid bent will be stressed only by the gravity live loads. This is possible in the example, since the bottom chord end members arc not required for stability with gravity loads on the truss.

Another special jointing problem is that which occurs when it is necessary to attach a truss to its supports in a manner that allows for some movement—possibly owing to the stretching of the chords then the truss is loaded or to thermal expansion. This problem is discussed in Section 11.4.

8.6 MANUFACTURED TRUSSES

The vast majority of trusses currently used in building construction in the United States are off-the-shelf, predesigned, patented industrial products that are marketed by various fabricators and suppliers. The two forms in greatest use are those shown in Figure 8.18.

The gable-form truss—typically made with 2 in. nominal thickness lumber and sheet steel gusset plates—is used as a combination roof rafter and ceiling joist for spans up to 30 ft or so. Spacing is usually a maximum of 2 ft, which permits the direct attachment of a thin plywood roof deck and a gypsum drywall ceiling surface. Roof overhangs are achieved simply by extending the ends of the top chords.

The parallel chord truss is used for both floor and flat roof construction. Roof pitch can be achieved by slightly tilting the whole truss—where no ceiling is required—or by tilting the top chord while keeping the bottom flat. The two most common types are the all steel, open web joist and the wood chord, steel web composite truss, called a trussed joist.

In many cases suppliers provide not only the products, but also the necessary engineering design and subcontracted field erection—an irresistible deal for architects and building owners. Caveat emptor! Where the use of these products is a feasible alternative, custom-designed trusses are seldom justifiable. Unless, of course, there is a need for special building form or detail, carrying of special loads, or some particular appearance.

Although the greatest use of these products is made for short to medium spans (20 to 40 ft for floors and 25 to 75 ft for roofs), various predesigned trusses are obtainable for larger spans. The availability and competitiveness of these products—as with

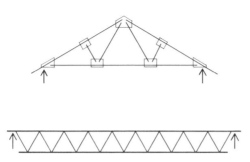

FIGURE 8.18 Common forms of light predesigned trusses. Upper: the W truss, used for framing of a sloped roof and flat ceiling in small wood structures. Lower: the common form of steel open-web joists and combination wood-chord and steel-web trusses.

any marketed materials or products—is regional in nature. Designers, therefore, should investigate the use of predesigned trusses on a local basis.

The upper illustration in Figure 8.19 shows the use of light open-web steel joists for an extensive flat roof for a one-story building. The support structure consists of steel beams and columns of rolled shapes. The roof surface is developed with a light formed sheet steel deck. This is a common use for this system, although the current availability of built-up elements of wood offers a highly competitive alternative for the short- to medium-span range shown here. However, the open form of the trusses may be an advantage, if extensive use must be made of the interstitial space between the ceiling and roof surfaces.

The lower illustration in Figure 8.19 shows the structure for a low-rise office building, in which light steel trusses are used for both the roof and floor systems. The structure here uses the combination of open-web joists and steel deck, but also uses spanning trusses for support of the open web joists. The support trusses are also available as manufactured products—called *joist girders*—from the same companies that produce open-web joists. Use of the support trusses, in place of beams with solid webs, extends even further the flexibility of use of the interstitial spaces.

FIGURE 8.19 Applications of open-web steel joists. Upper: An extensive flat roof with primary support framing of steel columns and beams and an infill with closely spaced joists. Lower: A low-rise office building using steel open web joists and supporting steel trusses, called *joist girders*, for an all truss spanning system.

8.7 TRUSSED BRACING FOR LATERAL FORCES

Although there are actually several ways to brace a frame against lateral loads, the term *braced frame* is used to refer to frames that utilize trussing as the primary bracing technique. In buildings, trussing is mostly used for the vertical bracing system in combination with the usual horizontal diaphragms. It is also possible, however, to use a trussed frame for a horizontal system, or to combine vertical and horizontal trussing in a truly three-dimensional trussed framework. The latter is more common for open tower structures, such as those used for large electrical transmission lines and radio and television transmitters.

Use of Trussing for Bracing

Post and beam systems, consisting of separate vertical and horizontal members, may be inherently stable for gravity loading, but they must be braced in some manner for lateral loads. The three basic ways of achieving this are through shear panels, moment-resistive joints between the members, or by trussing. The trussing, or triangulation, is usually formed by the insertion of diagonal members in the rectangular bays of the frame.

If single diagonals are used, they must serve a dual function: acting in tension for the lateral loads in one direction and in compression when the load direction is reversed (see Figure 8.20*a*). Because long tension members are more efficient than long compression members, frames are often braced with a crisscrossed set of diagonals (called X-bracing) to eliminate the need for the compression members. In any event the trussing causes the lateral loads to induce only axial forces in the members of the frame, as compared to the behavior of the rigid frame. It also generally results in a frame that is stiffer for both static and dynamic loading, having less deformation than the rigid frame.

While the stiffness of a truss represents an advantage in some regards (notably in less movement that causes damage to the nonstructural elements of the building) it means that the truss lacks the potential for resiliency and energy absorption that exists with more flexible structures. Thus significant deflection of the truss can occur only with buckling of compression members, tensile yielding of tension members, or major deformation of joints, none of which is really desirable. Joints in particular should be made to resist any loosening, brittle fracture, tearing, or other undesired forms of failure, and should preferably be stronger than the members they connect. This is all to say that the stiffness of a braced frame can be an advantage, but the structure pays a penalty to develop it. You have to accept the penalty to get the bonus.

Single-story, single-bay buildings may be braced as shown in Figure 8.20*a*. Single-story, multibay buildings may be braced by bracing less than all of the bays in a single plane of framing, as shown in Figure 8.20*b*. The continuity of the horizontal framing is used in the latter situation to permit the rest of the bays to tag along. Similarly, a single-bay, multistoried, towerlike structure, as shown in Fig. 8.20*c*, must have its frame fully braced, whereas the more common type of frame for the multistored building, as shown in Figure 8.20*d*, is usually only partly braced. Since either the single diagonal or the crisscrossed X-bracing causes obvious problems for interior circulation and for openings for doors and windows, building planning often makes the limited bracing a necessity.

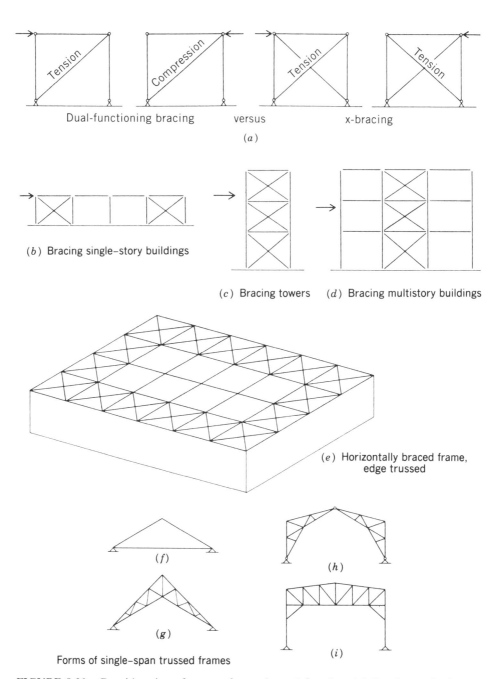

FIGURE 8.20 Considerations for use of truss-braced framing: (*a*) Bracing a single rectangular unit with dual-functioning single diagonals or X-bracing with a potential tension-only functioning. (*b*) Use of selected braced units to provide bracing for other, unbraced, units to which the braced units are linked by the framing system. (*c*) X-bracing for a multilevel tower. (*d*) Use of a selected braced bent within the whole system of a multilevel building column and beam frame. (*e*) Use of horizontal planes of trussing to provide lateral stability. (*f*) Defining the simple, single-unit truss as a tied rafter system. (*g*) Use of the scissors truss to create a rigid frame with pure truss action. (*h*) Creation of the gabled bent with what are, in effect, two scissors trusses connected to form a three-hinged structure. (*i*) Creation of a rigid, trussed bent, shown here with knee-braces added for additional stiffness or for reduction of the unbraced column height.

Just about any type of floor construction used for multistored buildings usually has sufficient capacity for diaphragm action in the lateral bracing system. Roofs, however, often utilize light construction or are extensively perforated by openings, so that the basic construction is not capable of the usual horizontal, planar diaphragm action. For such roofs or for floors with many openings, it may be necessary to use a trussed frame for the horizontal part of the lateral bracing system. Figure 8.20*e* shows a roof for a single-story building in which trussing has been placed in all the edge bays of the roof framing in order to achieve the horizontal structure necessary. As with vertical trussed frames, the horizontal trussed frame may be partly trussed, as shown in Figure 8.20*e*, rather than fully trussed.

For single-span structures, trussing may be utilized in a variety of ways for the combined gravity and lateral load resistive system. Figure 8.20*f* shows a typical gable roof with the rafters tied at their bottom ends by a horizontal member. The tie, in this case, serves the dual functions of resisting the outward thrust due to gravity loads and of one of the members of the single triangle, trussed structure that is rigidly resistive to lateral loads. Thus the wind force on the sloping roof surface, or the horizontal seismic force caused by the weight of the roof structure, is resisted by the triangular form of the rafter-tie combination.

The horizontal tie shown in Figure 8.20*f* may not be architecturally desirable in all cases. Some other possibilities for the single-span structure—all producing more openness beneath the structure—are shown in Figure 8.20*g*, *h*, and *i*. Figure 8.20*g* shows the so-called scissors truss, which can be used to permit more openness on the inside or to permit a ceiling that has a form reflecting that of the gable roof. Figure 8.20*h* shows a trussed bent that is a variation on the three-hinged arch.

The structure shown in Figure 8.20*i* consists primarily of a single-span truss that rests on end columns. If the columns are pin-jointed at the bottom chord of the truss,

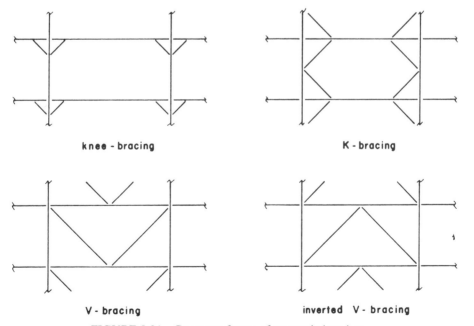

FIGURE 8.21 Common forms of eccentric bracing.

138   GENERAL CONSIDERATIONS FOR TRUSSED CONSTRUCTION

the structure lacks basic resistance to lateral loads and must be separately braced. If the column in Figure 8.20i is continuous to the top of the truss, it can be used in rigid frame action for resistance to lateral loads. Finally, if the knee-braces, shown in the figure are added, the column is further stiffened, and the structure has more load resistance and less deflection under lateral loading.

The knee-brace (Figure 8.21) is one form of diagonal bracing described as *eccentric bracing*, a name deriving from the fact that one or more of the bracing connections is off the column-to-beam joint. Other forms of eccentric bracing are the K-brace, V-brace, and inverted V-brace. V-bracing is also sometimes described as chevron bracing.

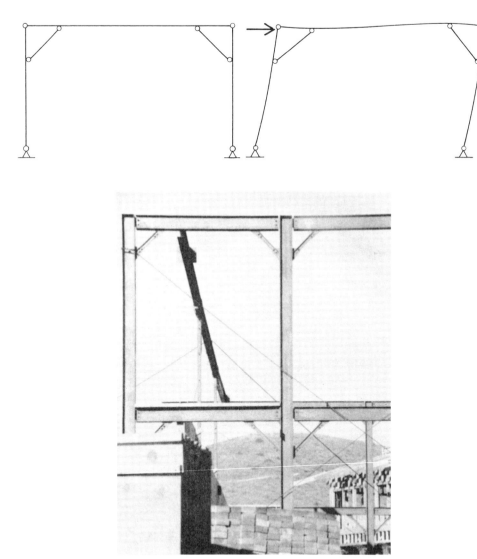

FIGURE 8.22 Knee-bracing used with steel frames using simple (not moment-resistive) connections. Upper figures show the assemblage and how it works in developing rigid frame action to resist lateral loads. Photograph shows installation in a light, multilevel frame.

Use of eccentric bracing results in some combined form of truss and rigid frame actions. The trussed form of the bracing produces the usual degree of stiffness associated with a braced frame, while the bending induced by the eccentricity of the bracing adds rigid frame deformations to the behavior (see Figure 8.22). For ultimate load failure, a significant event is the development of plastic hinging in the members, with the eccentric bracing joints.

Eccentric bracing was historically developed for use as core wind bracing in high-rise steel structures. It has more recently, however, become quite favored for bracing of steel frames for seismic resistance, where its combination of high stiffness and high energy capacity is advantageous. In zones of high seismic risk, braced frames are now most commonly achieved with either very heavy X-bracing or some form of eccentric bracing. Figure 8.23 shows the use of V-bracing for perimeter bracing in a low-rise office building, with diagonals of round steel pipe.

Trussing may be used for combined purposes of both the support for gravity loads and the development of lateral bracing. This occurs naturally when the trussing is formed as part of a *bent*, which is the general term for a planar framework in which the members interact to share the loads. In the multiple-story building, there may be several such bents, with each performing a particular task. Selected bents may be used for only gravity loads and others for combined gravity and lateral load resistance.

The structure in Figure 8.24 uses combinations of concentric and eccentric truss-

FIGURE 8.23 Use of eccentric bracing. Form shown here is described as *V-bracing* or *chevron bracing*.

140 GENERAL CONSIDERATIONS FOR TRUSSED CONSTRUCTION

FIGURE 8.24 Use of combined eccentric bracing (inverted V-bracing) and concentric, or ordinary, bracing with dual-functioning, single diagonals in a single plane.

ing in combination with the horizontal and vertical framing for development of complex bents. The high degree of structural indeterminacy of such a structure provides extensive redundancy that is an advantage for resistance to energy loads, particularly of the kind generated by earthquakes.

Planning of Bracing

Some of the problems to be considered in using braced frames are the following:

1. Diagonal members must be placed so as not to interfere with the action of the gravity-resistive structure or with other building functions. If the bracing members are designed essentially as axial stress members, they must be located and attached so as to avoid loadings other than those required for their bracing functions, they must also be located so as not to interfere with door, window, or roof openings or with ducts, wiring, piping, light fixtures, and so on.
2. As mentioned previously, the reversibility of the lateral loads must be considered. As shown in Figure 8.20a, such consideration requires that diagonal members be dual functioning (as single diagonals) or redundant (as X-bracing) with one set of diagonals working for load from one direction and the other set working for the reversal loading.
3. Although the diagonal bracing elements usually function only for lateral loading, the vertical and horizontal elements must be considered for the various possible combinations of gravity and lateral load. Thus the total frame must be analyzed for all the possible loading conditions, and each member must be designed for the particular critical combinations that represent its peak response conditions.
4. Long, slender bracing members, especially in X-braced systems, may have considerable sag due to their own dead weight, which requires that they be supported by sag rods or other parts of the structure.
5. The trussed structure should be "tight." Connections should be made in a manner to assure that they will be initially free of slack and will not loosen

under the load reversals or repeated loadings. This means generally avoiding connections that tend to loosen or progressively deform such as those that use nails, loose pins, and unfinished bolts.
6. To avoid loading on the diagonals, the connections of the diagonals are sometimes made only after the gravity-resistive structure is fully assembled and at least partly loaded by the building dead loads.
7. The deformation of the trussed structure must be considered, and it may relate to its function as a distributing element, as in the case of a horizontal structure, or to the establishing of its relative stiffness, as in the case of a series of vertical elements that share loads. It may also relate to some effects on nonstructural parts of the building, as was discussed for shear walls.
8. In most cases it is not necessary to brace every individual bay of the rectangular frame system. In fact, this is often not possible for architectural reasons. As shown in Fig. 8.20b, walls consisting of several bays can be braced by trussing only a few bays, or even a single bay, with the rest of the structure tagging along like cars in a train.

The braced frame can be mixed with other bracing systems in some cases. Figure 8.25a shows the use of a braced frame for the vertical resistive structure in one direction and a set of shear walls in the other direction. In this example the two systems act independently, except for the possibility of torsion, and there is no need for a deflection analysis to determine the load sharing.

Figure 8.25b shows a structure in which the end bays of the roof framing are X-braced. For loading in the direction shown, these braced bays take the highest shear in the horizontal structure, allowing the deck to be designed for a lower shear stress.

Figure 8.26 shows a low-rise office building in which X-braced steel bents are used in combination with wood-framed shear walls for the lateral bracing system. The detail shown in the lower illustration in Figure 8.26 illustrates the typical use of steel gusset plates welded to the vertical and horizontal framing members for attachment of the diagonal braces. In this case the diagonal members consist of single steel channel sections turned back-to-back to form the X-braces.

Although buildings and their structures are often planned and constructed in two dimensional components (horizontal floor and roof planes and vertical wall or framing bent planes), it must be noted that the building is truly three dimensional. Bracing against lateral forces is thus a three-dimensional problem, and although a single horizontal or vertical plane of the structure may be adequately stable and strong, the whole system must interact appropriately. While the single triangle is the basic unit for a planar truss, the three-dimensional truss may not be truly stable just because its component planes are braced.

In a purely geometric sense the basic unit for a three-dimensional truss is the four-sided figure called a tetrahedron. However, since most buildings consist of spaces that are rectangular boxes, the three-dimensional trussed building structure usually consists of rectangular units rather than multiples of the pyramidal tetrahedral form. When so used, the single planar truss unit is much the same as a solid planar wall or deck unit, and general reference to the box-type system typically includes both forms of construction.

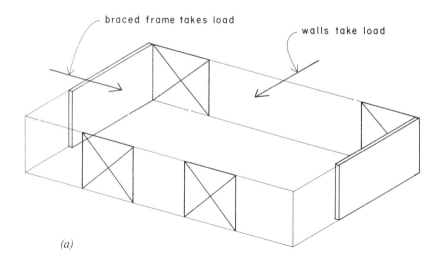

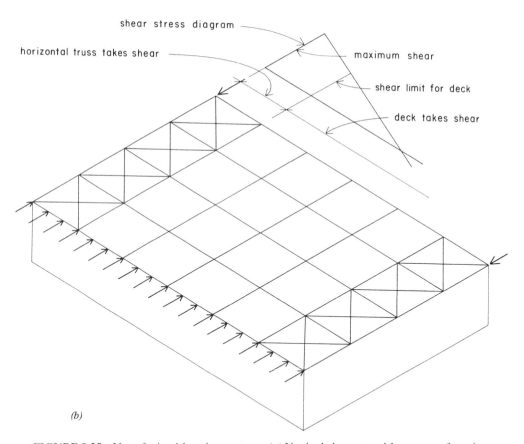

FIGURE 8.25 Use of mixed-bracing systems. (*a*) Vertical elements with separate functions. (*b*) Interactive horizontal structure with ends consisting of X-braced units and center portion consisting of a planar deck diaphragm.

FIGURE 8.26 X-braced bents of steel provide a stiff braced core for a building of mixed steel and wood construction. Upper photo shows the internal steel structure and exterior wood structure. Lower photo shows the use of single-channel shapes for the truss diagonals.

Typical Construction

Development of the details of construction for trussed bracing is in many ways similar to the design of spanning trusses. The materials used (generally wood or steel), the form of individual truss members, the type of jointing (nails, bolts, welds, etc.), and the magnitudes of the forces are all major considerations. Since many of the members of the complete truss serve dual roles for gravity and lateral loads, member selection is seldom based on truss action alone. Quite often trussed bracing is produced by simply adding diagonals (or X-bracing) to a system already conceived for the gravity loads and for the general development of the desired architectural forms and spaces.

9

DESIGN OF WOOD TRUSSES

Wood trusses include a range of possibilities in modern construction, including some with a close resemblance to ancient forms in heavy timber. The ones in most common use, however, are those of simple light wood elements, used with other light wood frame construction and manufactured trusses of composite materials (wood plus steel members). This chapter deals with the general use of wood for a variety of trussed construction.

9.1 BASIC CONSIDERATIONS FOR WOOD

Use of wood for structures is controlled by many practical considerations based on centuries of experience of carpenters and general wood workers. Usage for trussed construction begins with some elementary considerations for the truss forms and their basic stability, integration into the building construction, and so on. Building codes and industry standards govern the designs of members and joints once the internal force actions are determined. This section treats some of the general considerations for practical utilization of wood in modern trussed construction.

Types of Wood Trusses

The forms used most widely for wood trusses are those shown in Figure 9.1. Some of the considerations involved in their use are as follows.

Gable W Truss. This is a popular roof truss that uses a minimum of web members and joints. The web pattern effectively reduces the span of the rafters to one half of the distance from eave to ridge and the span of the ceiling joists to one third of the clear span of the truss. This permits the use of single 2 by 4 members for the chords for spans up to about 30 ft, resulting in a very light and economical structure.

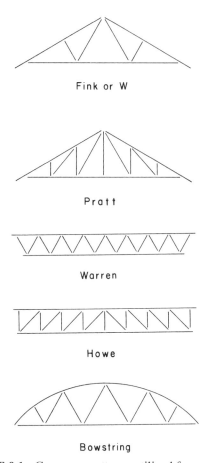

FIGURE 9.1 Common patterns utilized for wood trusses.

Gable Pratt Truss. For trusses with spans in excess of 30 ft, it is usually necessary to increase the number of web members from that used in the W truss. The most popular form for such a truss is the Pratt, primarily due to the fact that the long diagonals are in tension, which permits the use of lighter elements.

Flat Warren Truss. For parallel chord trusses of composite construction, the Warren is the most popular form, since it permits the use of a single size steel web member and a single joint detail throughout the length of the truss. This advantage is lost if the chords are not truly parallel.

Flat Howe Truss. If a parallel chord truss is all wood, the most popular form is the Howe. Although the diagonals are compression members, they tend to be shorter than in the gable truss and not as critical for slenderness effects. The Warren pattern is a reasonable alternative and may be more desirable if considerable space is required for the incorporation of ducts within the depth of the trusses. (See discussion in Section 8.2.)

Bowstring Truss. If an arched roof form is desired and use of the space immediately beneath the arch is not required, the wood bowstring truss may be a highly competitive alternative structure. Usually chords are made with glue laminated elements and web members with single or multiple elements of solid lumber. Spans of up to 200 ft and more are possible.

Design Considerations with Wood

From the initial decision to use a wood truss to the final development of fabrication details, there are numerous considerations involved in the design work. The major concerns are for the following.

1. *Decision to Use Wood.* Basic issues are cost, fire rating required, compatibility with other elements of the building construction, and any problems of exposure to the weather.
2. *Type of Wood.* This is a highly regional issue in terms of the availability of particular species of wood. Trusses usually use higher grades (stress ratings) of wood than the average for wood structures. The two most popular woods are Douglas fir and Southern pine, both for their high strength and their general availability.
3. *Form.* The most common truss profiles are those shown in Figure 9.2. Choice of profile is generally a matter of consideration of the problem of roof drainage as well as general architectural design. Truss patterns most commonly used are those shown in Figure 9.1.
4. *Truss Members.* For practicality truss members usually are selected from sizes available as standard products. Single pieces of solid wood are produced in the sizes given in Table B.11 (Appendix B). When large members, or higher stress capacities, are required, members may consist of glue laminated elements.

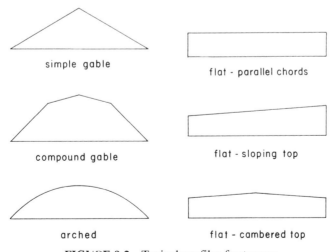

FIGURE 9.2 Typical profiles for trusses.

TABLE 9.1 Design Values for Visually Graded Structural Lumber of Douglas Fir–Larch[a] (Values in psi)

Species and Commercial Grade	Size and Use Classification	Extreme Fiber in Bending F_b		Tension Parallel to Grain F_t	Horizontal Shear F_v	Compression Perpendicular to Grain $F_{c\perp}$	Compression Parallel to Grain F_c	Modulus of Elasticity E
		Single Member Uses	Repetitive Member Uses					
		Dimension Lumber 2 to 4 in. Thick, 2 in. and Wider (Moisture Not Exceeding 19%)						
Select structural		1450	1668	1000	95	625	1700	1,900,000
No. 1 and better		1150	1323	775	95	625	1500	1,800,000
No. 1		1000	1150	675	95	625	1450	1,700,000
No. 2		875	1006	575	95	625	1300	1,600,000
		Timbers (Surfaced Green)						
Dense select structural	Beams and stringers	1900	—	1100	85	730	1300	1,700,000
Select structural		1600	—	950	85	625	1100	1,600,000
Dense No. 1		1550	—	775	85	730	1100	1,700,000
No. 1		1300	—	675	85	625	925	1,600,000
No. 2		875	—	425	85	625	600	1,300,000

Dense select structural	Posts and timbers	1750	—	1150	85	730	1350	1,700,000
Select structural		1500	—	1000	85	625	1150	1,600,000
Dense No. 1		1400	—	950	85	730	1200	1,700,000
No. 1		1200	—	825	85	625	1000	1,600,000
No. 2		750	—	475	85	625	700	1,300,000
Decking (Moisture Not to Exceed 19%)								
Select dex	Decking	1750	2013	—	—	625	—	1,800,000
Commercial dex		1450	1668	—	—	625	—	1,700,000

Source: Data adapted from *National Design Specification for Wood Construction*, 1991 edition (Ref. 3), with permission of the publishers, National Forest Products Association. The table in the reference document lists several other wood species and has extensive footnotes.

[a] Values listed are for normal duration loading with wood that is surfaced dry or green and used at 19% maximum moisture content.

150 DESIGN OF WOOD TRUSSES

5. *Jointing.* The methods used to achieve the truss joints depend mostly on the size of the truss and are discussed in the examples shown in this chapter.
6. *Assembly and Erection.* Consideration must be given to the production process—whether performed in a shop or at the job site. This may affect the methods used for jointing and involve problems of size for transportation.
7. *The Building Structure.* The trusses must be integrated into the general structural system for the building. Attachment of items to the trusses, supports for the trusses, and use of the trusses as part of the general lateral bracing system are some possible concerns.

Design Values

The general reference for allowable stresses and modulus of elasticity values to be used for design work is the publication that is prepared as a supplement to the NDS entitled *Design Values for Wood Construction.* Table 9.1 is adapted from this publication and presents values for one popular species of structural lumber: Douglas fir–larch. To obtain values from the table the following data must be determined for a particular piece of lumber.

1. *Species.* The NDS publication lists values for 26 different species, only one of which is included in Table 9.1.
2. *Moisture Condition at Time of Use.* The moisture condition assumed for the table values is given with the species designation in the table. Adjustments for other moisture conditions are described in the footnotes or in the various specifications in the NDS.
3. *Grade.* This is indicated in the first column of the table which is based on visual grading standards.
4. *Size or Use.* The second column of the table identifies size ranges or usage (e.g., beams and stringers) of the lumber. Note that for Douglas fir–larch the grade "select structural" appears five times for various sizes and uses.
5. *Structural Function.* Values are given for stresses in flexure, tension, shear, and compression, and for the modulus of elasticity.

In the reference document there are extensive footnotes to the table. Data from Table 9.1 will be used in various design examples in this book and some of the issues treated in the document footnotes will be explained. In many situations there are modifications to the design values, as described in the next section.

Bearing Stress

There are various situations in which a wood member may develop a contact bearing stress, essentially, a surface compression stress. Some examples are the following:

At the base of a wood column supported in direct bearing. This is a case of bearing stress that is in a direction parallel to the grain.
At the end of a beam that is supported by bearing on a support. This is a case of bearing stress that is perpendicular to the grain.

Within a bolted connection at the contact surface between the bolt and the wood at the edge of the bolt hole.

In a timber truss where a compression force is developed by direct bearing between the two members. This is frequently a situation involving bearing stress that is at some angle to the grain other than parallel or perpendicular.

For connections, the bearing condition is usually incorporated into the general assessment of the unit value of connecting devices. This is discussed in Chapter 11. The two critical dimensions that define the bearing contact area—bolt diameter and member thickness—are included in the data for determination of bolt values.

Limiting stress values for direct bearing are based only on the wood species and relative density. Ordinarily, a single value is given for dense grades and another single value for other (ordinary or not dense) grades. The value for compression perpendicular to the grain (at beam ends, for example) is given in Table 9.1. Taken from a separate table in the NDS, the value for bearing parallel to the grain, designated F_g, for Douglas fir-larch is as follows:

For all dense grades: 1730 psi

For other grades: 1480 psi

The situation of stress at an angle to the grain is discussed later in this section. It consists of finding a compromised value somewhere between the allowable values for the two limiting stress conditions.

Modifications of Design Values

The values given in Table 9.1 are basic references for establishing the allowable values to be used for design. These values are based on some defined norms, and in many situations the design values will be modified for actual use in structural computations. In some cases the form of the modification is a simple increase or decrease that is achieved by a percentage increase or reduction factor. In other situations the modification is more complex, such as the modification of allowable compression for slenderness effects. Some of the modifications are described in the footnotes to the table from which Table 9.1 is adapted. Other modifications are described in the various sections of the NDS that deal with specific types of problems. The following are some of the major types of modifications.

Moisture. The table gives the assumed moisture condition on which the table values are based. Increases may be permitted in some values for wood that is cured to a lower moisture value. If exposed to weather, a wet usage condition may require some reductions of values.

Load Duration. The table values are based on so-called normal duration load, which is actually rather meaningless. Increases in the design values are permitted for various degrees of short duration loading. A decrease in some values may be required for loading that is prolonged over the life of the structure (basically referring to dead load). Table 9.2 presents a summary of the NDS requirements for adjustment for load duration.

152 DESIGN OF WOOD TRUSSES

TABLE 9.2 Modification of Design Values for Load Duration[a]

Duration of Load and General Use	C_D Factor; Multiply by Design Values
Permanent: dead load	0.9
Ten years: occupancy live load	1.0
Two months: snow load	1.15
Seven days: construction load (used for roofs where no snow exists)	1.25
Ten minutes: wind and earthquakes	1.6
Impact: wheel bumps, braking of moving equipment, slamming of heavy doors[b]	2.0

[a] Adapted from the NDS, 1991 edition (Ref. 3), with permission of the publishers, National Forest Products Association.
[b] Not to be used for connections or for certain pressure-treated members.

Temperature. Where prolonged exposure to temperatures over 150°F exist, some values must be reduced.

Treatments. Impregnation with chemicals for enhanced resistance to rot, vermin, and insects, or fire may require reductions in some values.

Size. Effectiveness in flexure is reduced in beams of depths exceeding 12 in.

Buckling. Various modifications may be required for beams or columns with a tendency to fail in buckling.

Load Orientation to Grain. The table gives separate values for allowable compression with respect to the grain direction in the wood. In some situations the load direction may be other than parallel (0°) or perpendicular (90°) to the grain and a specific value for stress must be derived, as described later in this section.

Specific usage conditions must be carefully studied to ascertain the amount of modification required for a given design situation.

Modifications for Loading with Respect to the Grain Direction

Under the condition shown in Figure 9.3 the load from member B exerts a compressive stress on member A on a surface inclined to the grain. The compressive

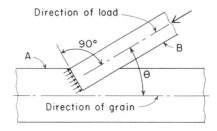

FIGURE 9.3 Condition of compression stress developed at an angle to the wood grain.

strength of wood is greatest parallel to the grain and is least perpendicular to the grain. The allowable unit compressive stress on an inclined surface is determined from the following expression, known as the Hankinson formula:

$$F_n = \frac{F_g \times F_{c\perp}}{F_g \times \sin^2 \theta + F_{c\perp} \times \cos^2 \theta}$$

in which F_n = allowable unit stress acting perpendicular to the inclined surface.

F_g = allowable unit stress in bearing parallel to the grain,

$F_{c\perp}$ = allowable unit stress in compression perpendicular to the grain,

θ = angle between the direction of the load and the direction of the grain.

When the load is applied parallel to the grain, θ is zero. When the load is applied perpendicular to the grain, θ is 90°. Table 9.3 gives values of $\sin^2 \theta$ and $\cos^2 \theta$ for various values of θ (theta).

Example 1. Two timbers 6 in. wide consisting of Douglas fir–larch, Dense grade, are framed together as indicated in Figure 9.2. The angle between the two pieces is 30°. Compute the allowable unit stress on the inclined bearing surface.

Solution: Referring to the previous discussion of bearing, we find that the allowable unit compressive stress parallel to grain is F_g = 1730 psi, and perpendicular to grain, $F_{c\perp}$ = 730 psi. Values of $\sin^2 \theta$ and $\cos^2 \theta$, when θ = 30°, are taken from Table 9.3. Then, substituting in the Hankinson formula the allowable unit compressive stress on the inclined surface is

$$F_n = \frac{1730 \times 730}{(1730 \times 0.25) + (730 \times 0.75)} = 1289 \text{ psi}$$

A similar modification is required with the use of some fasteners, such as bolts and split-ring connectors. As an alternative to the computations just illustrated, the functions of the Hankinson formula may be displayed in a graphical form and necessary modifications can be approximated from the graph. The use of such a method is illustrated in Section 9.2.

Wood Tension Members

Wood is relatively strong in resisting tension parallel to its grain. Design consists of finding the cross-section area required for the allowable stress. If bolted connections (as shown in Figure 9.4) or other details result in a reduction of the cross section, the net area is used for design analysis. When nails, screws, staples, or clinched steel plates are used for connections, the cross section is not reduced for stress calculation.

Although slenderness usually is not a critical factor for tension members, a problem sometimes occurs when long horizontal members are used. As shown in Figure

154 DESIGN OF WOOD TRUSSES

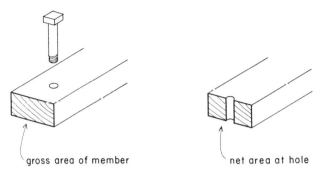

FIGURE 9.4 Reduction of the member cross section, resulting in the net area in tension.

9.5, a solution sometimes used in these situations is to add some vertical members to reduce the length of the horizontal spans of the tension members.

The following example illustrates the usual procedure for design of a wood tension member.

Example 2: Douglas fir, No. 1, used at normal moisture condition. Tension force of 6000 lb with load duration of 7 days. Bolted connections with $\frac{3}{4}$ in. bolts in single rows. Find the required size of wood member with nominal thickness of 2 in.

Solution: From Table 9.3 we note that the unmodified design value for F_t is 675 lb/in.² for all wood with nominal thickness of 2 in. The required area for tension is thus calculated as

$$A_{net} = \frac{\text{design force}}{\text{modified } F_t} = \frac{6000}{(1.25)(675)} = 7.11 \text{ in.}^2$$

From Table B.11 we note that a 2 by 6 member provides a gross area of 8.25 in.². However, this area must be reduced by an amount equal to the profile of the bolt hole. For this calculation the usual procedure is to assume the hole to be $\frac{1}{16}$ in. wider than the bolt. The area of the profile of the bolt hole is thus

TABLE 9.3 Values for Use in the Hankinson Formula

$\sin^2 \theta$	θ (Degrees)	$\cos^2 \theta$	$\sin^2 \theta$	θ (Degrees)	$\cos^2 \theta$
0.00000	0	1.00000	0.50000	45	0.50000
0.00760	5	0.99240	0.58682	50	0.41318
0.03015	10	0.96985	0.67101	55	0.32899
0.06698	15	0.93302	0.75000	60	0.25000
0.11698	20	0.88302	0.82140	65	0.17860
0.17860	25	0.82140	0.88302	70	0.11698
0.25000	30	0.75000	0.93302	75	0.06698
0.32899	35	0.67101	0.96985	80	0.03015
0.41318	40	0.58682	0.99240	85	0.00760
0.50000	45	0.50000	1.00000	90	0.00000

$$A = \left(\text{bolt diameter} + \frac{1}{16}\right) (\text{thickness of piece})$$
$$= (0.75 + 0.0625)(1.5)$$
$$= 1.22 \text{ in.}^2$$

and the actual net area for tension stress is thus

$$A_{net} = A_{gross} - (\text{area of hole})$$
$$= 8.25 - 1.22$$
$$= 7.03 \text{ in.}^2$$

which indicates that the 2 × 6 is not adequate so a 2 × 8 must be used.

Note that the design value for the tension stress was modified in the calculation by the appropriate load duration factor from Table 9.2.

Wood Compression Members

Design of wood compression members is considerably more complicated than that for tension members because of the influence of slenderness as a modifying condition. Figure 9.6 illustrates the typical form of the relationship between axial compression capacity and slenderness for a linear compression member (column). The two limiting conditions are those of the very short member and the very long member. The short member (such as a block of wood) fails in crushing, which is limited by the mass of material and the stress limit in compression. The very long member (such as a yardstick) fails in elastic buckling, which is determined by the stiffness of the member; stiffness is determined by a combination of geometric property (shape of the cross section) and material stiffness property (modulus of elasticity). Between these two extremes—which is where most wood compression members fall—the behavior is indeterminate as the transition is made between the two distinctly different modes of behavior.

Over the years, several methods have been employed to deal with this situation in the design of wood columns (or any columns for that matter). The 1986 edition of the NDS used three separate formulas to deal with the full range for L/d, reflecting the three distinct regions of the graph in Figure 9.6. In the 1991 edition, however, a single formula is employed, effectively covering the whole range of the graph. The formula

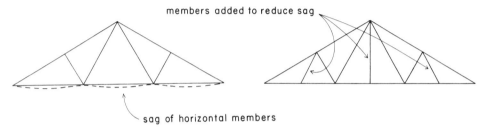

FIGURE 9.5 Functioning of web members to reduce sag of the bottom chord.

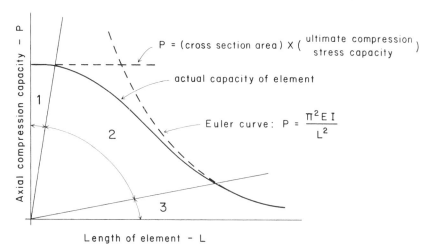

FIGURE 9.6 Relation of member length to axial compression capacity.

and its various factors are complex, and its use involves considerable computation; nevertheless, the basic process is essentially simplified through the use of a single, defined relationship.

In practice, use is now commonly made of either tabulated data (derived from the formula for specific species and grades of wood) or CAD procedures, which operate the solving of the formula in the computer. The following discussion presents the somewhat laborious use of the new NDS procedure with hand calculations (not computer assisted); the basic purpose being to illuminate the process and not to demonstrate practical design procedures.

Column Load Capacity: 1991 NDS

The basic formula for determination of the capacity of a wood column is

$$P = (F_c^*)(C_p)(A)$$

where A = area of the column cross-section

F_c^* = the allowable design value for compression parallel to the grain, as modified by applicable factors

C_p = the column stability factor

P = the allowable column axial compression load

The column stability factor is determined as follows:

$$C_p = \frac{1 + (F_{cE}/F_c^*)}{2c} - \sqrt{\left[\frac{1 + (F_{cE}/F_c^*)}{2c}\right]^2 - \frac{(F_{cE}/F_c^*)}{c}}$$

where F_c^* = the stress value defined above

F_{cE} = the Euler buckling stress, as determined by the formula below

c = 0.8 for sawn lumber, 0.85 for round poles, 0.9 for glued-laminated timbers

For the buckling stress:

$$F_{cE} = \frac{(K_{cE})(E)}{(L_e/d)^2}$$

where K_{cE} = 0.3 for visually graded lumber and machine-evaluated lumber, 0.418 for machine-stress rated lumber and glued-laminated timber

E = modulus of elasticity for the species and grade

L_e = the effective length (unbraced column height as modified by any factors for support conditions)

d = the cross-section dimension (column width) measured in the direction that buckling occurs

For a basic reference, the accounting for the buckling phenomenon typically uses a member that is pinned at both ends and prevented from lateral movement only at the ends. This condition may exist in some cases in actual construction; however, other situations also occur. Thus it may become necessary to adjust for other situations in the computation. This is ordinarily done by considering an altered—or modified—buckling length.

The following example illustrates the use of the NDS formulas for determination of the safe axial load for wood columns.

Example 3. A wood compression member consists of a 6 × 6 of Douglas fir–larch, No. 1 grade. Find the safe axial compression load for unbraced lengths of: (1) 2 ft, (2) 8 ft, (3) 16 ft.

Solution: From Table 9.1 find values of F_c = 1000 psi and E = 1,600,000 psi. With no basis for adjustment given, the F_c value is used directly as the F_c^* value in the column formulas.

(1) $\quad L/d = 2(12)/5.5 = 4.36$

$$F_{cE} = \frac{(K_{cE})(E)}{(L_e/d)^2} = \frac{(0.3)(1,600,000)}{(4.36)^2} = 25,250 \text{ psi}$$

$$\frac{F_{cE}}{F_c^*} = \frac{25,250}{1000} = 25.25,$$

$$C_p = \frac{1 + 25.25}{1.6} - \sqrt{\left[\frac{1 + 25.25}{(1.6)}\right]^2 - \frac{25.25}{0.8}} = 0.993$$

Then, the allowable compression is

$$P = (F_c^*)(C_p)(A) = (1000)(0.993)(5.5)^2 = 30{,}038 \text{ lb}$$

(2) $\quad L/d = 8(12)5.5 = 17.45$

$$F_{cE} = \frac{(0.3)(1{,}600{,}000)}{(17.45)^2} = 1576 \text{ psi}$$

$$\frac{F_{cE}}{F_c^*} = \frac{1576}{1000} = 1.576$$

$$C_p = \frac{2.576}{1.6} - \sqrt{\left[\frac{2.576}{1.6}\right]^2 - \frac{1.576}{0.8}} = 0.821$$

$$P = (1000)(0.821)(5.5)^2 = 24{,}835 \text{ lb}$$

(3) $\quad L/d = 16(12)/5.5 = 34.9$

$$F_{cE} = \frac{(0.3)(1{,}600{,}000)}{(34.9)^2} = 394$$

$$\frac{F_{cE}}{F_c^*} = \frac{394}{1000} = 0.394$$

$$C_p = \frac{1.394}{1.6} - \sqrt{\left[\frac{1.394}{1.6}\right]^2 - \frac{0.394}{0.8}} = 0.355$$

$$P = (1000)(0.3355)(5.5)^2 = 10{,}736 \text{ lb}$$

Design of Wood Compression Members

The design for compression is complicated by the relationships in the column formulas. The allowable compression stress for the column is a function of the actual column dimensions, which are not known at the beginning of the design process. A direct design process therefore becomes one of trial and error. For this reason, designers typically use various design aids: graphs, tables, or computer-aided processes.

A problem arises because of the large number of different species and grades of wood, resulting in many different combinations of allowable compression and modulus of elasticity. Nevertheless, aids are commonly used for at least a preliminary design choice, even when the wood properties do not quite match the design conditions. An advantage lies in the relatively small choice for column sizes if standard lumber dimensions are used.

Table 9.4 gives the axial compression capacity for a range of standard lumber sizes and specific values of unbraced length. Note that the design values for mem-

TABLE 9.4 Safe Loads for Wood Compression Members[a]

Column Section		Unbraced Length (ft)									
Nominal Size	Area (in.²)	6	8	10	12	14	16	18	20	22	24
2 × 4	5.25	1.26	0.70								
2 × 6	8.25	1.98	1.07								
2 × 8	10.875	2.6	1.41								
2 × 10	13.875	3.3	1.80								
2 × 12	16.875	4.0	2.2								
3 × 4	8.75	5.4	3.2	2.1							
3 × 6	13.75	8.5	5.0	3.3							
3 × 8	18.125	11.2	6.6	4.3							
3 × 10	23.125	14.2	8.4	5.5							
3 × 12	28.125	17.3	10.3	6.7							
4 × 4	12.25	12.6	8.2	5.5	3.9	2.9					
4 × 6	19.25	19.8	12.9	8.7	6.2	4.6					
4 × 8	25.375	26	17.0	11.5	8.2	6.1					
4 × 10	32.375	33	22	14.6	10.4	7.7					
4 × 12	39.375	40	26	17.8	12.7	9.4					
4 × 14	46.375	48	31	21	14.9	11.1					
4 × 16	53.375	55	36	24	17.2	12.8					
6 × 6	30.25	30	26	22	17.4	13.7	10.9	8.8	7.2	6.0	
6 × 8	41.25	41	36	30	24	18.6	14.8	12.0	9.8	8.2	
6 × 10	52.25	52	46	38	30	24	18.7	15.2	12.5	10.4	
6 × 12	63.25	63	55	46	36	28	23	18.4	15.1	12.6	
6 × 14	74.25	73	65	54	43	33	27	21	17.7	14.8	
6 × 16	85.25	84	75	62	49	38	30	25	20	17.0	
8 × 8	56.25	54	51.5	48.1	43.5	38.0	32.3	27.4	23.1	19.7	16.9
8 × 10	71.25	68.4	65.3	61.0	55.1	48.1	41.0	34.7	29.3	24.9	21.4
8 × 12	85.25	82.8	79.0	73.8	66.7	58.2	49.6	42.0	35.4	30.2	26.0
8 × 14	101.25	106	101	93	83	71	60	50	42	36	30
8 × 16	116.25	122	116	107	95	82	69	58	48	41	35

[a]Load capacity in kips for solid-sawn sections of select structural grade Douglas fir-larch under normal moisture and load duration conditions.

160 DESIGN OF WOOD TRUSSES

bers with nominal dimensions smaller than 5 in. are not obtained from the size classification group of "posts and timbers" in Table 9.1. Thus different values for F_c and E must be used for the two groups of sizes in Table 9.4, even though the grade is the same.

Spaced Columns

A type of element sometimes used in wood structures is the spaced column. This consists of two or more wood members of the same size that are fastened together to share a compressive load as a single unit (see Figure 9.7). One occurrence of this is the compression member in a truss that consists of members of layered wood elements. The following example shows the general procedure for analysis of such a compression member.

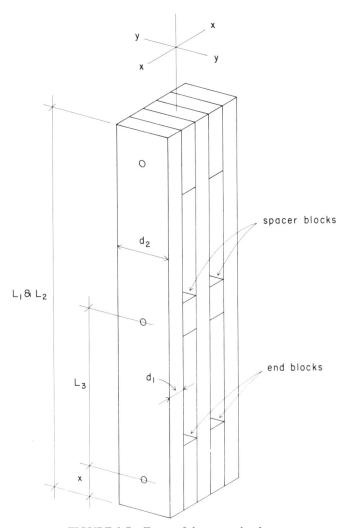

FIGURE 9.7 Form of the spaced column.

BASIC CONSIDERATIONS FOR WOOD 161

Example 4. A spaced column of the form shown in Figure 9.7 consists of three 3 × 12 pieces of Douglas fir–larch, No. 1 grade. Dimension L_1 is 11 ft-8 in. and x is 6 in. Find the axial compression capacity.

Solution: There are two separate conditions to be investigated for the *spaced column*. These relate to the effects of relative slenderness in the two directions, as designated by the x- and y- axes in Figure 9.7. In the y direction, the column behaves simply as a set of solid-sawn columns. Thus the stress is limited by the dimensions d_2 and L_2 and their ratio. For this condition we determine the following for the example.

$$\frac{L_2}{d_2} = \frac{(11.67)(12)}{11.25} = 12.5$$

Using this value for the slenderness ratio, we determine the capacity of a solid-sawn column with a cross section having an area three times that for one 3 × 12. This computation is the same as that presented in the preceding discussion of the solid-sawn column. Application of the example data will yield a value of 0.884 for the factor C_p. The capacity for this condition is then determined by multiplying this factor times the product of the tabulated value for compression stress and the column cross-section area. In this example, there is a second condition to consider, so we will determine the C_p factor for that condition before determining which is the critical condition for the column capacity.

For the condition of behavior with regard to buckling in the x direction, we first check for conformance with two limitations:

1. Maximum value for $L_3/d_1 = 40$.
2. Maximum value for $L_1/d_1 = 80$.

Thus, using the example data,

$$70/2.5 = 28 \quad \text{(less than 40)}$$

and

$$140/2.5 = 56 \quad \text{(less than 80)}$$

so the limits are not exceeded.

The capacity for this condition depends on the value of L_1/d_1 and is determined in a manner similar to that for the solid-sawn column, except that a modified form is used for the factor F_{cE} as follows:

$$F_{cE} = \frac{K_{cE} \, K_x \, E}{(L/d)^2}$$

The value for K_x is based on the conditions of the end blocks in the column. In the illustration in Figure 9.7 the distance x indicates the distance from the end of the column to the centroid of the connectors that are used to fasten the end blocks into the column. Two values for K_x are given, based on the relation of the x distance to the overall length of the column (L_1 in Figure 9.7). Thus

1. K_x = 2.5 when x is equal to or less than $L_1/20$.
2. K_x = 3.0 when x is between $L_1/20$ and $L_1/10$.

For our example, with x = 6 in., $L_1/20$ = 140/20 = 7. Thus K_x = 2.5, and we determine the value for F_{cE} as

$$F_{cE} = \frac{K_{cE} \; K_x \; E}{(L/d)^2} = \frac{(0.3)(2.5)(1,700,000)}{(56)^2} = 406 \text{ psi}$$

This value is used in the formula for C_p, as given in the discussion for the solid-sawn column. For our example, the computation is as follows:

$$C_p = \frac{1 + 0.28}{1.6} - \sqrt{\left[\frac{1 + 0.28}{1.6}\right]^2 - \frac{0.28}{0.8}} = 0.26$$

As this yields a value for C_p, that is less than that for the condition of y direction buckling, the capacity is limited by this condition. Thus the capacity is determined as

$$P = (F_c)(C_p)(A) = (1450)(0.26)(3)(28.13) = 31,815 \text{ lb}$$

Compression Plus Bending

There are a number of situations in which structural members are subjected to combined effects of axial compression and bending. Stresses developed by these two actions are both of the direct type (tension and compression) and can be combined for consideration of a net stress condition. However, the basic actions of a column and a bending member are essentially different in character, and it is therefore customary to consider this combined activity by what is called *interaction*.

The classic form of interaction is represented by the graph in Figure 9.8. Referring to the notation on the graph:

The maximum axial load capacity of the column (without bending) is P_0.
The maximum bending capacity of the member (without compression) is M_0.
At some applied compression load below P_0 the column has some capacity for bending in combination with the load. This combination is indicated as P_n and M_n.

The classic form of the interaction relationship is expressed in the formula

$$\frac{P_n}{P_0} + \frac{M_n}{M_0} = 1$$

BASIC CONSIDERATIONS FOR WOOD   163

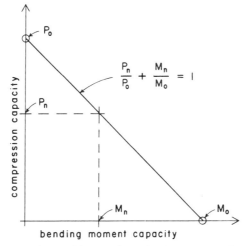

FIGURE 9.8 Classic form of interaction of axial compression and bending.

The plot of this equation is the straight line connecting P_0 and M_0 as shown on Figure 9.8.

A graph similar to that in Figure 9.8 can be produced using stresses rather than loads and moments. This is the procedure generally used in wood and steel design, with the graph taking a form expressed as

$$\frac{f_a}{F_a} + \frac{f_b}{F_b} \leq 1$$

in which f_a = computed stress due to the actual load,

F_a = allowable column action stress,

f_b = computed stress due to bending,

f_b = allowable bending stress.

For solid-sawn columns, the 1991 NDS provides the following formula for investigation of a column with combined axial compression and bending about one axis.

$$\left(\frac{f_c}{F'_c}\right)^2 + \frac{f_b}{F_b(1 - (f_c/F_{cE}))} \quad \text{equal to or less than 1}$$

in which f_c = computed compressive stress

F'_c = tabulated design value for compressive stress multiplied by the C_p factor

f_b = computed bending stress

164 DESIGN OF WOOD TRUSSES

F_b = tabulated design value for bending stress

F_{cE} = the value determined for a solid-sawn column

The following example will serve to demonstrate an application for the procedure.

Example 5. Figure 9.9 shows a loading condition for a top chord member of a truss. The proposed member is that given in Example 4 in the preceding discussion of the spaced column. Investigate the member for the combined column action and bending.

Solution: From Example 4 we determine the following:

The member is a spaced column consisting of three 3 × 12 pieces of Douglas fir–larch, No. 1 grade. A = 84.39 in.^{2}

$$F_c = 1450 \text{ psi}, F_{cE} = 406 \text{ psi}, C_p = 0.26$$

(Note: the truss chord is actually sloped, so its true length is 11.7 ft, as given in the preceding example. However, its *horizontal projection* is 10 ft, which is what we used for the span for the bending investigation).

For the interaction condition we determine the following:

$$f_c = P/A = 26{,}300/84.39 = 312 \text{ psi}$$
$$F_c' = C_p F_c = 0.26(1300) = 338 \text{ psi}$$

For this situation the bending takes place about the main axis of the members, and the d dimension used for determination of the bending factor in the interaction formula is that of the member of the direction of bending; for this example the 11.25 in. dimension. Thus

$$L/d = (10 \times 12)/11.25 = 10.67$$

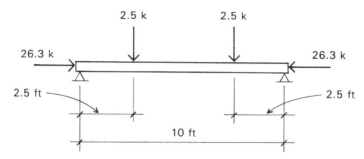

FIGURE 9.9 Example of a truss member subjected to combined compression and bending.

Then

$$F_{cE} = 0.3E/(L/d)^2 = 0.3(1{,}600{,}000)/(10.67)^2 = 4216 \text{ psi}$$

$$f_c/F_{cE} = 312/4216 = 0.074$$

Observing Figure 9.9, the maximum bending is

$$M = (2.5)(2.5) = 6.25 \text{ k-ft}$$
$$= (6.25)(12)(1000) = 75{,}000 \text{ in.-lb}$$

From Table 9.1: $S = 52.73$ in.^{3} for one 3×12, and thus S for the three element member is $3(52.73) = 158.19$ in.^{3}. Then

$$f_b = M/S = 75{,}000/158.19 = 474 \text{ psi}$$

Then, assuming a stress increase factor of 1.25 for the roof load (Table 9.2), the interaction investigation is as follows:

$$\left[\frac{312}{1.25(338)}\right]^2 + \frac{474}{1.25(1000)(1 - 0.074)} = 0.545 + 0.410 = 0.955$$

and the member is adequate.

9.2 JOINTS IN WOOD TRUSSES

As stated previously, most wood trusses are factory produced; thus their joints are achieved by methods yielding to economical, mass-production technology, not to means achievable by hand labor. However, when only a few, specially designed trusses are to be built, their joints must usually be achieved by the ordinary methods available for fastening of wood framing members in general. This section treats primarily the latter situation.

Bolted Joints

When steel bolts are used to connect wood members, there are several design considerations. The principal concerns are the following.

1. *Net Stress in Member.* Holes drilled for the placing of bolts reduce the member cross section. For this analysis the hole is assumed to have a diameter one sixteenth of an inch larger than that of the bolt. The most common situations are those shown in Figure 9.10a. When bolts are staggered, it may be necessary to make two investigations, as shown in the illustration.
2. *Bearing of the Bolt on the Wood and Bending in the Bolt.* When the members are thick and the bolt thin and long, the bending of the bolt will cause a concentration of stress at the edges of the member. The bearing on the wood is further

166 DESIGN OF WOOD TRUSSES

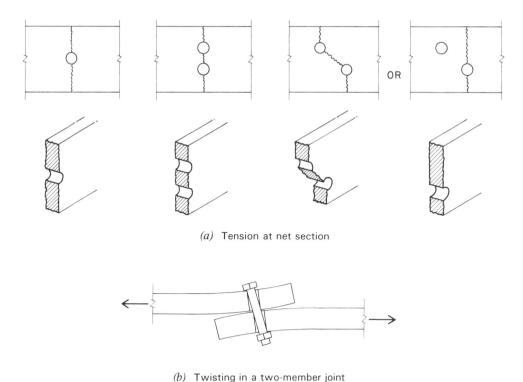

FIGURE 9.10 Critical behaviors in tension members: (*a*) Effect of bolt holes on reduction of cross section. (*b*) Twisting in a single lapped joint (called a two-member joint).

limited by the angle of the load to the grain, since the wood is much stronger in the grain direction.

3. *Number of Members Bolted.* The worst case, as shown in Figure 9.10*b*, is that of the two-member joint. In this case the lack of symmetry in the joint produces considerable twisting. This situation is referred to as single shear, since the bolt is subjected to shear on a single plane. When more members are joined, this twisting effect is reduced.

4. *Ripping Out the Bolt When Too Close to an Edge.* This problem, together with that of the minimum spacing of the bolts in multiple bolt joints, is dealt with by using the criteria given in Figure 9.11. Note that the limiting dimensions involve the consideration of the bolt diameter D; the bolt design length L; the type of force—tension or compression; and the angle of load to the grain of the wood.

5. *Length of the Bolt.* The bolt length is one part of the definition of the bearing contact surface with the wood. It may also be of concern if the length is relatively large in proportion to the bolt diameter, resulting in considerable bending of the bolt. For design control, bolt length is defined in terms of the thicknesses of the connected wood members.

The 1991 NDS presents materials for design of bolted joints in layered stages. The first stage involves general concerns for the basic wood construction; notably,

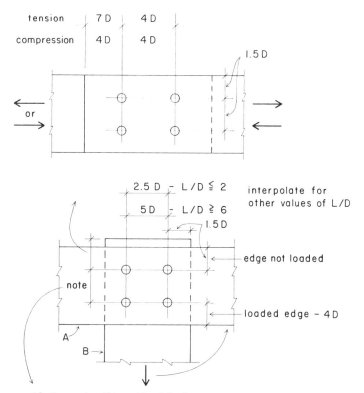

FIGURE 9.11 Edge, end, and spacing dimensions for bolts in wood structures.

adjustments for load duration, moisture condition, and the particular species and grade of the wood.

The second layer of concern has to do with some considerations for structural fastenings in general. These are addressed in NDS Part VII, which contains provisions for some of the general concerns for the form of connections, arrangements of multiple fasteners, and the eccentricity of forces in connections.

The third layer of concern involves considerations for the particular type of fastener. NDS Part VII contains provisions for bolted connections, while six additional parts and some appendix materials treat particular fastening methods and special problems. All in all, the number of separate concerns are considerable.

For a single-bolted joint, the capacity of a bolt may be expressed as

$$Z' = Z(C_D)(C_M)(C_t)(C_g)(C_\Delta)$$

where Z' = the adjusted bolt design value

Z = the nominal bolt design value, based on one of many possible modes of failure

C_D = the load duration factor (from Table 9.3)

168 DESIGN OF WOOD TRUSSES

C_M = the factor for any special moisture condition

C_t = the temperature factor for extreme climate conditions

C_g = the group action factor for joints with more than one bolt in a row in the direction of the load

C_Δ = a factor related to the geometry (general form) of the connection (such as single shear, double shear, eccentrically loaded, etc.)

For all of this complexity, additional adjustments may be necessary. For bolted joints, two additional concerns have to do with the angle of the load to the grain direction in the connected members and the adequacy of dimensions in the bolt layout.

For the direction of loads with respect to the grain, there are two major positions: (1) that with the load parallel to the grain (see Figure 9.12a) and that with the load perpendicular to the grain (Figure 9.12b). Between these is a so-called angle-to-grain loading (Figure 9.12c), for which an adjustment is made using the Hankinson formula. Figure 9.13 illustrates the application of the Hankinson formula in the form of a graph. For the illustration the bolt capacity parallel to the grain is expressed as P and the capacity perpendicular to the grain is expressed as Q. Considering the P direction as zero and the Q direction as 90°, angles between zero and 90 are expressed by the specific value, designated by the Greek letter theta in the illustration. An example of the use of the graph is shown in Figure 9.13.

Bolt layout becomes critical when a number of bolts must be used in members of relatively narrow dimension. Major dimensional concerns are illustrated in Figure 9.11. The NDS generally establishes two limiting dimensions. The first limit is the minimum required for design use of the full value of the bolt capacity. The second limit is an absolute minimum, for which some reduction of capacity is specified. In general, any dimensions falling between these two limits in real situations may be used to establish bolt capacity values by direct interpolation between the limiting capacity values.

Using all of the NDS requirements in a direct way in practical design work is a

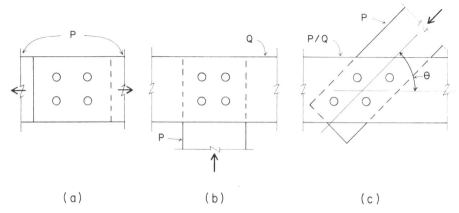

FIGURE 9.12 Relation of direction of load to orientation of wood grain in joints with wood members.

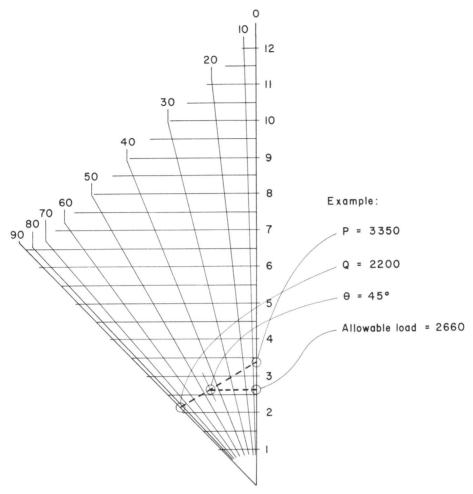

FIGURE 9.13 Hankinson graph for determination of load values with loadings at an angle to the wood grain. See also Fig. 9.3 and Table 9.3.

mess. Consequently, the NDS, or others, provide some shortcuts. One such aid is represented by a series of tables that allow the determination of a bolt capacity value by direct use of the table. Care must be used in using the various values for data that are incorporated in the table. However, even more care must be used in paying attention to what is not incorporated into the tables; notably, first layer concerns described above for load duration and moisture condition.

Table 9.5 presents a sampling from much larger tables in the 1991 NDS; data being given here only for Douglas fir–larch, while data for several species of wood are included in the reference source. Table 9.5 is compiled from two separate tables in the NDS, where one table provides data only for single shear joints and another table provides data only for double shear joints. These tables provide values for joints achieved with all wood members; other NDS tables also provide data for joints achieved with combinations of wood members and steel splice or gusset plates.

TABLE 9.5 Bolt Design Values for Wood Joints with Douglas Fir–Larch (lb/bolt)

Thickness			Douglas Fir–Larch					
Main Member	Side Member	Bolt Diameter	Single Shear			Double Shear		
t_m inches	t_s inches	D inches	$Z_\parallel$ lbs.	$Z_{s\perp}$ lbs.	$Z_{m\perp}$ lbs.	$Z_\parallel$ lbs.	$Z_{s\perp}$ lbs.	$Z_{m\perp}$ lbs.
1-1/2	1-1/2	1/2	480	300	300	1050	730	470
		5/8	600	360	360	1310	1040	530
		3/4	720	420	420	1580	1170	590
		7/8	850	470	470	1840	1260	630
		1	970	530	530	2100	1350	680
2-1/2	1-1/2	1/2	610	370	370	1230	730	790
		5/8	850	520	430	1760	1040	880
		3/4	1020	590	500	2400	1170	980
		7/8	1190	630	550	3060	1260	1050
		1	1360	680	610	3500	1350	1130
3	1-1/2	1/2	610	370	420	1230	730	860
		5/8	880	520	480	1760	1040	1050
		3/4	1190	590	550	2400	1170	1170
		7/8	1390	630	610	3180	1260	1260
		1	1590	680	670	4090	1350	1350
3-1/2	1-1/2	1/2	610	370	430	1230	730	860
		5/8	880	520	540	1760	1040	1190
		3/4	1200	590	610	2400	1170	1370
		7/8	1590	630	680	3180	1260	1470
		1	1830	680	740	4090	1350	1580
	3-1/2	1/2	720	490	490	1430	970	970
		5/8	1120	700	700	2240	1410	1230
		3/4	1610	870	870	3220	1750	1370
		7/8	1970	1060	1060	4290	2130	1470
		1	2260	1230	1230	4900	2580	1580
4-1/2	1-1/2	5/8	880	520	590	1760	1040	1190
		3/4	1200	590	750	2400	1170	1580
		7/8	1590	630	820	3180	1260	1890
		1	2050	680	890	4090	1350	2030
	3-1/2	5/8	1120	700	730	2240	1410	1460
		3/4	1610	870	1000	3220	1750	1760
		7/8	2190	1060	1160	4390	2130	1890
		1	2610	1290	1290	5330	2580	2030
5-1/2	1-1/2	5/8	880	520	590	1760	1040	1190
		3/4	1200	590	790	2400	1170	1580
		7/8	1590	630	1060	3180	1260	2030
		1	2050	680	1270	4090	1350	2480
	3-1/2	5/8	1120	700	730	2240	1410	1460
		3/4	1610	870	1030	3220	1750	2050
		7/8	2190	1060	1260	4390	2130	2310
		1	2660	1290	1390	5330	2580	2480

JOINTS IN WOOD TRUSSES   171

TABLE 9.5 *(Continued)*

Thickness			Douglas Fir–Larch					
Main Member	Side Member	Bolt Diameter	Single Shear			Double Shear		
t_m inches	t_s inches	D inches	$Z_\parallel$ lbs.	$Z_{s\perp}$ lbs.	$Z_{m\perp}$ lbs.	$Z_\parallel$ lbs.	$Z_{s\perp}$ lbs.	$Z_{m\perp}$ lbs.
7-1/2	1-1/2	5/8	880	520	590	1760	1040	1190
		3/4	1200	590	790	2400	1170	1580
		7/8	1590	630	1010	3180	1260	2030
		1	2050	680	1270	4090	1350	2530
	3-1/2	5/8	1120	700	730	2240	1410	1460
		3/4	1610	870	1030	3220	1750	2050
		7/8	2190	1060	1360	4390	2130	2720
		1	2660	1290	1630	5330	2580	3380

Source: Adapted from data in the *National Design Specification for Wood Construction* (Ref. 3), with permission of the publishers, the National Forest Products Association.

The following examples illustrate the use of the materials presented here from the NDS.

Example 1. A three-member (double shear) joint is made with members of Douglas fir–larch, select structural grade lumber (see Figure 9.14). The joint is loaded as shown, with the load parallel to the grain direction in the members. The tension force is 9 kips. The middle member (designated main member in Table 9.5) is a 3 × 12 and the outer members (side members in Table 9.5) are each 2 × 12. Is the joint adequate for the required load if it is made with four $\frac{3}{4}$ in. bolts as shown? No

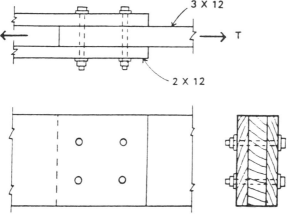

FIGURE 9.14 Example 1.

adjustment is required for moisture or load duration and it is assumed that the layout dimensions are adequate for use of the full design values for the bolts.

Solution: From Table 9.5, the allowable load per bolt is 2400 lb. With four bolts, the total joint capacity of the bolts is thus

$$T = 4(2400) = 9600 \text{ lb}$$

based only on the bolt actions in bearing and shear in the joint.

For tension stress in the wood, the critical condition is for the middle member, since its thickness is less than the total of the thicknesses of the side members. With the holes typically being considered as $\frac{1}{16}$-in. larger than the bolts, the net section through the two bolts across the member is thus

$$A = (2.5)[11.25 - (2)(13/16)] = 24.06 \text{ in.}^2$$

From Table 9.1, the allowable tension stress for the wood is 1000 psi. The maximum tension capacity of the middle member is thus

$$T = \text{(allowable stress) (net area)}$$
$$= (1200)(24.06) = 24,060 \text{ lb}$$

and the joint is adequate for the load required.

It should be noted that the NDS provides for a reduction of capacity in joints with multiple connectors. However, the factor becomes negligible for joints with as few as only two bolts per row and other details as given for the preceding example.

Example 2. A bolted two-member joint consists of two 2 X 10 members of Douglas fir–larch, select structural grade lumber, attached at right angles to each other, as shown in Figure 9.15. What is the maximum capacity of the joint if it is made with two $\frac{7}{8}$-in. bolts?

Solution: This is a single-shear joint with both members of 1.5-in. thickness and no concern for tension stress on a net section. The load limit is thus that for the perpendicular-to-grain loading on the 1.5-in.-thick side piece. From Table 9.5, this value is 470 lb per bolt, and the total joint capacity is thus

$$C = (2)(470) = 940 \text{ lb}$$

Example 3. A three-member joint consists of two outer members, each 2 X 10, bolted to a 4 X 12 middle member. The outer members are arranged at an angle to the middle member, as shown in Figure 9.16. Wood of all members is Douglas fir–larch, select structural grade. Find the maximum compression force that can be transmitted through the joint by the outer members.

Solution: In this case we must investigate both the outer and middle members. The load is parallel to the grain in the outer members and at an angle of 45° to the grain in the middle member.

For the outer members, Table 9.5 yields a value of 2400 lb per bolt. (Main member 3.5-in., side members 1.5-in. thickness, double shear, load parallel to grain.)

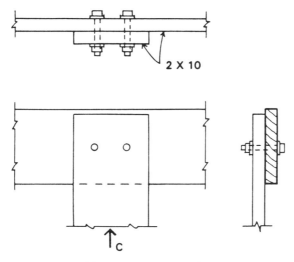

FIGURE 9.15 Example 2.

For the main member we find the two limiting values of 2400 lb and 1370 lb for the directions parallel and perpendicular to the grain. Entering these on the graph in Figure 9.13, we find a load of approximately 1740 lb per bolt for the 45-degree loading. Since this is less than the limit for the outer members, the total capacity of the joint is thus

$$C = (2)(1740) = 3480 \text{ lb}$$

Nailed Joints

Nails are available in a wide range of sizes and forms for many purposes. They range in size from tiny tacks to huge spikes. Most nails are driven by someone pounding a hammer—more or less as they have been for hundreds of years. For

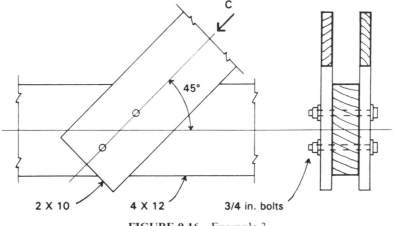

FIGURE 9.16 Example 3.

174 DESIGN OF WOOD TRUSSES

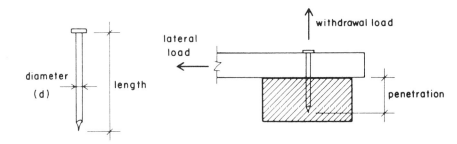

FIGURE 9.17 Typical form of the common wire nail and its basic loading considerations.

situations where many nails must be driven, however, we now have a variety of mechanical driving devices. We also have other types of fasteners, such as screws, staples, and toothed plates, which have replaced nails for many connections.

For structural fastening in light wood framing, the nail most commonly used is called, appropriately, the *common wire nail*, or simply the common nail. Basic concerns for structural use of nails, as shown in Figure 9.17, are the following:

Nail Size. Critical dimensions are the diameter and length. Sizes are specified in pennyweight units, designated as 4d, 6d, and so on, and referred to as four penny, six penny, and so on.

Load Direction. Pull-out loading in the direction of insertion is called *withdrawal*; shear loading perpendicular to the nail shaft is called *lateral load*.

Penetration. Nailing is typically done through some element and into another. The resistance capacity is based on the length of the embedment in the second

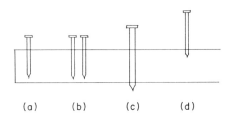

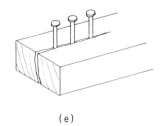

FIGURE 9.18 Poor nailing practices: (*a*) Nail too close to member edge. (*b*) Nails too closely spaced. (*c*) Nail too large for wood piece. (*d*) Too little penetration of nail into holding piece. (*e*) Too many nails in a single row parallel to the wood grain.

member. This length is called the penetration. A minimum amount of penetration is required to develop the full lateral load capacity.

Species and Grade of the Wood. The harder, tougher, and more dense the wood, the more the load resistance capability.

Design of good nail joints requires a little engineering and a lot of good carpentry. Some obvious situations to avoid are those shown in Figure 9.18.

Withdrawal load capacities of common nails are given in units of load capacity per inch of penetration. In general, it is best not to rely on withdrawal for the development of structural joints.

Lateral load capacities for common nails are given in pounds of total force capacity per nail. Values for commonly used sizes are given in Table 9.6. Note that orientation of the load to the grain in the wood is not a concern, except that it is assumed that the nails are driven at right angles to the grain in general.

Nails are not ideal fasteners for truss joints because of the potential for movement in the joints and possible loosening of the joints due to initial shrinkage of the wood or of repeated swelling and shrinkage over time due to fluctuations of moisture and temperature. In light trusses with plywood gusset plates, it is common to glue the plates in place, using the nails to clamp the glued joint, although the nails will also actually be designed for the joint loads. It is also common to use wood screws in place of nails for their greater resistance to loosening.

The following example illustrates the design of a typical nailed joint.

Example 4. A structural joint is formed as shown in Figure 9.19, with the wood members connected by 16d common wire nails. Wood is Douglas fir–larch. What is the maximum value for the compression force in the two side members?

Solution: From Table 9.6, we read a value of 141 lb per nail. (Side member thickness of 1.5-in., 16d nails.) The total joint load capacity is thus

$$C = (10)(141) = 1410 \text{ lb}$$

No adjustment is made for direction of load to the grain. However, the basic form of nailing assumed here is so-called side grain nailing, in which the nail is inserted at 90° to the grain direction and the load is perpendicular (lateral) to the nails.

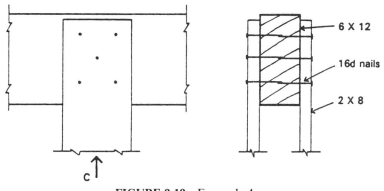

FIGURE 9.19 Example 4.

TABLE 9.6 Lateral Load Capacity of Common Wire Nails (lb/nail)

Side Member Thickness t_s inches	Nail Length L inches	Nail Diameter D inches	Penny-Weight	G = 0.50 Douglas Fir–Larch Z lbs.
1/2	2	0.113	6d	59
	2-1/2	0.131	8d	76
	3	0.148	10d	90
	3-1/4	0.148	12d	90
	3-1/2	0.162	16d	105
	4	0.192	20d	124
	4-1/2	0.207	30d	134
	5	0.225	40d	147
	5-1/2	0.244	50d	151
	6	0.263	60d	171
5/8	2	0.113	6d	66
	2-1/2	0.131	8d	82
	3	0.148	10d	97
	3-1/4	0.148	12d	97
	3-1/2	0.162	16d	112
	4	0.192	20d	130
	4-1/2	0.207	30d	140
	5	0.225	40d	151
	5-1/2	0.244	50d	155
	6	0.263	60d	175
3/4	2-1/2	0.131	8d	90
	3	0.148	10d	105
	3-1/4	0.148	12d	105
	3-1/2	0.162	16d	121
	4	0.192	20d	138
	4-1/2	0.207	30d	147
	5	0.225	40d	158
	5-1/2	0.244	50d	162
	6	0.263	60d	181
1	3	0.148	10d	118
	3-1/4	0.148	12d	118
	3-1/2	0.162	16d	141
	4	0.192	20d	159
	4-1/2	0.207	30d	167
	5	0.225	40d	177
	5-1/2	0.244	50d	181
	6	0.263	60d	199
1-1/4	3-1/4	0.148	12d	118
	3-1/2	0.162	16d	141
	4	0.192	20d	170
	4-1/2	0.207	30d	186
	5	0.225	40d	200
	5-1/2	0.244	50d	204
	6	0.263	60d	222

TABLE 9.6 (*Continued*)

Side Member Thickness	Nail Length	Nail Diameter	Penny-Weight	G = 0.50 Douglas Fir–Larch
t_s inches	L inches	D inches		Z lbs.
1-1/2	3-1/2	0.162	16d	141
	4	0.192	20d	170
	4-1/2	0.207	30d	186
	5	0.225	40d	205
	5-1/2	0.244	50d	211
	6	0.263	60d	240

Source: Adapted from data in the *National Design Specification for Wood Construction* (Ref. 3), with permission of the publishers, the National Forest Products Association.

Adequate penetration of the nails into the supporting member is also a problem, but use of the combinations given in Table 9.6 assures adequate penetration if the nails are fully buried in the members.

Shear Developers

A problem for joints in wood frameworks is that of the potential for deformation, or lack of tightness, in the joints. This is especially critical for some situations; truss joints are one such case. The degree of the problem is affected by the form of the joint and the type of fastening used.

Bolted joints generally lack enduring tightness for a number of reasons. To reduce slippage (shearing movement) in a joint, it is sometimes possible to insert a device that is specifically used to reduce the slipping of the wood members at their contact faces. Toothed or ridged elements of metal are sometimes used for large timber joints; these are simply placed between the members and the tightening of the bolts causes them to bite into both members.

Two shear-developing devices used for both timber and light frame elements are the *split-ring connector* and the *shear plate*. The split-ring connector consists of a steel ring that is installed by cutting matching circular grooves in the faces of the lapped pieces of wood. When the ring is inserted into the grooves and the bolt is tightened, the ring is squeezed into the grooves and the resulting connection has both a degree of tightness and a shear-resisting strength greater than that with the bolt alone.

The shear plate is a round device that is also sunk in a circular groove. Two plates are used at each contact face; one in each member. The bolts are inserted through close matching holes in the shear plates. The plates bite into and grab the wood, but the basic nature of the joint is more that of a connection of steel members. A single-shear plate can also be used for splice joints using steel side plates; a principal use for this device.

The following discussion illustrates the use of split-ring connectors. The form of data and the general design process for shear plates is quite similar. Extensive data and criteria for both devices is provided in the NDS.

Split-Ring Connectors

The ordinary form of the split-ring connector and the method of installation are shown in Figure 9.20. Design considerations for this device include the following:

1. *Size of the Ring.* Rings are available in the two sizes shown in the figure with nominal diameters of 2.5 and 4 in.
2. *Stress on the Net Section of the Wood Member.* As shown in Fig. 9.20, the cross section of the wood piece is reduced by the ring profile (A in the figure) and the bolt hole. If rings are placed on both sides of a wood piece, there will be two reductions for the ring profile.
3. *Thickness of the Wood Piece.* If the wood piece is too thin, the cut for the ring will bite excessively into the cross section. Rated load values reflect concern for this.
4. *Number of Faces of the Wood Piece Having Rings.* As shown in Figure 9.21, the outside members in a joint will have rings on only one face, whereas the inside members will have rings on both faces. Thickness considerations therefore are more critical for the inside members.

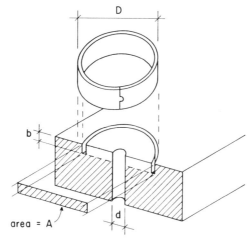

FIGURE 9.20 Split-ring connectors for shear development in bolted joints.

JOINTS IN WOOD TRUSSES   179

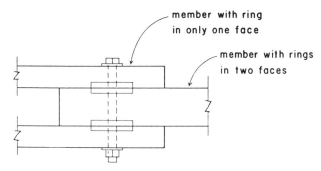

FIGURE 9.21 Determination of the number of faces of a member with split-ring connectors.

5. *Edge and End Distances.* These must be sufficient to permit the placing of the rings and to prevent splitting out from the side of the wood piece when the joint is loaded. Concern is greatest for the edge in the direction of loading—called the loaded edge (see Figure 9.22).
6. *Spacing of Rings.* Spacing must be sufficient to permit the placing of the rings and the full development of the ring capacity of the wood piece.

Figure 9.23 shows the four placement dimensions that must be considered. The limits for these dimensions are given in Table 9.7. In some cases, two limits are given. One limit is that required for the full development of the ring capacity (100% in the table). The other limit is the minimum dimension permitted for which some reduction factor is given for the ring capacity. Load capacities for dimensions between these limits can be directly proportioned.

Table 9.8 gives capacities for split-ring connections for both dense and regular grades of Douglas fir–larch and Southern pine. As with bolts, values are given for load directions both parallel to and perpendicular to the grain of the wood. Values for loadings at some angle to the grain can be determined with the use of the Hankinson formula given in Section 9.1 or the graph shown in Figure 9.13.

The following example illustrates the procedures for the analysis of a joint using split-ring connectors.

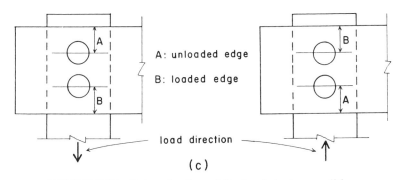

FIGURE 9.22 Determination of the loaded edge condition.

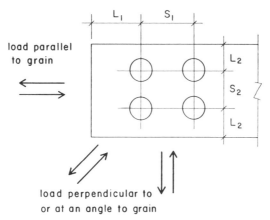

FIGURE 9.23 Reference figure for the edge, end, and spacing distances in Table 9.7.

TABLE 9.7 Spacing, Edge Distance, and End Distance for Split-Ring Connectors

Load Direction with Respect to Grain		Distances (in Inches) and Corresponding Percentages of Design Values from Table 9.8			
		Parallel		Perpendicular or Angle	
Ring Size (in.)		2.5	4	2.5	4
L_1	Tension	5.5, 100% 2.75 min., 62.5%	7, 100% 3.5 min., 62.5%	5.5, 100% 2.75 min., 62.5%	7, 100% 3.5 min., 62.5%
	Compression	4, 100% 2.5 min., 62.5%	5.5, 100% 3.25 min., 62.5%	5.5, 100% 2.75 min., 62.5%	5.5, 100% 3.25 min., 62.5%
L_2	Unloaded	1.75 min., 100%	2.75 min., 100%	1.75 min., 100%	2.75 min., 100%
	Loaded[a]	1.75 min., 100%	2.75 min., 100%	2.75, 100% 1.75 min., 83%	3.75, 100% 2.75 min., 83%
S_1		6.75, 100% 3.5 min., 50%	9, 100% 5 min., 50%	3.5 min., 100%	5 min., 100%
S_2		3.5 min.	5 min.	4.25, 100% 3.5 min., 50%	6, 100% 5 min., 50%

Reference for Table is Figure 9.23.
[a] See Table 9.8 and Figure 9.22.

TABLE 9.8 Design Values for Split-Ring Connectors

Ring Size (in.)	Bolt Diameter (in.)	Faces with Connectors[a]	Actual Thickness of Piece (in.)	Load Parallel to Grain Design Value/Connector (lb)		Distance to Loaded Edge[c] (in.)	Load Perpendicular to Grain Design Value/Connector (lb)	
				Group A Woods[b]	Group B Woods[b]		Group A Woods[b]	Group B Woods[b]
2.5	$\frac{1}{2}$	1	1 min.	2630	2270	1.75 min.	1580	1350
						2.75 or more	1900	1620
			1.5 or more	3160	2730	1.75 min.	1900	1620
						2.75 or more	2280	1940
		2	1.5 min.	2430	2100	1.75 min.	1460	1250
						2.75 or more	1750	1500
			2 or more	3160	2730	1.75 min.	1900	1620
						2.75 or more	2280	1940
4	$\frac{3}{4}$	1	1 min.	4090	3510	2.75 min.	2370	2030
						3.75 or more	2840	2440
			1.5 or more	6020	5160	2.75 min.	3490	2990
						3.75 or more	4180	3590
		2	1.5 min.	4110	3520	2.75 min.	2480	2040
						3.75 or more	2980	2450
			2	4950	4250	2.75 min.	2870	2470
						3.75 or more	3440	2960
			2.5	5830	5000	2.75 min.	3380	2900
						3.75 or more	4050	3480
			3 or more	6140	5260	2.75 min.	3560	3050
						3.75 or more	4270	3660

[a] See Figure 9.21.
[b] Group A includes dense grades, and Group B regular grades of Douglas fir-larch.
[c] See Figure 9.23.

182 DESIGN OF WOOD TRUSSES

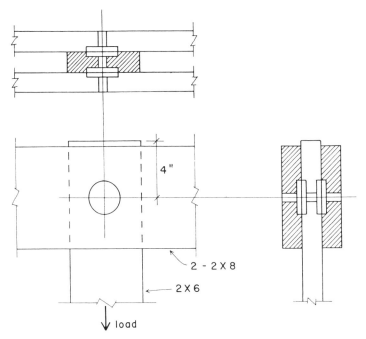

FIGURE 9.24 Example 5.

Example 5. The joint shown in Figure 9.24, using $2\frac{1}{2}$ in. split-rings and wood of Douglas fir–larch, No. 1 grade, sustains the load indicated. Find the limiting value for the load.

Solution: Separate investigations must be made for the members in this joint. For the 2 × 6,

Load is parallel to the grain.

Rings are in two faces.

Critical dimensions are member thickness of 1.5 in. [38 mm] and end distance of 4 in.

From Table 9.7 we determine that the end distance required for use of the full capacity of the rings is 5.5 in. and that, if the minimum distance of 2.75 in. is used, the capacity must be reduced to 62.5% of the full value. The value to be used for the 4 in.

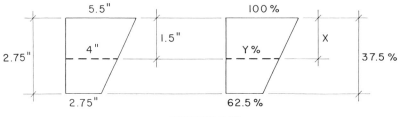

FIGURE 9.25

end distance must be interpolated between these limits, as shown in Figure 9.25. Thus

$$\frac{1.5}{x} = \frac{2.75}{37.5} \quad x = \frac{1.5}{2.75}(37.5) = 20.45\%$$

$$y = 100 - 20.45 = 79.55\%, \text{ or approximately } 80\%$$

From Table 9.8 we determine the full capacity to be 2100 lb per ring. Therefore the usable capacity is

$$(0.80)(2100) = 1680 \text{ lb/ring } [7.47 \text{ kN/ring}]$$

For the 2 × 8,

Load is perpendicular to the grain.
Rings are in only one face.
Loaded edge distance is one-half of 7.25 in., or 3.625 in. [92 mm].

For this situation the load value from Table is 9.8 is 1940 lb [8.63 kN]. Therefore the joint is limited by the conditions for the 2 × 6, and the capacity of the joint with the two rings is

$$T = (2)(1680) = 3360 \text{ lb } [14.9 \text{ kN}]$$

It should be verified that the 2 × 6 is capable of sustaining this load in tension stress on the net section at the joint. As shown in Figure 9.26, the net area is

$$A = 8.25 - (2)(1.10) - \left(\frac{9}{16}\right)(0.75) = 5.63 \text{ in.}^2 \ [3632 \text{ mm}^2]$$

From Table 9.1 the allowable tension stress is 675 psi [4.65 MPa], and therefore the capacity of the 2 × 6 is

$$T = (675)(5.63) = 3800 \text{ lb } [16.9 \text{ kN}]$$

and the member is not critical in tension stress.

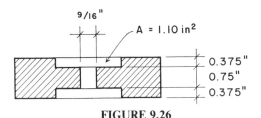

FIGURE 9.26

184 DESIGN OF WOOD TRUSSES

FIGURE 9.27 Use of metal connectors for light wood trusses.

Miscellaneous Jointing Methods

Most of the jointing methods treated in the preceding discussions are used only for custom-designed trusses. Trusses produced in factories or assembly shops generally use some form of proprietary jointing methods that employ woodworking and hardware that is not developed for ordinary on-site carpenter work.

Light wood trusses now often employ some form of perforated metal plate connectors, of a form described in the illustration in Figure 9.27. The sheet metal plate connector is pressed into the sides of the joined wood pieces, functioning both as a gusset plate and a multiple fastening device. The specific details of such connectors are sometimes peculiar to individual fabricators as proprietary products.

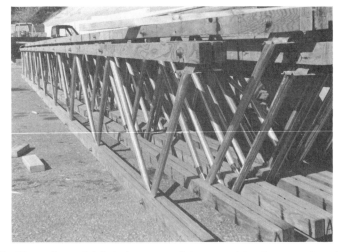

FIGURE 9.28 Formation of joints in a proprietary combination wood and steel truss.

Manufactured trusses of various forms and sizes often employ some proprietary jointing techniques. Composite trusses with wood chords and steel-web members are in this category, an example being that shown in Figure 9.28. Accepted load capacities for such trusses are based on load tests of sample trusses, rather than reviewed computations of the truss designs.

Jointing methods used for manufactured trusses may appear complicated when compared to the simpler forms used for ordinary wood joints achieved by hand labor. However, they would not be used, if they were not indeed truly easy to achieve and quite economical for the particular factory production processes. In general, both hand and factory methods have their applications, and both are likely to continue to be used, with neither generally appropriate for the other's particular situation.

While truss construction has always involved considerable use of crafted jointing and the development of some elaborate hardware, the assemblage of light wood frames was in the past achieved with little sophisticated hardware. Now, this has changed considerably, and jointing of wood frame structures in general is achieved with an extensive inventory of connecting devices. Thus the achieving of truss joints is somewhat less special in the general development of today's wood structures.

9.3 LIGHT WOOD TRUSSES WITH SINGLE MEMBERS

The following example illustrates the analysis and design of a light wood truss that is utilized as a combined roof rafter and ceiling joist for a gable-form roof. The truss configuration and design loads are as shown in Figure 9.29. Wood used will be Douglas fir–larch, select structural grade. Trusses are 2 ft on center.

Although this truss will be very light, its weight should be added to the given superimposed loads. If we use the approximate value from Table 7.1, the truss weight is 10 lbs/ft. Thus

$$\text{total truss weight} = 10(30) = 300 \text{ lb}$$

$$\text{load per joint} = \frac{300}{5} = 60 \text{ lb}$$

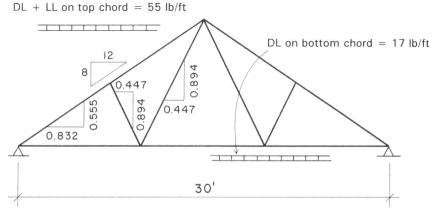

FIGURE 9.29 Truss form and loading for the design example.

186   DESIGN OF WOOD TRUSSES

Thus the joint design loads are as follows:

Top chord interior: roof load = (55)(7.5) = 413
                     truss weight =   60
                     total =   473 lb
                                           (say 480)

Bottom chord interior: ceiling load = (17)(10) = 170
                        truss weight =   60
                        total =   230 lb

Ends: roof load = (55)(4.33) = 240
       ceiling = (17)(5)   = 85
       total =   = 325 lb

This loading is shown on the truss space diagram in Figure 9.30, together with the truss reactions. The Maxwell diagram and the separated joint diagram for this loading are shown in Figure 9.31. We now proceed to design the members and joints for the truss with the following assumptions:

1. Members are to be single elements.
2. Joints will be made with nails and gusset plates.
3. Roof and ceiling construction provides adequate lateral bracing for the chords.
4. Roof live load duration permits an increase of 1.25 in the allowable stresses.

It is likely that the top chord will be made with a single member on each side of the gable roof as shown in Figure 9.32. The critical design force, therefore, is the higher of the two values, that given for member *BJ*. Thus this member will be

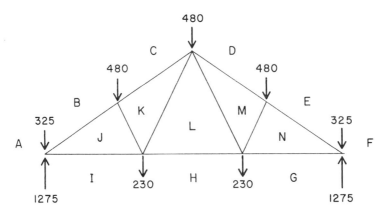

FIGURE 9.30 Truss loads and reactions.

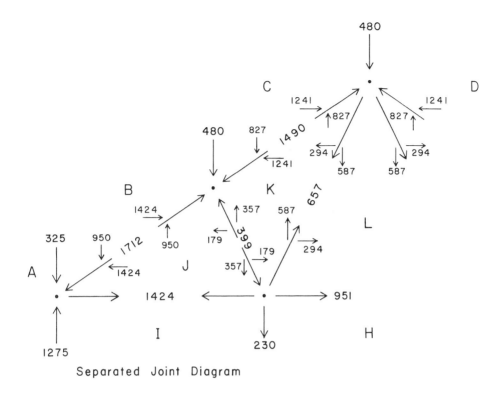

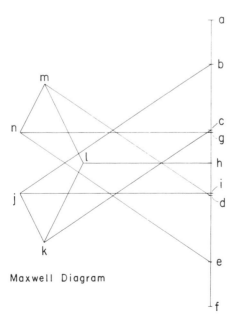

FIGURE 9.31 Analysis for internal forces. The upper figure shows the separated joint diagram for the algebraic analysis and the lower figure shows the graphical analysis, with the Maxwell diagram for member forces.

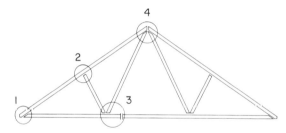

FIGURE 9.32 Layout form for the members and joints.

designed for the axial force in compression of 1712 lb and the bending due to the direct loading. For the compression force, we determine that

$$\text{member length is } \frac{1}{0.832} (7.5 \times 12) = 108 \text{ in.}$$

If we select a 2 × 4, we obtain from Table 5.1:

$$A = 5.25 \text{ in.}^2, \quad S = 3.063 \text{ in.}^3$$

With the member braced by the roof deck on its weak axis, the slenderness ratio is

$$\frac{L}{d} = \frac{108}{3.5} = 30.86$$

Assuming a member with 2 in. nominal thickness, we determine from Table 9.1 that

$$F_c = 1700 \text{ lb/in.}^2$$
$$E = 1,900,000 \text{ lb/in.}^2$$

From Table 9.1,

$$F_b = 1450 \text{ lb/in.}^2$$

$$M = \frac{wL^2}{8} = \frac{(55)(7.5)^2}{8} = 387 \text{ lb-ft} \quad \text{(bending moment)}$$

$$f_b = \frac{M}{S} = \frac{(387)(12)}{3.063} = 1515 \text{ lb/in.}^2 \quad \text{(actual bending stress)}$$

$$f_c = \frac{1712}{5.25} = 326 \text{ lb/in}^2 \quad \text{(actual axial stress)}$$

Using this data, an investigation may now be made for the allowable axial compression capacity and the condition of combined compression and bending for the top chord, as discussed in Section 7.7, using the procedures for wood members as illustrated in Section 9.1. This investigation will show the 2 × 4 to be considerably overstressed. Increasing the size to a 2 × 6 considerably increases the bending resistance (section modulus property) and reduces the slenderness (lower L/d ratio), so that the combined action will be found not to be critical.

It is doubtful that the bottom chord would be made of a single piece for its entire 30 ft length. We therefore assume it to be spliced with an arrangement of members as shown in Figure 9.32. Thus the bottom chord will be designed for the axial tension force of 1424 lb in member JI, although the splice need only be designed for the force of 829 lb in member LH.

The bottom chord must be designed for the combined bending and axial tension action as follows:

Assuming a 2 × 4 member, for which A = 5.25 in.^{2} and S = 3.063 in.^{3},

$$f_t = \frac{1424}{5.25} = 271 \text{ lb/in}^2 \quad \text{(actual axial stress)}$$

$$M = \frac{wL^2}{8} = \frac{(20)(10)^2}{8} = 250 \text{ lb-ft} \quad \text{(bending moment)}$$

$$f_b = \frac{M}{S} = \frac{(250)(12)}{3.063} = 979 \text{ lb/in}^2 \quad \text{(actual bending stress)}$$

From Table 9.1,

$$F_t = 1000 \text{ lb/in}^2, \quad F_b = 1450 \text{ lb/in}^2$$

Then

$$\frac{f_t}{F_t} + \frac{f_b}{F_b} = \frac{271}{(1.25)1000} + \frac{979}{(1.25)1450}$$

$$= 0.217 + 540$$

$$= 0.757$$

which indicates that the 2 × 4 is quite adequate.

Investigation of the web members will show that 2 × 4 members are quite adequate. In fact, it would be possible to use 2 × 3 members to save weight, although consideration should be given to the jointing method which may benefit from a wider member.

For the truss joints we will assume the use of nailed plywood gusset plates. Nailing options, as shown in Figure 9.33, consist of 6d nails driven from both sides or 12d nails driven through from one side and clinched. For these conditions we determine the following:

6d nails   nail length = 2 in. (Table 9.6)

actual penetration = 2 − 0.5 = 1.5 in.

190 DESIGN OF WOOD TRUSSES

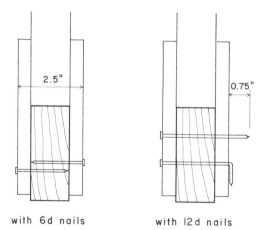

FIGURE 9.33 Nailing of the joints with plywood gusset plates, showing both double-sided nailing and single, driven-through nails with clinched ends.

$$\text{lateral capacity} = 59 \text{ lb}$$
$$\text{usuable capacity} = 1.25(59) = 74 \text{ lb/nail}$$

12d nails

$$\text{nail length} = 3.25 \text{ in.}$$
$$\text{protrusion} = 3.25 - 2.5 = 0.75 \text{ in.}$$
$$\text{lateral capacity} = 90 \text{ lb}$$

Although the penetration requirement for the 12d nail is not quite met, the double shear action of the clinched nail is much less critical for this relationship, and most designers will accept the full value for this nailing. If this is acceptable, the value of a single clinched nail becomes

$$\text{usable capacity} = 1.25(90)(2) = 225 \text{ lb/nail}$$

The four truss joints that must be designed are those labeled one through four in Figure 9.32. Possible layouts for these joints are shown in Figure 9.34. The nailing shown in the illustration is that determined for the single-sided nailing with the 6d nails. Some of the considerations in the joint designs are as follows.

Joint 1. There are several options for this joint. That shown is usually preferred since it permits extension of the top chord to form an overhang at the roof edge. For the nail design, the critical force is the horizontal force of 1424 lb in the bottom chord, since the vertical component in the top chord is taken in direct bearing on the support. With the 6d nails, the required number is determined as

$$\text{no.} = \frac{\text{force in member}}{\text{nail capacity}} = \frac{1424}{74} = 19.2$$

which can be satisfied with ten nails on each side of the joint.

LIGHT WOOD TRUSSES WITH SINGLE MEMBERS   191

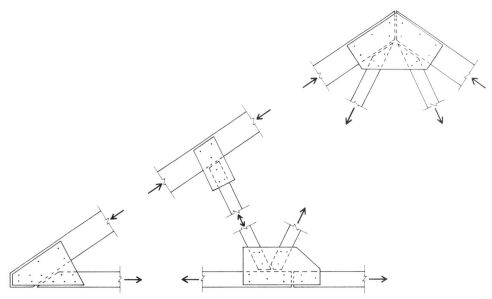

FIGURE 9.34 Form of joints for the truss with single members and gusset plates.

Joint 2. The load transfer is essentially made through direct bearing of the end of the web member on the side of the top chord. The gusset plates and nails, therefore, function primarily to hold the pieces together, although they also help to provide lateral bracing for the top chord.

Joint 3. This joint also functions as the splice for the bottom chord. Nailing is determined independently for all four members as follows:

$$\text{Member } IJ: \quad \text{no.} = \frac{1424}{74} = 19.2, \text{ use 10 per side}$$

$$\text{Member } JK: \quad \text{no.} = \frac{399}{74} = 5.39, \text{ use 3 per side}$$

$$\text{Member } KL: \quad \text{no.} = \frac{657}{74} = 8.88, \text{ use 5 per side}$$

$$\text{Member } LH: \quad \text{no.} = \frac{951}{74} = 12.85, \text{ use 7 per side}$$

Joint 4. As with gabled rafters, with a purely symmetrical load, the chords transfer force by simply direct bearing against their ends. Thus, although force is involved in the attachment of the web members, the gusset functions essentially only to hold the chords together. However, with wind load, or other unsymmetrical loading, there will be some stress transfer from chord to chord through the joint. Therefore, it is desirable to use a reasonable number of nails between the gussets and the top chords to provide for the unsymmetrical loading possibility. For the web members, the nailing is the same as at their opposite ends, as determined at Joint 3.

In most cases it is likely that trusses of this type will be provided as prefabricated

192 DESIGN OF WOOD TRUSSES

products using stock patterns and some form of patented jointing devices. However, they may also be quite economically produced as hand-built, site-fabricated elements in the form illustrated here.

9.4 HEAVY TIMBER TRUSSES

When the loading or span conditions produce internal forces of a magnitude in excess of that which can be developed by light wood elements and nailed or glued gusset plate joints, one option is the so-called heavy timber truss. This truss is similar to that developed in the example in Section 9.3 in that the truss members consist of single elements assembled in a single plane. However, because of the size of the members and the forces required, the joints usually are made with heavy steel plate gussets and steel bolts. The following example illustrates the design of such a truss.

We will assume the truss has a configuration and span the same as that shown in Figure 9.29 for the light wood truss. However, instead of being 2 ft on center, the trusses are 12 ft on center. Although this is likely to result in a slightly heavier roof construction, we will assume the truss loading to be six times that for the truss in Section 9.3. The internal forces will therefore be six times those shown in Figure 9.31. These increased forces are shown in Figure 9.35.

Because of the jointing technique used, it is necessary that all of the truss members have one common dimension. For practicality it is also usually desirable that the joints use a single bolt size. These requirements demand some coordinated design effort which may involve several tries to find the ideal sizes of members and the best bolt size and layout patterns.

Design of the single-piece timber members can be achieved with the procedures described in Section 9.1. It is desirable to use a high grade of wood to assure material reasonably free of defects and a reduction of dimensional changes due to shrinkage. Selection of member sizes must be made to have both the truss member design and joint design work as a package effort. This may even affect reconsideration of the truss form and layout dimensions. Some of the procedures for an optimized design in this regard are discussed in Chapter 16.

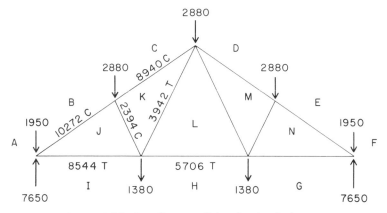

FIGURE 9.35 Loading condition for the timber truss.

Selection of a common thickness for members, and design for the other dimension required for accommodation of minimum jointing, may result in a minimum truss member of considerable size; possibly with considerable redundant strength. This would produce an overly wasteful, heavy truss, and all the design factors—truss spacing, form, height/span ratio, wood grade, and so on—would need to be reconsidered.

Figure 9.36 shows the form of the joints for the truss. Side plates consist of elements of steel plates that are welded together to produce the forms shown. Some of the considerations in the design of such joints are as follows:

1. Steel plates should have a width slightly less than that of the wood members.
2. Thickness of the steel plates should be adequate for the stress on the net section of the plates.
3. Edge and end distances must satisfy the requirements for both the wood and

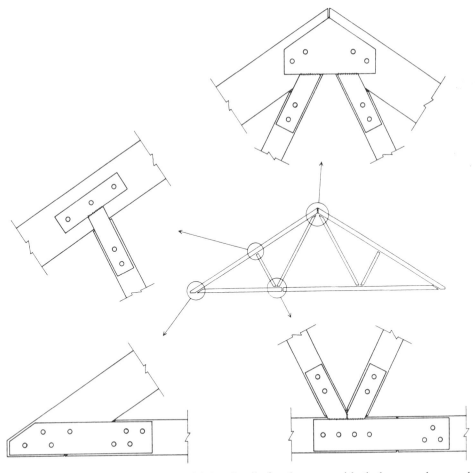

FIGURE 9.36 Member layout and joint details for the truss with timber members and bolted joints.

194 DESIGN OF WOOD TRUSSES

steel members. (See discussion for edge and end distance in wood in Section 9.2 and for steel in Section 10.2.)

4. Maximum diameter of bolts should be one fourth of the width of the wood members and one-third the width of the steel plates.
5. A minimum of two bolts should be used in each member at a joint.

Figure 9.37 shows a structure for the roof over a swimming pool. The main spanning elements are heavy timber trusses with single-piece, solid-sawn timber members and bolted joints with welded steel gusset plates. While this is a structure with a clear line of heritage from outdated construction methods, the technology for achieving jointing is undoubtedly quite current. The overall effect is not notably different from much ancient construction, although it fits well into the general building design here that otherwise uses very little historic detail.

Ancient Timber Trusses

Use of timber trusses in the past is discussed in Part 1. A principal change from those times has mostly to do with the development of jointing. Precise cutting and fitting of the wood members to form wood-to-wood, direct-bearing compression joints is

FIGURE 9.37 Timber structure using trusses with bolted joints and steel gusset plates.

HEAVY TIMBER TRUSSES   195

generally not feasible in present construction. This has partly to do with the lack of available craft work, but also relates to better and stronger jointing methods now available. More notable is the possibility for strong wood-to-wood tension joints that reduce the need for use of steel tension members.

Figure 9.38 shows some details for trusses taken from the 1931 edition of *Architects and Builders Handbook* (Ref. 14). While some very interesting forms are displayed here, most of the elaborate jointing is really not required with the present availability of jointing methods. These joints could still be made, of course, but only with

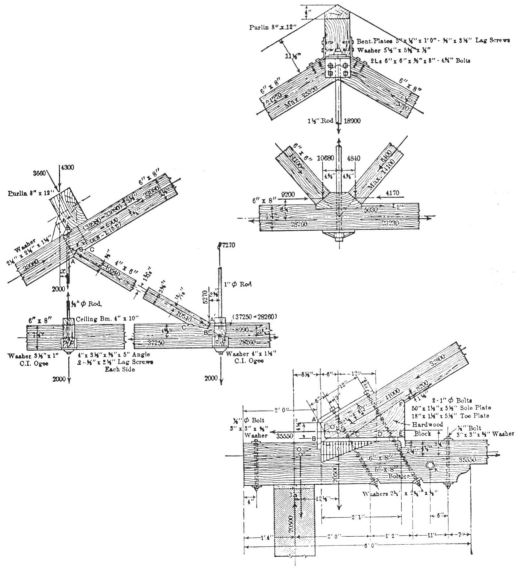

FIGURE 9.38 Joint details for early twentieth century timber trusses. (From *Architects and Builders Handbook*, 1931, (Ref. 14), reproduced with permission of the publishers, John Wiley and Sons.)

196 DESIGN OF WOOD TRUSSES

a considerable expenditure for hand-crafted work, and—in the end—a structure with some less security than that possible with current methods.

9.5 WOOD TRUSSES WITH MULTIPLE-ELEMENT MEMBERS

The trusses described in the two preceding sections used members consisting of single pieces, with individual pieces all of the same thickness so as to facilitate jointing with gusset plates. Another basic means for achieving joints with wood members is to lap the members over each other and use some form of connector that develops shear to resist the slipping away of the lapped parts. In order not to have the two-member joint with its twisting failure (see Figure 9.10b), this means that some members will be of more than a single piece.

Figure 9.39 shows an alternative form for the truss developed in the preceding two sections. In this case, the connections are achieved as lapped-member joints and the selection of individual members is made to develop symmetrical jointing. In some cases, blocking is used in the joints to achieve splicing (in the bottom chord, for example) or to make the spaced arrangement of the multiple members work for the whole assemblage (see the joints at the support and at the midpoint of the top chord).

In the truss with multiple-piece members, the compression members will take the form of *spaced columns*, as discussed in Section 9.1. This makes it more feasible to use relatively thin pieces, as the slenderness limit is less for the spaced column.

FIGURE 9.39 Form of the wood truss with multiple-piece members and lapped joints.

WOOD TRUSSES WITH MULTIPLE-ELEMENT MEMBERS **197**

The selection of individual member sizes will be constrained by the following considerations:

Magnitude and type (tension or compression) of the member force.

Length of the member for compression members. (And maybe for sag of the horizontal bottom chord.)

General overall size of the truss, effecting whether the members are in the light framing or heavy timber category. Logical development of the truss member pattern and the match up of lapping, member thicknesses, and number of pieces for each member to make the joints work.

Type of shear-resisting fasteners and the number required at each joint, as it relates to minimum thickness or necessary width of the members.

Because of the additional total material required for the multiple-piece members, this form is not likely to be used for closely spaced trusses, such as that developed in Section 9.3. As with any truss used in multiples, the feasible range for truss spacing will be a factor to consider with others in the logical design develop-

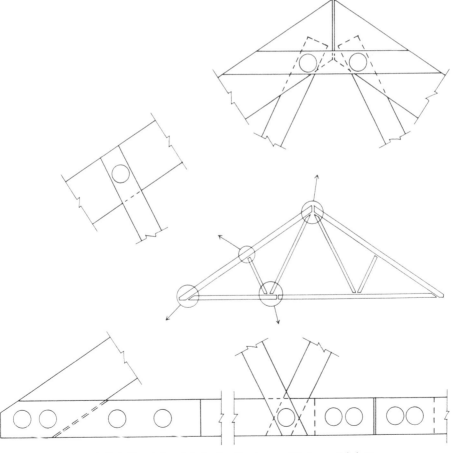

FIGURE 9.40 Joint details for the truss with lapped joints.

ment. Spacing will mostly affect the total load and thus the specific magnitudes of member forces, which will in turn affect member sizes and joint loads for the fasteners. Some amount of trial and error is to be expected in developing a logical combination of truss form, truss spacing, fastening type and size, and so on.

One type of fastener used frequently for this situation (lapped members) is that using shear developers, such as the split-ring connector described in Section 9.2. The advantages in using such a connector, over a joint with only a bolt, include its increased load capacity and its very low deformation during loading (translating to lack of slip between the members and a lot less deflection for the truss). Figure 9.40 shows a layout of the joints for the truss in Figure 9.39 with indication of placement for split-ring connectors. As shown in Figure 9.40, some joints are achieved with a single connector and others with two. However, as discussed in Sec. 9.2, it must also be noted that some members will have a connector in one face and others in two. Thus at the lower chord joint in Figure 9.39, the double-member diagonal has a connector in one face of each of the two members, while the single diagonal has connectors in two faces.

A form of truss often achieved with multiple element members is the bowstring truss (see Figure 9.1). This truss is usually used for spans long enough to make it possible to use straight pieces for the arched top chord. The individual pieces are wide enough to permit them to be cut to the arch profile on one side. The arch is thus made up of several separate pieces in its length and typically of multiple element members. Splicing of the arched top chord and the horizontal bottom chord and achieving the lapped joints, is subject to all the preceding discussion about development of the truss with multiple element members.

Figure 9.41 shows the use of a bowstring truss with multiple-element members and lapped joints for a warehouse-type building. The span is approximately 60 ft here, and trusses are 8 ft on center. While there is undoubtedly some actual "truss" action here, the diagonals function mostly to brace the arch and hold the bottom chord from sagging. The number of bolts in the joints seems to indicate that the structure is truly designed primarily as a tied arch.

FIGURE 9.41 Wood bowstring trusses used for a roof structure.

9.6 COMBINATION TRUSSES: WOOD AND STEEL

As steel elements—particularly steel rods with threaded ends—were developed in the nineteenth century, construction with wood trusses began to incorporate some steel rods as tension members. Wood has considerable capacity in axial tension along the grain, but the problem of developing tension connections in all-wood truss is a limiting factor. An example of an early combination truss with steel members is shown in Figure 9.42. This is still a valid form of structure for a truss using major compression members of solid timber. However, the use of steel side plates and side gussets of welded steel plates has made it possible to develop wood tension members (see Figure 9.37), so the usage shown in Figure 9.42 has less significance for the heavy timber truss.

Combination trusses today consist mainly of manufactured trusses using wood chords and some or all web members of steel. An example of the product of one manufacturer is shown in Figure 9.28. The feasibility of using the steel members depends considerably on the ability to develop a simple and economical jointing method, as well as a form of steel truss member.

Modern jointing methods generally permit the use of wood tension members in more situations than in the past. This includes the light all-wood truss with metal plate connectors and larger trusses with multiple-element members using joints with shear developers.

It is part of the modern spirit, however, to use mixtures of materials and elements,

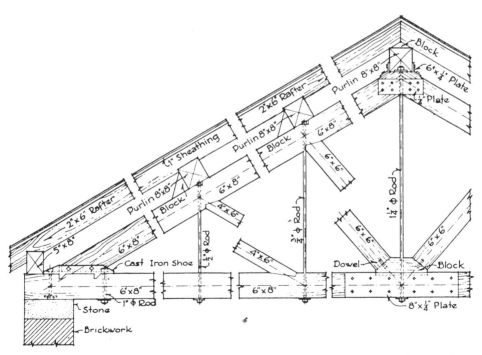

FIGURE 9.42 Form of early twentieth century timber truss with steel rod tension members. (From *Materials and Methods of Construction*, 1932, (Ref. 16), reproduced with permission of the publishers, John Wiley and Sons.)

200 DESIGN OF WOOD TRUSSES

each appropriate to their tasks. There is nothing sacred about a structure that is all wood, all steel, all concrete, and so on. Truss chords are now frequently made of laminated elements, permitting much greater lengths of single pieces and thus eliminating some expensive jointing. As with wood construction in general, mixtures are likely to become more common and more complex, with various combinations of elements of solid-sawn lumber, laminated wood, plywood, and even wood fiber products, as well as steel.

9.7 TRUSSED BRACING FOR WOOD STRUCTURES

The trussed frame is quite commonly used in steel frameworks, but is now less often used for wood frameworks. However, it remains a possibility for bracing of a wood frame if other options are not reasonable. A type of construction used frequently in

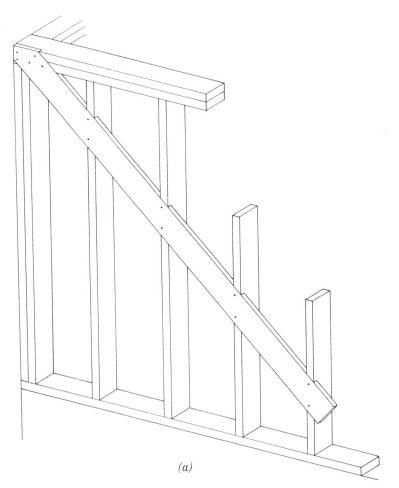

(a)

FIGURE 9.43 Trussed bracing for wood structures: (a) Let-in bracing for stud wall construction. (b) X-bracing with steel rods.

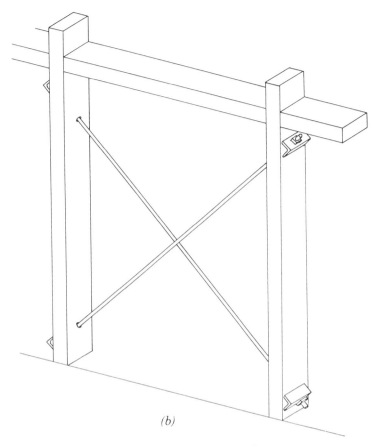

FIGURE 9.43 (*Continued*)

the past employed thin boards (1-in. nominal thickness) placed diagonally across the studs of a light frame structure, as shown in Figure 9.43a. In order not to interfere with placement of surfacing materials, these diagonal members were placed in notches cut into the stud faces (called *letting in*, thus the terms used to describe the bracing was *let-in bracing*).

Let-in bracing is now less used due to the rated capacities established for many forms of wall surfacing for use as shear walls. It is still used in some cases to hold the framework in position before the surfacing is applied. A special construction sometimes used is one in which stucco is applied without a sheathing backup; using only a backup paper attached to a wire mesh. Figure 9.44 shows the use of let-in bracing with such a directly applied stucco finish.

Timber structures built without some useable shear wall construction are sometimes braced by trussing. One form used is that shown in Figure 9.43b, consisting of steel tie rods arranged as X-bracing. This is not a highly popular bracing method, as it is difficult to assure the lasting tightness of the rods during the shrinkage of the timber frame over time.

If trussing is used to brace wood frames today, it is more often developed as a trussed steel frame incorporated into the wood framework.

FIGURE 9.44 Let-in bracing for wood stud construction. Used here for construction in which stucco is applied directly to the studs with no wall sheathing.

10

DESIGN OF STEEL TRUSSES

Steel is used for trusses of a great range of sizes. Manufactured open-web joists are made in depths as small as 8 in. for use as short-span rafters or joists where noncombustible construction is required. At the opposite end of the size range are the huge trusses used for bridges and for some very long span roofs. Trussed bracing was used for the Eiffel Tower in the last century, for the Empire State Building in the 1930s, and for the John Hancock Building in Chicago in the 1970s. This chapter deals with the problems of analysis and design of steel trusses, with an emphasis on relatively light truss construction for roofs and floors of short to medium span.

10.1 BASIC CONSIDERATIONS FOR STEEL

Steel truss construction for buildings is frequently achieved with trusses and truss systems that are the proprietary products of various manufacturers. This includes systems with open-web joists and joist girders and various systems for three-dimensional, two-way spanning trusses or space frame structures.

Custom-designed trusses are mostly achieved with standard, hot-rolled products (shapes, pipe, tubes, etc.) and jointing methods typically using welding in the fabricating shop and high-strength bolts at the erection site. This section presents a short summary of basic considerations for use of steel in these situations. More extensive treatment of steel structures is available from various sources, including *Simplified Design of Steel Structures* (Ref. 6) and *Steel Buildings: Analysis and Design* (Ref. 7)

Types of Steel Trusses

Some of the commonly used forms for steel trusses are those shown in Figure 10.1. Some of the considerations of their use are as follows.

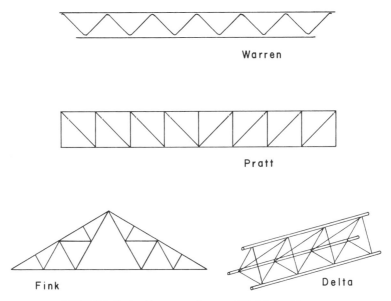

FIGURE 10.1 Common forms of light steel trusses.

Gable Truss. The W form truss—popular in wood—is less used in steel, primarily because of the desire to reduce the length of the truss members in steel trusses. Because of the high stress capacity of steel, it is often possible to use very small elements for the truss members. This results in increased concern for the problems of slenderness of compression members, of span length for members subjected to bending, and of the sag of horizontal tension members. This often makes it more practical to use some pattern that produces shorter members, such as the Compound Fink shown in Figure 10.1.

Warren Truss. The Warren is the form used for light open web joists, in which the entire web sometimes consists of a single round steel rod with multiple bends. When a large scale is used, the Warren offers the advantage of providing a maximum of clear open space for the inclusion of building service elements that must pass through the trusses (ducts, piping, catwalks, etc.).

Flat Pratt Truss. For the parallel-chorded truss, the Pratt offers the advantage of having the longest web members in tension and the shorter vertical members in compression.

Delta Truss. Another popular form of truss is the three-dimensional arrangement referred to as a delta truss. This truss derives its name from the form of its cross section—an equilateral triangle resembling the capital Greek letter delta (Δ). Where lateral bracing is not possible—or is not desired—for ordinary planar trusses, it may be possible to use the delta truss, which offers resistance to both vertical and horizontal loading. The delta form is also one used for trussed columns, as discussed in Section 12.3.

Design Considerations with Steel

Various factors may influence the basic decision to use steel as well as the design of truss members and joints. Some of the major concerns are the following.

Decision to Use Steel. Basic issues are cost, availability of fabricators, fire rating required (steel qualifies as a noncombustible material), and compatibility with other elements of the construction.

Type of Steel. Most steel structural elements are produced in the common grade known as A36 steel, short for ASTM A36, with a design yield stress of 36,000 lb/in.^{2}. Other steels are usually used only where design forces are exceptionally high or when some special property is required, such as resistance to exposure conditions. Prefabricated open web joists are produced in two steel stress classifications, called ordinary joists and high strength joists.

Form. Common truss patterns are those shown in Figure 10.1. Choice of truss profile also depends on the general architectural design and the problems of roof drainage.

Truss Members. These usually consist of elements from the standard forms produced as rolled products—rods, bars, plates, tubes, and shapes: I, T, C, Z, and so on. T forms are most commonly produced by splitting I- or H-shaped elements. The design properties for the various standard steel products available are given in the tables in the *AISC Manual* (Ref. 8). Tables B2 through B.10 in Appendix B give the properties for a limited number of elements of the size range and type most frequently used for steel trusses of short to medium span.

Jointing. Steel elements may be joined by riveting, bolting, or welding; they may be connected directly to each other or to intermediate splice or gusset plates. As with steel structures in general, the currently favored technique is to use welded connections in the fabricating shop and bolted connections for site erection. Choice of the connectors depends on the size of the truss, the magnitudes of forces, and the shapes of truss members. Examples of ordinary connection details are given in the designs in Sections 10.3 through 10.6.

Assembly and Erection. Small trusses are ordinarily fabricated in a single piece in the shop, with field connections limited to those at the supports and for bracing. For large trusses—where problems of transportation and erection require it—it is usually necessary to design the truss for shop fabrication in units. This may affect the choice of the truss pattern, of member shapes, and of connecting methods.

The Building Structure. The trusses must be integrated into the general structural system for the building. Some possible concerns are the need for attachment of items to the truss, the supports for the truss, need for temporary bracing during construction, and the use of the trusses as part of the general lateral bracing system for the building.

Allowable Stresses in Steel Structures

Stresses used for design of elements of steel structures are usually those specified in the current edition of the *Manual of Steel Construction*, published by the American Institute of Steel Construction, and commonly referred to simply as the *AISC Manual*. (See Ref. 2.) Most building codes require that steel structures be designed and fabricated in accordance with the specifications given in the *AISC Manual*.

The most common grade of steel used for ordinary structural steel elements—bars, rods, plates, and rolled sections—is ASTM A36, usually referred to simply as A36. This is a highly ductile steel that functions well for riveted, bolted and welded connections, and whose most significant property is its yield stress value of 36,000 lb/in.^{2}. Most allowable stresses are based on the yield strength, designated as F_y, or on the ultimate strength, designated as F_u. A36 steel will be used for all the examples of design work in this book.

Allowable stresses for steel, as specified in the eighth edition of the *AISC Manual* (Ref. 8), are summarized in Table 10.1. A brief discussion of the various types of stresses follows.

Tension. When welded connections are used, the critical tension stress is usually that on the gross cross-section area of elements. When bolts or rivets are used, the stress on the net area at the holes in the element must also be considered. Where groups of connectors occur, it may be necessary to investigate more than one chain of holes to find the critical net section, as shown in Figure 10.2. When round rods are threaded at their ends and used as tension elements, the tension stress must be investigated at the root of the threads. When steel gusset plates or splice plates are used, the various possibilities of tearing of the plates must be investigated; this may involve some combination of tension and shear in the plates, as illustrated in Figure 10.2.

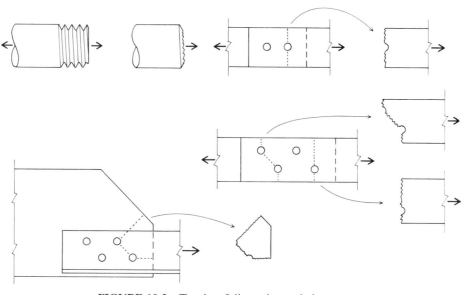

FIGURE 10.2 Tension failures in steel elements.

TABLE 10.1 Allowable Stresses on Steel Truss Members of A36 Steel

Type of Stress	Stress Limits[a,b]	Values for A36 Steel (kips/in.^{2})
Tension	On gross area:	
F_t	$F_t = 0.60 F_y$	21.60
	On effective net area:	
	$F_t = 0.50 F_u$	29.00
	On threaded rods:	
	$F_t = 0.33 F_u$	19.00
Compression[a]	For members completely braced against buckling ($L/r = 0$):	
F_a	$F_a = 0.60 F_y$	21.60
Shear	On a net plane through a row of fasteners[c]:	
F_v	$F_v = 0.30 F_u$	17.40
Bending[b]	For rolled shapes, bent on their major axis, adequately braced against buckling:	
F_b	$F_b = 0.60 F_y$	21.60

[a] For members subject to buckling, see Table 10.3.
[b] For other situations, see Section 1.5.1.4 of the *AISC Manual* (Ref. 8).
[c] See Figure 10.2.

The specifications refer to a condition of stress in tension at the location of pins. Although rivets and bolts are sometimes referred to as pin-type connectors, the code reference is generally not to this form of connection.

Shear. Concern for shear stress in truss design is usually limited to two situations. The first is the shear that occurs in a bolt or rivet—which is usually dealt with for design purposes by using the given rated capacities for specific types of connectors, as discussed in Section 10.2. The other situation involves the possibility of shear tearing in gusset plates, which is also discussed in Section 10.2.

Compression. In truss design the most common compression stress situation is that of the column action of the compression elements of the truss. Allowable compression varies as a function of the length of the element, as illustrated in Figure 9.5. For the very short element, the stress is limited by the actual stress capacity of the material—as indicated by its F_y value—while for the very long element the condition is one of elastic buckling. If buckling is the predominant action, the significant properties become the stiffness of the material (E) and the stiffness of the element as indicated by its L/r ratio. For the intermediate zone between these two extremes, an empirical formula is used to effect the transition between the two types of behavior.

The graph in Figure 10.3 consists of a plot of the allowable compression stress for a steel column of A36 steel as a function of its L/r ratio. Specific values for the allowable stress for a range of L/r values from 40 to 200 are given in Table 10.2. The lower limit of 40 is used here simply for brevity, since truss members are typically quite slender and seldom have a value below this. The upper limit of 200 is the maximum permitted by the specifications.

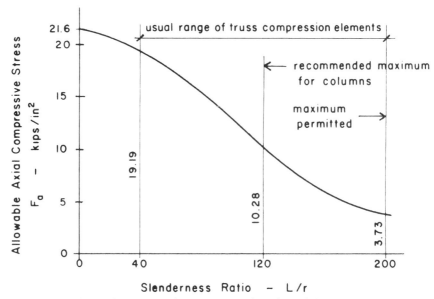

FIGURE 10.3 Allowable compression stress as a function of slenderness of the compression member. (From AISC requirements for A36 steel.)

Bending. In trusses, bending is usually limited to that which occurs when chord members are directly loaded, as illustrated in Figure 7.10. Thus the condition is one of combined bending and axial load, rather than bending alone. In some situations the chord member may be a zero stress member (no truss load), in which case it would be investigated for the bending alone. In the current specifications the allowable bending stress is dependent on the conditions of lateral bracing of the element, although when bending occurs with truss members they are usually quite adequately braced by the elements that apply the bending loading.

Bearing. Critical bearing stresses in trusses are usually those that occur at the edges of holes in riveted and bolted connections. This situation is discussed in Section 10.2.

Combined Stress. This refers primarily to the situations of the combined actions of axial load and bending as they occur in truss chord members with directly applied loading. The problems of analysis and design for these conditions are discussed later in this section. Although the potential problems of analysis are quite complex, the conditions that ordinarily exist in truss structures generally permit a reasonably simplified approach.

Design of Steel Tension Members

When the cross section of the member is not reduced, the stress permitted for design is simply

$$F_t = 0.6F_y = 0.6(36,000) = 21,600 \text{ lb/in.}^2 \text{ for A36 steel}$$

TABLE 10.2 Allowable Axial Compression Stress (F_a) for A36 Steel

$\dfrac{L}{r}$	F_a (k/in.²)	$\dfrac{L}{r}$	F_a (k/in.²)	$\dfrac{L}{r}$	F_a (k/in.²)	$\dfrac{L}{r}$	F_a (k/in.²)
40	19.19	81	15.24	121	10.14	161	5.76
41	19.11	82	15.13	122	9.99	162	5.69
42	19.03	83	15.02	123	9.85	163	5.62
43	18.95	84	14.90	124	9.70	164	5.55
44	18.86	85	14.79	125	9.55	165	5.49
45	18.78	86	14.67	126	9.41	166	5.42
46	18.70	87	14.56	127	9.26	167	5.35
47	18.61	88	14.44	128	9.11	168	5.29
48	18.53	89	14.32	129	8.97	169	5.23
49	18.44	90	14.20	130	8.84	170	5.17
50	18.35	91	14.09	131	8.70	171	5.11
51	18.26	92	13.97	132	8.57	172	5.05
52	18.17	93	13.84	133	8.44	173	4.99
53	18.08	94	13.72	134	8.32	174	4.93
54	17.99	95	13.60	135	8.19	175	4.88
55	17.90	96	13.48	136	8.07	176	4.82
56	17.81	97	13.35	137	7.96	177	4.77
57	17.71	98	13.23	138	7.84	178	4.71
58	17.62	99	13.10	139	7.73	179	4.66
59	17.53	100	12.98	140	7.62	180	4.61
60	17.43	101	12.85	141	7.51	181	4.56
61	17.33	102	12.72	142	7.41	182	4.51
62	17.24	103	12.59	143	7.30	183	4.46
63	17.14	104	12.47	144	7.20	184	4.41
64	17.04	105	12.33	145	7.10	185	4.36
65	16.94	106	12.20	146	7.01	186	4.32
66	16.84	107	12.07	147	6.91	187	4.27
67	16.74	108	11.94	148	6.82	188	4.23
68	16.64	109	11.81	149	6.73	189	4.18
69	16.53	110	11.67	150	6.64	190	4.14
70	16.43	111	11.54	151	6.55	191	4.09
71	16.33	112	11.40	152	6.46	192	4.05
72	16.22	113	11.26	153	6.38	193	4.01
73	16.12	114	11.13	154	6.30	194	3.97
74	16.01	115	10.99	155	6.22	195	3.93
75	15.90	116	10.85	156	6.14	196	3.89
76	15.79	117	10.71	157	6.06	197	3.85
77	15.69	118	10.57	158	5.98	198	3.81
78	15.58	119	10.43	159	5.91	199	3.77
79	15.47	120	10.28	160	5.83	200	3.73
80	15.36						

When the cross section is reduced—typically by holes for rivets or bolts—the maximum stress on the net area is

$$F_t = 0.6F_u = 0.5(58,000) - 29,000 \text{ lb/in.}^2 \text{ for A36 steel}$$

Another situation that must be considered occurs when the holes are small, in which case the design may be critical for the gross area.

Some additional considerations for tension members are the following

1. *Minimum Size for Connections.* If rivets or bolts are used, the member must have sufficient width to permit drilling of holes and minimum thickness for bearing on the connectors.
2. *Minimum Stiffness.* The specifications require a minimum L/r of 240 for a structural tension member. For bracing elements a minimum L/r of 300 is permitted. Although this will usually result in some minimal bending resistance, the possibility of sag of long horizontal members should be considered also.
3. *Development of the Full Cross Section.* For truss members that consist of angles or tees, it is often not possible to connect the whole element at the joints. Thus, as shown in Figure 10.4, the connectors are placed in only one leg of the angle or only in the web or flange of the tee. When this occurs, the specifications require that the cross-section area used in design be limited to that of the connected portion only.
4. *Effective Net Area.* The specifications give reduction factors to be used to establish the so-called effective net area A_e. This area rather than the true net area A_n, is to be used in actual design calculations. The effective area is thus determined as

$$A_e = C_t A_n$$

and some of the typical reduction factors are

$C_t = 0.75$ when only two connectors are used at the end of a member (the minimum number permitted).

$C_t = 0.90$ for **W**, **M** or **S** shapes connected by their flanges.

The effective net area for gusset or splice plates is limited to 85% of the gross area.

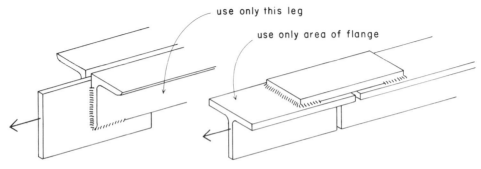

FIGURE 10.4 Influence of form of connections on the effectiveness of cross sections of steel tension members.

BASIC CONSIDERATIONS FOR STEEL 211

The following examples illustrate the procedures for design of simple tension elements for trusses.

Given: Steel tension member, 14 ft long, axial tension force of 48,000 lb, welded joints with no reduction of cross section.

Required: Select steel elements consisting of a round rod, single angle, double angle, tee, round pipe, and square tube; all of A36 steel.

We will assume that the full section is developed at the joints, although this may be questionable for the angles and the tee. For the gross area, the allowable stress is

$$F_t = 0.6F_y = 0.6(36,000) = 21,600 \text{ lb/in.}^2$$

We thus require a gross area as follows:

$$A = \frac{\text{tension force}}{\text{allowable stress}} = \frac{48,000}{21,600} = 2.22 \text{ in.}^2$$

If the element is not a brace, the maximum allowable L/r is 240. Thus the minimum r for the member is

$$r = \frac{L}{240} = \frac{(14)(12)}{240} = 0.70$$

Based on these two considerations alone, some possible choices are those shown in Table 10.3. For the rod and single angle, the minimum r value is a critical concern. Otherwise the problem is limited to finding a member with the least cross-section area.

Let us now consider the problem when the cross section is reduced at the joints.

Given: Same as preceding, $T = 48,000$ lb, $L = 14$ ft.

Required: Select a round rod with threaded end, a pair of angles with a single row of $\frac{7}{8}$ in. bolts, and a tee with a double row of in. bolts.

For the rod, the allowable stress on the gross section is the F_t value of 21,600 lb/in.². For the stress on the net section at the thread, the allowable stress from Table 10.1 is

$$F_t = 0.33F_u = 0.33(58,000) = 19,000 \text{ lb/in.}^2 \text{ (rounded off)}$$

Since the stress on the net section is obviously more critical, we thus determine the required area at the thread to be

$$A = \frac{48,000}{19,000} = 2.53 \text{ in.}^2$$

From Table B.2 this area is provided by a rod with diameter of 2.25 in. If the L/r criteria is applied, however, a larger size would be required.

212 DESIGN OF STEEL TRUSSES

TABLE 10.3 Choices for the Unreduced Tension Member

Type of Element and Size of Choice	Area	r	$\dfrac{L}{r}$
Round Rod			
1.75 in. dia.[a]	2.405	0.44	382
3.00 in. dia.[b]	7.069	0.75	224
Single Angle			
$4 \times 3\frac{1}{2} \times \frac{5}{16}$ [c]	2.25	0.730	230
Double Angle			
Two $3 \times 2 \times \frac{1}{4}$ [c]	2.38	0.891	189
T			
WT4 × 7.5[c]	2.22	0.876	192
Round Pipe			
3 in. nominal dia.	2.23	1.16	145
Square Tube			
$3 \times 3 \times \frac{1}{4}$	2.59	1.10	153

[a] Minimum member based on area requirement.
[b] Minimum member if maximum L/r is 240.
[c] Assumes use of full cross section. (See Figure 10.4.)

For the pair of angles, we reduce the cross section for two effects: the hole for the bolt and the ineffectiveness of the outstanding legs. For an approximate choice, therefore, we should look for a pair of angles with a total area about twice that required at the net section. We must also consider the need for a minimum leg size to accommodate the $\frac{7}{8}$ in. diameter bolt and a minimum thickness to develop the bearing of the bolt on the side of the hole.

We can select the angles from the information given in Table B.10 or Table B.9, in which r values are given for the pairs of angles on their y-axes. (See Figure 10.5.) Try an angle with a $\frac{3}{8}$ in. thickness. The area of the holes to be deducted thus becomes

$$A = \tfrac{15}{16}(0.375)(2) = 0.703 \text{ in.}^2$$

$$A = \frac{T}{F_t} = \frac{48{,}000}{29{,}000} = 1.66 \text{ in.}^2 \quad \text{(required net area)}$$

$$A = 1.66 + 0.703 = 2.363 \text{ in.}^2 \quad \text{(required gross area)}$$

$$L = \frac{2.363}{0.375} = 6.30 \text{ in.} \quad \text{(total leg width required)}$$

which indicates that we need a $3\frac{1}{2}$ in. leg as a minimum.

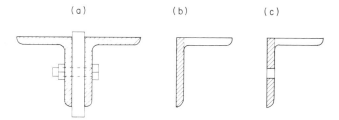

FIGURE 10.5 Effective area of a double-angle steel tension member: (*a*) Gross area of cross section. (*b*) Effective portion of area due to form of connection. (*c*) Effective area as reduced to net section at bolt hole.

As a starting point, this indicates that we will need a wider leg if the angle is thinner, but that we could possibly use a narrower one if it is thicker. On this basis, two possible choices are shown in Table 10.4.

For the tee, the situation is essentially similar to that for the angles. In this case we use only the area of the flange, and we would need a flange width of at least 6.30 in. if it is $\frac{3}{8}$ in. thick, as determined in the previous calculation. Two possibilities for the tee are given in Table 10.5.

Design of Steel Compression Members

When completely unbraced the member will buckle on its weak axis. Figure 10.6 shows the various elements typically used for truss members in steel trusses. Except for the round pipe and the square tube, the elements typically have a major axis (usually designated as the *X–X* axis) and a minor, or weaker, axis (usually designated as the *Y–Y* axis). For the single angle, the weak axis is the diagonal *Z–Z* axis. Allowable axial compression stress is established on the basis of the critical slenderness ratio L/r, with L being the full length of the member and r being that for the weak axis.

When a member is braced, and the bracing is with respect to the weak axis, it is sometimes necessary to consider separate slenderness ratios for both axes: L_x/r_x and L_y/r_y. The greater of these two ratios will establish the allowable stress for design.

When subjected to bending, members usually will be braced by the elements of the construction that apply the bending load to the member. Thus they will be braced in a direction perpendicular to the direction of the bending loading and will have the bending and buckling both occurring with respect to the same axis.

TABLE 10.4 Choices for the Reduced Tension Member—Double Angles

Angle Size and Thickness (in.)	Gross Area (in.²)	Area of the Connected Legs Only (in.²)	Area of One $\frac{15}{16}$ in. Hole (in.²)	Net Area with Two $\frac{15}{16}$ in. Holes (in.²)
$3\frac{1}{2} \times 2\frac{1}{2} \times \frac{3}{8}$	4.22	$2(3.5 \times \frac{3}{8}) = 2.625$	$(\frac{15}{16} \times \frac{3}{8}) = 0.3516$	1.922
$4 \times 3 \times \frac{5}{16}$	4.18	$2(4 \times \frac{5}{16}) = 2.50$	$(\frac{15}{16} \times \frac{5}{16}) = 0.2930$	1.914

TABLE 10.5 Choices for the Reduced Tension Member—Structural Tees

Designation	Gross Area (in.²)	Area of the Flanges Only (in.²)	Area of One $\frac{15}{16}$ in. Hole (in.²)	Net Area of Flange with Two $\frac{15}{16}$ in. Holes (in.²)
WT5 × 13	3.81	(5.770 × 0.440) = 2.539	($\frac{15}{16}$ × 0.440) = 0.4125	1.714
WT6 × 13	3.82	(6.490 × 0.380) = 2.466	($\frac{15}{16}$ × 0.380) = 0.3563	1.754

When the width-to-thickness ratio of unstiffened elements, such as tee stems and outstanding legs of angles, exceeds certain limits, the allowable compression stress must be reduced for the effects of localized buckling. Existence of this condition with an F_y of 36 ksi is indicated by the presence of a value for Q_s, in Tables B.6 through B.11. Reference is made to the specifications in the AISC Manual (Ref. 2) for the proper design of such elements. An approximate design stress may be obtained by simply multiplying the usual allowable stress by the reduction factor Q_s, as shown in the examples.

The following example illustrates the problem of selecting a compression member using the data available in the tables.

Given: Compression member of A36 steel; 9 ft long; axial force of 48 kips

Required: Select members of each type listed in Tables B.3 through B.9, except for the round, solid rod.

The selection process is most easily done by using tables with allowable compression loads predetermined for specific lengths of various elements, such as those available in Part 3 of the *AISC Manual* (Ref. 2). Otherwise the process is one of trial and error, since the allowable stress is a function of L/r, which cannot be known until the member is chosen. However, with a finite number of choices, it is not as difficult as it may seem. One approach is to first find the allowable stress, r value, and corresponding area for a range of L/r values, and then to look for elements with the r and area combinations that are associated. Using data from Table 10.2, we can determine

$$\frac{L}{r} = 50, F_a = 18.35 \text{ k/in.}^2,$$

required

$$A = \frac{48}{18.35} = 2.62 \text{ in.}^2$$

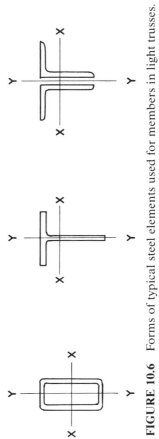

FIGURE 10.6 Forms of typical steel elements used for members in light trusses.

216 DESIGN OF STEEL TRUSSES

and the corresponding

$$r = \frac{L}{50} = \frac{(9)(12)}{50} = 2.16 \text{ in.}$$

If

$$\frac{L}{r} = 100, F_a = 12.98,$$

$$A = \frac{48}{12.98} = 3.70 \text{ in.}^2,$$

$$r = \frac{108}{100} = 1.08 \text{ in.}$$

If

$$\frac{L}{r} = 150, F_a = 6.64,$$

$$A = \frac{48}{6.64} = 7.23 \text{ in.}^2,$$

$$r = \frac{108}{150} = 0.72 \text{ in.}$$

With these combinations of r and area values in mind, we can now scan the data in the tables and make a reasonable first guess. Thus if we try the following member

$$4 \times 4 \text{ square tube, wall } t = 0.3125 \text{ in.}$$

$$A = 4.36 \text{ in.}^2, r = 1.48 \text{ in.}$$

$$L/r = 108/1.48 = 73;$$

from Table 10.2, $F_a = 20.38$ k/in.^{2}

then

$$\text{allowable load} = F_a \times A = (20.38)(4.36) = 88.9 \text{ k}$$

This indicates that the member is stronger than necessary, and it would pay to look for a lighter element. Although there may be other criteria for selecting a particular shape or a specific dimension, the usual search is for the member with the least area in its cross section, which generally produces lower cost in terms of the volume of steel used.

Table 10.6 gives a number of possible choices for the member in this example. These have been taken from the sections listed in the properties tables, which does not include all of the sections that may be available. For real design situations, it

BASIC CONSIDERATIONS FOR STEEL   217

TABLE 10.6   Choices for the Compression Member

Type of Section and Size	Area (in.²)	Weight per ft (lb)	Least r (in.)	$\dfrac{L}{r}$	F_a (k/in.²)	Q_s	Allowable Load $F_a \times$ Area (kips)
Round Pipe							
4 in. nominal	3.17	10.79	1.51	72	16.22		51
Square Tube							
$4 \times 4 \times \frac{1}{4}$	3.59	12.21	1.51	72	16.22		58
Tee							
WT5 $\times$ 11	3.24	11.0	1.33	81	15.24	0.999	49
Single Angle							
$6 \times 6 \times \frac{3}{8}$	4.36	14.9	1.19	91	14.09	0.911	56
Double Angle							
$2(4 \times 3\frac{1}{2} \times \frac{1}{4})$	3.63	12.4	1.54	70	16.43	0.911	54

may be desirable to determine if some of the less used sections are available in a particular area.

When a compression member is braced on one axis, it may be necessary to investigate both axes. If the bracing is continuous, as it virtually is when a deck is directly attached to a top chord, then the member need be investigated only on the other axis.

Design for Combined Bending and Tension

When bottom chords are directly loaded—as by an attached ceiling—they must be investigated for the combined actions of tension and bending. This requires satisfying the following formula:

$$\frac{f_a}{0.6F_y} + \frac{f_b}{F_b} \leq 1.0$$

in which f_a is the axial tension stress, f_b is the actual bending stress, and F_b is the allowable bending stress.

The following example illustrates the procedure for design and analysis of such a member.

Given: Truss member of A36 steel with conditions as shown in Figure 10.7.

Required: Choose a double angle element for the member.

This is a trial and error process. As a guide it is sometimes useful to find the area and section modulus required if the actions of tension and bending occur separately. It is then understood to be necessary to find values for both properties that are slightly higher. Thus

For *T* alone:

$$\text{required } A = \frac{T}{0.6F_y} = \frac{10,000}{21,600} = 0.463 \text{ in.}^2$$

For *M* alone:

$$\text{Assume } M = \frac{wL^2}{8} = \frac{(150)(8)^2}{8} = 1200 \text{ lb/ft}$$

$$\text{Required } S = \frac{M}{F_b} = \frac{(1.2)(12)}{24} = 0.60 \text{ in.}^3$$

Try: Two $3 \times 2 \times \frac{1}{4}$ angles

From Table B.10: $A = 2.38$ in.^{2}, $S = 1.08$ in.^{3}

$$f_a = \frac{10}{2.38} = 4.20 \text{ k/in.}^2$$

$$f_b = \frac{M}{S} = \frac{(1.2)(12)}{1.08} = 13.33 \text{ k/in.}^2$$

$$\frac{f_a}{0.6F_y} + \frac{f_b}{F_b} = \frac{4.20}{21.60} + \frac{13.33}{21.60}$$

$$= 0.194 + 0.617$$

$$= 0.811$$

This indicates that the member is quite adequate; therefore it may be possible to use a smaller angle. However, this member will sustain considerable deflection on the 8 ft span. This, plus considerations of the connection design, may make it desirable not to reduce the member size further.

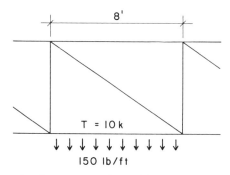

FIGURE 10.7 Example of a truss member subjected to combined tension and bending.

Design for Combined Bending and Compression

When the top chord of a truss is directly loaded, the typical situation that occurs is one of combined bending and axial compression with the chord braced against buckling in the direction perpendicular to the plane of the truss. The following example illustrates the procedures for the analysis and design of such an element.

Given: The top chord of a truss with the loading condition shown in Figure 10.8.

Required: Select a structural tee of A36 steel.

We assume that the construction is capable of bracing the chord on its weak axis. Therefore, the r value to be used for determination of the allowable compression stress (F_a) is the one with respect to the same axis as that about which the bending occurs. Thus it can be expected that the L/r ratio will be quite low, since the element must be reasonably stiff about this axis. A procedure for a first approximation is to assume a low L/r, determine the corresponding value for F_a, and find required values for the area and the section modulus of the compression and bending each occur alone. (See also the procedure for combined bending and tension.) Thus

$$L = \frac{1}{0.894} (8 \times 12) = 107 \text{ in.}$$

For $L/r = 50$, $F_a = 18.35$ k/in.² (from Table 10.2.)

$$\text{required } A = \frac{C}{F_a} = \frac{18,000}{18.35} = 0.981 \text{ in.}^2$$

$$\text{Assume critical } M = \frac{wL^2}{8} = \frac{(300)(8)^2}{8} = 2400 \text{ lb-ft}$$

$$\text{required } S = \frac{M}{F_b} = \frac{(2400)(12)}{21,600} = 1.33 \text{ in.}^3$$

For a first try we double both of these values and look for a steel section with

$$A = 2(0.981) = 1.962 \text{ in.}^2$$
$$S = 2(1.33) = 2.66 \text{ in.}^3$$

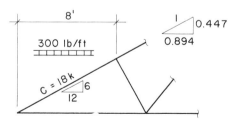

FIGURE 10.8 Example of a truss member subjected to combined compression and bending.

From a scan of Table B.6,

WT5 × 13 – $A = 3.81$ in.^{2}, $S = 1.91$ in.^{3}
WT6 × 9.5 – $A = 2.79$ in.^{2}, $S = 2.28$ in.^{3}, $Q_s = 0.797$
WT7 × 11 – $A = 3.25$ in.^{2}, $S = 2.91$ in.^{3}, $Q_s = 0.621$

Trying the lighter member, we determine the following.

For the WT6 × 9.5,

$$f_a = \frac{C}{A} = \frac{18{,}000}{2.79} = 6452 \text{ lb/in.}^2$$

Assuming the tee to be braced on its y-axis and using $r_x = 1.90$ in.,

$$\frac{L}{r} = \frac{107}{1.90} = 56.3 \quad \text{(From Table 10.2, } F_a = 17{,}810 \text{ lb/in.}^2.)$$

$$f_b = \frac{M}{S} = \frac{(2400)(12)}{2.28} = 12{,}632 \text{ lb/in.}^2$$

Then, for the combined action,

$$\frac{f_a}{F_a Q_s} + \frac{f_b}{(1 - f_a/F'_e)F_b Q_s} \leq 1$$

in which

$$F'_e = \frac{5.15E}{(L/r)^2} = \frac{5.15(29{,}000{,}000)}{(56.3)^2}$$

$$= 48{,}122 \text{ lb/in.}^2$$

and

$$\frac{f_a}{F'_e} = \frac{6452}{48{,}122} = 0.137$$

so that

$$\frac{f_a}{F_a Q_s} + \frac{f_b}{(1 - f_a/F'_e)F_b Q_s} = \frac{6452}{17{,}810(0.797)} + \frac{12{,}632}{0.863(21{,}600)(0.797)}$$

$$= 0.454 + 0.851$$

$$= 1.305$$

This indicates that the WT6 × 9.5 is not quite adequate and a heavier tee, or one with a higher value for Q_s must be used. However, other considerations of the truss construction, such as the details for joints and chord splices, for attachment of the roof deck, for support details, and for bracing, may provide additional criteria for the selection of the member.

10.2 JOINTS IN STEEL TRUSSES

For light steel trusses of short to medium span, the preferred joint connections are those that utilize either welding or high-strength, friction-type bolts. The friction-type bolted connection is preferred over the bearing type because of the resulting tightness (lack of slipping) in the connections. The typical procedure is to use welds only for shop connections and bolts for field connections, although bolts may also be used for shop connections. Bearing-type bolts may be used for the connection of the truss to its supports or for attachment of bracing or other framing elements to the truss.

Bolted Joints

The three basic loading situations for a bolted connection are shown in Figure 10.9. For truss joints, the conditions are generally limited to those of the shear loadings. With shear loading, the primary design concerns are the following.

1. Shear on the bolt is calculated as a stress on the bolt cross-section area.
2. Bearing on the edge of the hole is calculated as compression on the bearing area, the area being the product of the bolt diameter and the thickness of the connected part.
3. Direct tension or tension/shear tearing is calculated on the net cross section of the connected parts.

A critical aspect of the design of bolted connections is the development of the geometric layout of the joint. The various considerations that typically must be dealt with are those shown in Figure 10.10 and described as follows.

1. *Minimum Edge Distance* (*b*). Measured from the center of the bolt to the nearest edge of the connected part.
2. *Spacing* (*s*). Center-to-center of bolts; in a single row, between parallel rows, or in any direction when the bolts are in staggered rows.

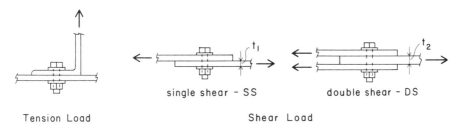

FIGURE 10.9 Loading conditions for bolted joints in steel structures.

222 DESIGN OF STEEL TRUSSES

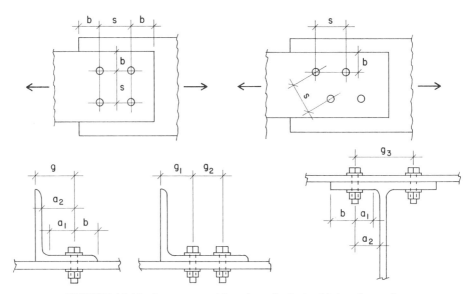

FIGURE 10.10 Layout considerations for bolted joints in steel.

3. *Clearance* (*a*). Limited by the fillet (rounded inside corner) of rolled shapes or by the proximity of adjacent parts; a_1 or a_2, respectively, as shown in the illustration.
4. *Gage Distance* (*g*). Which is the usual location for fasteners in legs of angles or flanges of rolled shapes.

Design data pertaining to the two standard types of high-strength bolts—A325 and A490—is given in Table 10.7. The following additional requirements should be considered in the design of bolted connections.

1. Connections should be designed for a minimum load of 50% of the full capacity of the connected elements or 6 kips, whichever is larger.
2. Except for bracing elements, connections should have at least two bolts in the end of each element.
3. For stress calculation, holes should be assumed to be $\frac{1}{16}$ in. larger than the bolt diameter.

Examples of the design of bolted connections are given in the joint designs for the truss in Section 10.3.

Welded Joints

For light steel trusses, welded joints usually are achieved with fillet welds that are placed along the edges of the connected parts. The limits on the size of this type of weld are given in Figure 10.11. Load capacities of fillet welds are based on their

TABLE 10.7 Design Data for High-Strength Bolts in Friction-Type Connections

Type of Bolt	A325				A490			
Nominal Diameter (in.)	$\frac{5}{8}$	$\frac{3}{4}$	$\frac{7}{8}$	1	$\frac{5}{8}$	$\frac{3}{4}$	$\frac{7}{8}$	1
Gross Area of Section (in.²)	0.3068	0.4418	0.6013	0.7854	0.3068	0.4418	0.6013	0.7854
Tension Capacity (kips)	13.5	19.4	26.5	34.6	16.6	23.9	32.5	42.4
Shear Capacity (kips)								
SS	5.4	7.7	10.5	13.7	6.7	9.7	13.2	17.3
DS	10.7	15.5	21.0	27.5	13.5	19.4	26.5	34.6
Minimum Thickness of Connected Parts of A36 Steel to Develop Full DS Capacity (in.)								
(t_2 in Figure 10.9)	$\frac{3}{16}$	$\frac{1}{4}$	$\frac{5}{16}$	$\frac{3}{8}$	$\frac{1}{4}$	$\frac{5}{16}$	$\frac{3}{8}$	$\frac{1}{2}$
Minimum Edge Distance (in.)								
At cut edge	$1\frac{1}{8}$	$1\frac{1}{4}$	$1\frac{1}{2}$	$1\frac{3}{4}$	$1\frac{1}{8}$	$1\frac{1}{4}$	$1\frac{1}{2}$	$1\frac{3}{4}$
At rolled edge	$\frac{7}{8}$	1	$1\frac{1}{8}$	$1\frac{1}{4}$	$\frac{7}{8}$	1	$1\frac{1}{8}$	$1\frac{1}{4}$
Spacing (in.)								
Preferred	3.00	3.00	3.00	3.00	3.00	3.00	3.00	3.00
Minimum	2.00	2.00	2.33	2.67	2.00	2.00	2.33	2.67
Minimum Width of Connected Element (in.)								
Angle leg, single row	2	$2\frac{1}{2}$	$2\frac{1}{2}$	3	2	$2\frac{1}{2}$	$2\frac{1}{2}$	3
Angle leg, two rows, not staggered	5	5	6	7	5	5	6	7
Tee flange	$5\frac{1}{4}$	$5\frac{1}{2}$	$5\frac{3}{4}$	6	$5\frac{1}{4}$	$5\frac{1}{2}$	$5\frac{3}{4}$	6

224 DESIGN OF STEEL TRUSSES

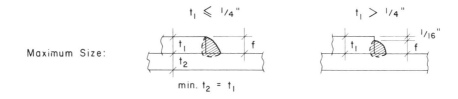

Minimum Size: when t =   up to 1/4"   1/4" to 1/2"   1/2" to 3/4"   over 3/4"

min. f =   1/8"   3/16"   1/4"   5/16"

FIGURE 10.11 Considerations for limited size of fillet welds.

TABLE 10.8 Allowable Loads on Fillet Welds

Size of Weld (in.)	Allowable Load (Kips per Linear Inch of Weld)	
	With EE 60 XX Electrodes	With EE 70 XX Electrodes
$\frac{3}{16}$	2.4	2.8
$\frac{1}{4}$	3.2	3.7
$\frac{5}{16}$	4.0	4.6
$\frac{3}{8}$	4.8	5.6
$\frac{1}{2}$	6.4	7.4
$\frac{5}{8}$	8.0	9.3

designated nominal size (f) and are quoted in terms of pounds per inch of the weld length. Capacities of ordinary fillet welds are given in Table 10.8.

It is generally desirable that welded joints be made symmetrical and otherwise formed so as to minimize potential tearing and twisting. Some good and bad practices in joint detailing are shown in Figure 10.12.

Examples of the design of welded joints are given in joint designs for the trusses in Sections 10.4 and 10.6.

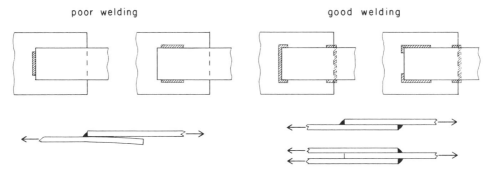

FIGURE 10.12 Good and poor welding practices.

10.3 DESIGN OF A LIGHT STEEL TRUSS WITH BOLTED JOINTS

The following example illustrates the procedures for the analysis and design of a light steel truss for a gable-form roof. The roof construction, truss configuration, and design loads are as shown in Figure 10.13. Truss construction is to be of steel with truss members consisting of double angles of A36 steel. Joints will use high-strength A325 bolts and gusset plates of A36 steel.

Trusses are to be spaced at 8 ft on center and the value for the truss weight from Table 7.1 is thus 48 lb/ft. This is probably a bit heavy for the light steel truss as the table somewhat favors weights for wood trusses. We will assume a modest weight of 25 lb/ft here, although the true weight can always be verified once some design

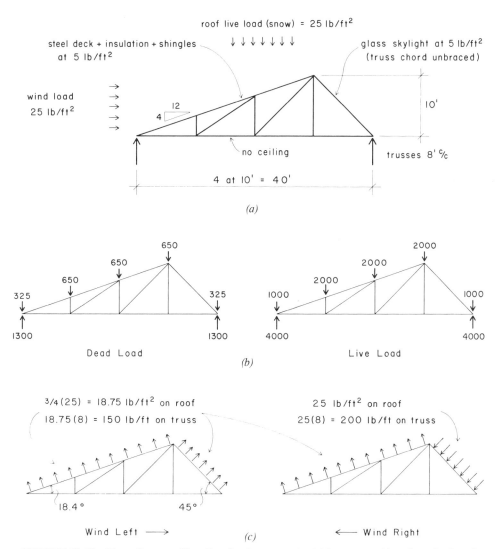

FIGURE 10.13 Truss form and loading for the example: (*a*) Layout and loading. (*b*) Gravity loads and reactions for truss analysis. (*c*) Wind loads for truss analysis.

226 DESIGN OF STEEL TRUSSES

decisions for member sizes have been made. For the same conditions, a truss with bolted joints and gusset plates will be somewhat heavier than one with members directly welded to each other at joints. Thus the assumption for weight at the very early stages of design investigation is unavoidably pretty much of a wild guess.

Using the data from Figure 10.13, the design gravity loads will be

$$DL = 25 + (8)(5) = 65 \text{ lb/ft} \quad \text{(dead load)}$$
$$LL = (8)(25) = 200 \text{ lb/ft} \quad \text{(live load)}$$

and the truss loadings will be as shown in Figure 10.13*b*.

Using the wind design criteria discussed in Section 3.5, the wind loadings will be as shown in Figure 10.13*c*. In order to find the reaction forces at the truss supports, we will use the procedure of finding individual reaction forces for the loads on each roof surface and then adding the results for the appropriate combinations shown in Figure 10.13*c*. The individual loading conditions and resulting reactions are shown in Figure 10.14 and are determined as follows.

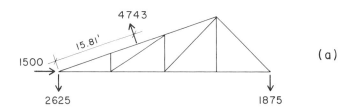

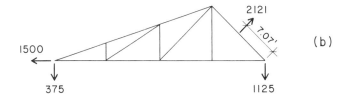

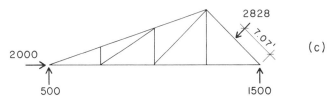

FIGURE 10.14 Determination of reactions due to wind loads.

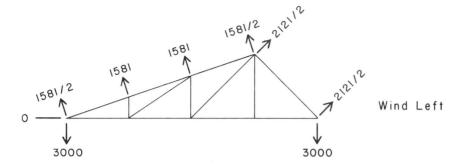

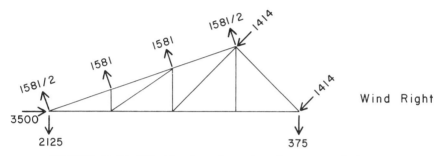

FIGURE 10.15 Wind loads and reactions for truss analysis.

For loading (*a*),

$$\text{total load on surface} = (150)\left(\frac{1}{0.9487}\right)(30) = 4743 \text{ lb}$$

$$R_{2v} = \frac{(4743)(15.81)}{40} = 1875 \text{ lb} \quad \text{(down as shown)}$$

$$R_{1v} = 4500 - 1875 = 2625 \text{ lb} \quad \text{(down as shown)}$$

$$R_{1h} = 1500 \text{ lb} \quad \text{(acting toward the right as shown)}$$

For loading (*b*),

$$\text{total load on surface} = (150)\left(\frac{1}{0.707}\right)(10) = 2121 \text{ lb}$$

$$R_{1v} = \frac{(2121)(7.07)}{40} = 375 \text{ lb} \quad \text{(down as shown)}$$

$$R_{2v} = 1500 - 375 = 1125 \text{ lb} \quad \text{(down as shown)}$$

$$R_{1h} = 1500 \text{ lb} \quad \text{(acting toward the left as shown)}$$

For loading (*c*),

$$\text{total load on surface} = (200)\left(\frac{1}{0.707}\right)(10) = 2828 \text{ lb}$$

228 DESIGN OF STEEL TRUSSES

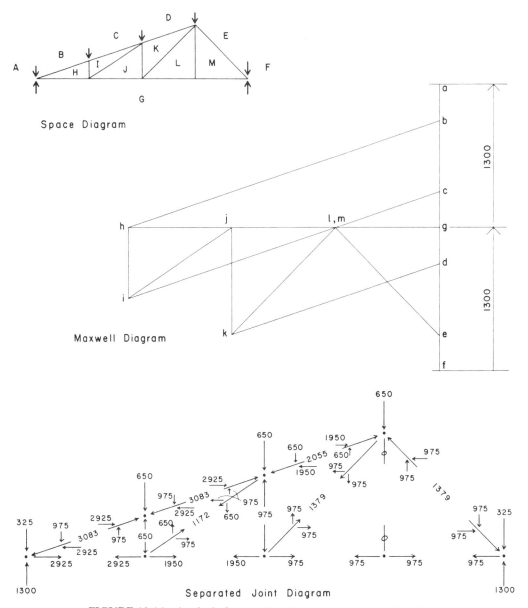

FIGURE 10.16 Analysis for member forces due to gravity loads.

$$R_{1v} = \frac{(2828)(7.07)}{40} = 500 \text{ lb} \quad \text{(up as shown)}$$

$$R_{2v} = 2000 - 500 = 1500 \text{ lb} \quad \text{(up as shown)}$$

$$R_{1h} = 2000 \text{ lb} \quad \text{(acting toward the right as shown)}$$

For the actual wind load conditions, we now combine these individual results as follows. For wind left, combine (*a*) + (*b*).

$$R_{1v} = 2625 + 375 = 3000 \text{ lb} \quad \text{(down)}$$
$$R_{2v} = 1875 + 1125 = 3000 \text{ lb} \quad \text{(down)}$$
$$R_{1h} = 1500 - 1500 = 0 \quad \text{(believe it or not)}$$

For wind right, combine $(a) + (c)$.

$$R_{1v} = 2625 - 500 = 2125 \text{ lb} \quad \text{(down)}$$
$$R_{2v} = 1875 - 1500 = 375 \text{ lb} \quad \text{(down)}$$
$$R_{1h} = 1500 + 2000 = 3500 \text{ lb} \quad \text{(toward the right)}.$$

Thus the total wind loadings will be as shown in Figure 10.15.

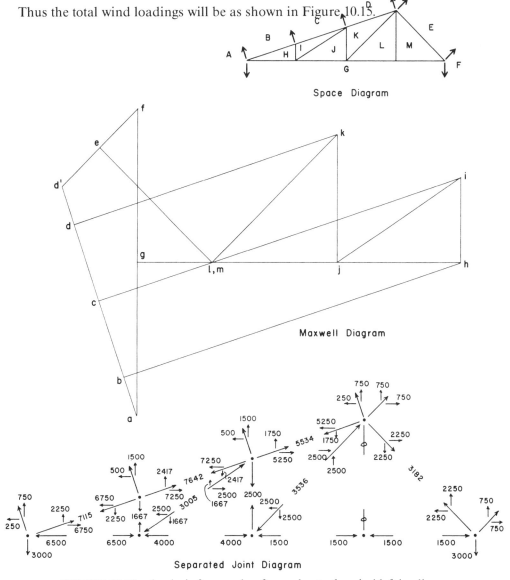

FIGURE 10.17 Analysis for member forces due to the wind left loading.

230 DESIGN OF STEEL TRUSSES

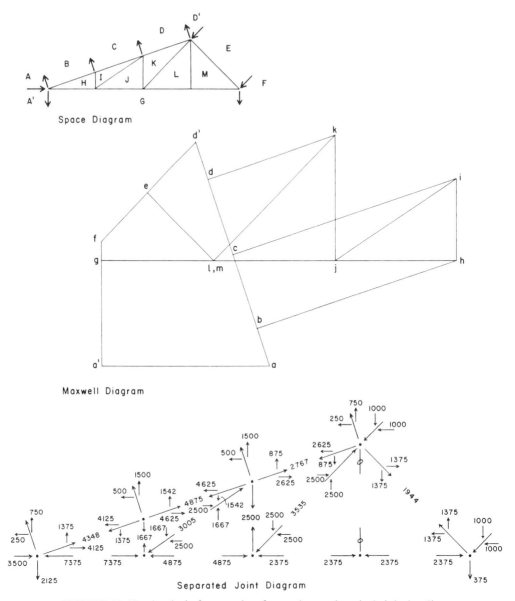

FIGURE 10.18 Analysis for member forces due to the wind right loading.

The determination of the internal forces for the gravity load is shown in Figure 10.16. This is actually the analysis for the dead load only. However, since the live load is dispersed in a similar manner on the structure, the values for the live loading may be simply proportioned from those obtained for the dead loading.

The determination of the internal forces for the two wind loadings is shown in Figures 10.17 and 10.18. A summary of the design forces for the individual truss members is given in Table 10.9. The member numbering used in the table refers to the assumed member configuration shown in Figure 10.19. It is assumed that the

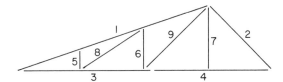

FIGURE 10.19 Configuration of the truss members.

two top chords will each be a single length, but that the 40 ft long bottom chord is too long and will require a splice.

The combined loadings to be used for design of the truss members are given in Table 10.10. The following three combinations have been considered:

1. Gravity load only—DL + LL. $A + B$ from Table 10.9.
2. Dead load + $\frac{1}{2}$ live load + wind load. $A + \frac{1}{2}B + C$ and $A + \frac{1}{2}B + D$ from Table 10.9.
3. Dead load alone + wind load. $A + C$ and $A + D$ from Table 10.9.

The maximum force and the reversal force, if any, for each member are selected from the resulting five combinations. The design of the individual members then proceeds as follows.

General Considerations:

Use $\frac{5}{8}$ in. A325 high-strength bolts.
From Table 10.7: Double shear capacity (DS) = 10.7 k/bolt
    Minimum angle leg = 2 in.

The minimum angle size of 2 in. is based on the required edge distance and the usual angle gage. For reasonable clearance of the bolt and washer a minimum leg of 2.5 in. is recommended. We thus consider the minimum double angle to be $2\frac{1}{2} \times 2 \times \frac{3}{16}$ for which

gross area = 1.62 in.² (from Table B.10)

$T = 21.6(1.62) = 35.0\text{k}$ (maximum T on gross area)

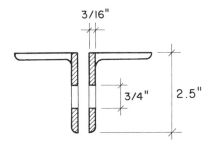

FIGURE 10.20 Net effective cross section of the minimum truss member.

232 DESIGN OF STEEL TRUSSES

TABLE 10.9 Internal Forces for the Truss

	Internal Forces[b] (kips)			
Member[a]	A Dead Load	B[c] Live Load	C Wind Left	D Wind Right
1	−3083	−9486	+7642	+4875
2	1379	−4242	+3182	+1944
3	+2925	+9000	−6500	−7375
4	+975	+3000	−1500	−2375
5	−650	−2000	+1667	+1667
6	−975	−3000	+2500	+2500
7	0	0	0	0
8	+1172	+3606	−3005	−3005
9	+1379	+4242	−3536	−3536

[a] See Figure 10.19.
[b] + = tension; − = compression.
[c] $\frac{200}{65}$ (dead load).

If used as a tension member, we consider the stress on the net effective area shown in Figure 10.20, which is determined as follows:

$$A = 2(2.5 \times 0.1875) = 0.9375 \text{ in.}^2 \quad \text{(area of angle legs)}$$

$$A = 2(0.1875 \times 0.75) = 0.2812 \text{ in.}^2 \quad \text{(area of hole profile)}$$

$$A = 0.9375 - 0.2812 = 0.6563 \text{ in.}^2 \quad \text{(effective net area)}$$

and

$$T = 29.0(0.6563) = 19.03\text{k} \quad \text{(allowable } T \text{ on net area)}$$

which is the maximum tension capacity of our minimum member.

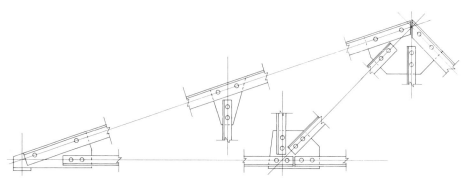

FIGURE 10.21 Joint details for the bolted truss.

TABLE 10.10 Design Forces for the Truss

Truss Member[a]	Design Forces[b] (kips)					Design Forces for Members		Design Forces for Connections[c]
	DL + LL $A + B$	DL + LL + WL$_t$ $\frac{3}{4}(A + B + C)$	DL + LL + WL$_r$ $\frac{3}{4}(A + B + D)$	DL + WL$_t$ $\frac{3}{4}(A + C)$	DL + WL$_r$ $\frac{3}{4}(A + D)$	Maximum	Minimum (Reversal)	
1	−12,569	−3,695	−5,771	+3,419	+1,344	−12,569	+3,419	−12,569
2	−5,621	−1,829	−2,758	+1,352	+424	−5,621	+1,352	−6,000
3	+11,925	+4,069	+3,413	−2,681	−3,338	+11,925	−3,338	+11,925
4	+3,975	+1,856	+1,200	−394	−1,050	+3,975	−1,050	+6,000
5	−2,650	−737	−737	+763	+763	−2,650	+763	+6,000
6	−3,975	−1,106	−1,106	+1,144	+1,144	−3,975	+1,144	+6,000
7	0	0	0	0	0	0	0	+6,000
8	+4,778	+1,330	+1,330	−1,375	−1,375	+4,778	−1,375	+6,000
9	+5,621	+1,564	+1,564	−1,618	−1,618	+5,621	−1,618	+6,000

[a] See Figure 10.19.
[b] + = tension; − = compression.
[c] Minimum load of 6 kips; also joint must develop 50% of member capacity.

Member 1

axial force $- C = 12{,}569$ lb, $T = 3419$ lb,

length $= \dfrac{120}{0.9487} = 126.5$ in.

$M = \dfrac{wL^2}{8} = \dfrac{(265)(10)^2}{8} = 3312$ lb-ft   (bending moment with $DL + LL$)

For a first guess, if $L/r = 60$, $F_a = 17.43$ k/in.² (from Table 10.2)

$A = \dfrac{C}{F_a} = \dfrac{12{,}569}{17{,}430} = 0.72$ in.²   (required area for axial load alone)

$S = \dfrac{M}{F_b} = \dfrac{(3312)(12)}{21{,}600} = 1.84$ in.³   (required S for bending alone)

As a rough guide we look for a double angle with approximately twice these values.

Try: $4 \times 3 \times \tfrac{3}{8}$, $A = 4.97$ in.², $S = 2.92$ in.³, $r_x = 1.26$ in.

$f_a = \dfrac{C}{A} = \dfrac{12.569}{4.97} = 2.53$ k/in.²

$\dfrac{L}{r} = \dfrac{126.5}{1.26} = 100$, $F_a = 12.98$ k/in.², $F'_e = 14.93$ k/in.²

$f_b = \dfrac{M}{S} = \dfrac{(3.312)(12)}{2.92} = 13.61$ k/in.²

Then, for the combined action,

$$\dfrac{f_a}{F_a} + \dfrac{f_b}{(1 - f_a/F'_e)F_b} = \dfrac{2.53}{12.98} + \dfrac{13.61}{(1 - 2.53/14.93)(21.6)}$$

$$= 0.195 + 0.759$$

$$= 0.954$$

Since this is quite close, but slightly less than one, we will consider the member to be an adequate choice, although other ones are possible. Note that the tension load is not a consideration since the member will easily sustain this force.

Member 2

axial force $- C = 5621$ lb, $T = 1352$ lb,

length $= \dfrac{120}{0.707} = 169.7$ in.

$DL + LL\ M = \dfrac{(265)(10)^2}{8} = 3312$ lb-ft

$$\text{wind } M = \frac{(200)(14.14)^2}{8} = 4998 \text{ lb-ft}$$

(WL$_r$ loading—see Figure 10.15)

Reduced for comparison, $M = \frac{3}{4}(4998) = 3749$ lb-ft

We thus consider two different loading conditions, as follows:

With DL + LL only,

$$C = 5621 \text{ lb and } M = 3312 \text{ lb-ft}$$

With DL + WL$_r$,

$$T = 424 \text{ lb and } M = 3749 \text{ lb-ft}$$

It should be evident that the DL + LL condition will be the more critical, so we will proceed with the design for that combination. Note that this member is unbraced on both axes; therefore, we will use the least radius of gyration, for whichever axis it occurs. If $L/r = 100$, $F_a = 12.98$ k/in.^{2} (from design for member 1),

$$A = \frac{5621}{12{,}980} = 0.433 \text{ in.}^2 \quad \text{(required area for } C \text{ alone)}$$

$$S = \frac{(3312)(12)}{21{,}600} = 1.84 \text{ in.}^3 \quad \text{(required } S \text{ for } M \text{ alone)}$$

$$r = \frac{170}{100} = 1.70 \text{ in.} \quad \left(\text{required } R \text{ for } \frac{L}{r} = 100\right)$$

Try: $4 \times 3 \times \frac{3}{8}$, $A = 4.97$ in.^{2}, $S = 2.92$ in.^{3}, $r = 1.26$ in.

$$f_a = \frac{C}{A} = \frac{5.621}{4.97} = 1.131 \text{ k/in.}^2$$

$$\frac{L}{r} = \frac{170}{1.26} = 135, \; F_a = 8.19 \text{ k/in.}^2, \; F'_e = 8.19 \text{ k/in.}^2$$

$$f_b = \frac{M}{S} = \frac{(3.312)(12)}{2.92} = 13.61 \text{ k/in.}^2$$

and for the combined action,

$$\frac{f_a}{F_a} + \frac{f_b}{(1 - f_a/F'_e)F_b} = \frac{1.131}{8.19} + \frac{13.61}{1 - 1.131/8.19)(21.6)}$$

$$= 0.138 + 0.731$$

$$= 0.869$$

236 DESIGN OF STEEL TRUSSES

This indicates that the element chosen is adequate. However, other choices are possible and consideration for construction detailing may indicate a better selection.

Member 3

$$\text{axial force} - C = 3338 \text{ lb}, \; T = 11925 \text{ lb}$$

$$\text{length} = 10 \text{ ft}$$

With no ceiling, there is no directly applied load and the bottom chord is thus designed only for the axial forces. Even though the tension force is the higher, it is not the critical design factor since the minimum member and minimum connection capacities are both greater than the internal tension force. The chord must be designed as a compression member and thus has a limit of 200 for the L/r ratio.

With no ceiling construction, we must consider the problem of lateral bracing for the bottom chord. We will assume that lateral bracing is provided only at the truss midspan, in which case the bottom chord has laterally unsupported length of 10 ft on its x-axis and 20 ft on its y-axis. With this condition the minimum required r values for the chord are

$$L_x = 10 \text{ ft},$$

$$\text{required } r_x = \frac{(10)(12)}{200} = 0.60 \text{ in.} \quad \text{(on the } x\text{-axis)}$$

$$L_y = 20 \text{ ft},$$

$$\text{required } r_y = \frac{(20)(12)}{200} = 1.20 \text{ in.} \quad \text{(on the } y\text{-axis)}$$

The minimum member, as established previously, is

$$2\tfrac{1}{2} \times 2 \times \tfrac{3}{16}, \; A = 1.62 \text{ in.}^2, \; r_x = 0.793 \text{ in.}, \; r_y = 0.923 \text{ in.}$$

The minimum member is thus not sufficient unless the truss is braced at each bottom chord panel point. If we stay with the single bracing at the center of the truss, we must find a member with a higher value for r_y.

Try: $3 \times 3 \times \tfrac{1}{4}, \; A = 3.13 \text{ in.}^2, \; r_x = 1.11 \text{ in.}, \; r_y = 1.33 \text{ in.}$

Then,

$$\frac{L}{r_y} = \frac{240}{1.33} = 181, \; F_a = 4.56 \text{ k/in.}^2 \quad \text{(from Table 10.2)}$$

$$C = (F_a)(A) = (4.56)(3.13) = 14.27 \text{k} \quad \text{(allowable axial compression)}$$

This indicates that the member is adequate. This member would also be the minimum size for member 4 unless additional lateral bracing is provided.

DESIGN OF A LIGHT STEEL TRUSS WITH BOLTED JOINTS

Members 5 and 6. Although these are also compression members, they are quite short and we will try the minimum member. Using the longer length for member 6,

$$L = 80 \text{ in.}, \quad \frac{L}{r} = \frac{80}{0.793} = 101, \quad F_a = 12.85 \text{ k/in.}^2$$

$$C = (F_a)(A) = (12.85)(1.62) = 20.8 \text{k} \quad \text{(allowable axial compression)}$$

Member 7. Since there is no actual design load for this member, it is designed as a minimum tension member, with the direct functions of supporting and bracing the bottom chord. We thus consider only the minimum r value required for a tension member, as follows:

$$L = 120 \text{ in.}, \quad \text{required } r = \frac{L}{240} = \frac{120}{240} = 0.50 \text{ in.}$$

for which the minimum member is adequate.

Member 8

$$\text{axial force} - T = 4778 \text{ lb}, \quad C = 1375 \text{ lb}$$

$$\text{length} = \frac{1}{0.832}(120) = 144 \text{ in.}$$

With the minimum member,

$$\frac{L}{r} = \frac{144}{0.793} = 182, \quad F_a = 4.51 \text{ k/in.}^2$$

$$C = (F_a)(A) = (4.51)(1.62) = 7.31 \text{k} \quad \text{(allowable axial compression)}$$

Member 9

$$\text{axial force} - T = 5621 \text{ lb}, \quad C = 1618 \text{ lb}$$

$$\text{length} = \frac{1}{0.707}(120) = 170 \text{ in.}$$

$$\text{required } r = \frac{L}{200} = \frac{170}{200} = 0.850 \text{ in.}$$

Try: $3 \times 2 \times \frac{3}{16}$, $A = 1.80 \text{ in.}^2$, $r_y = 0.879 \text{ in.}$, $r_x = 0.966 \text{ in.}$

$$\frac{L}{r} = \frac{170}{0.879} = 193, \quad F_a = 4.01 \text{ k/in.}^2$$

$$C = (F_a)(A) = (4.01)(1.80) = 7.22 \text{k} \quad \text{(allowable axial compression)}$$

Since the minimum two-bolt connection has a capacity in excess of all of the internal forces, the joints will all be developed with only two bolts per truss member. The form of the joints will be as shown in Figure 10.21. From Table 10.7 we find that the minimum thickness for the gusset plate is $\frac{3}{16}$ in. Although this size is acceptable in terms of stress calculation, many designers prefer to use a slightly stiffer plate and would probably select one with $\frac{1}{4}$ in. thickness. In addition to the problem of bearing stress on the side of the bolt holes, the gusset plate should be investigated for the problem of tearing. We will investigate this for the highest tension force—that in Member 3 at the left support point. As shown in Figure 10.22, the tearing action involves a combination of shearing along the horizontal rupture plane and tension on the vertical rupture plane. Assuming the thinner $\frac{3}{16}$ in. plate and a hole diameter of $\frac{3}{4}$ in. for the $\frac{5}{8}$ in. bolts, the calculations are as follows:

For shear,

$$\text{allowable stress} = F_v = 0.30 F_u = 0.30(58) = 17.4 \text{ k/in.}^2$$
$$\text{shear area} = \tfrac{3}{16}[4.5 - 1.5(0.75)] = 0.6328 \text{ in.}^2$$

For tension,

$$\text{allowable stress} = F_t = 0.50 F_u = 0.50(58) = 29.0 \text{ k/in.}^2$$
$$\text{tension area} = \tfrac{3}{16}[1.5 - 0.5(0.75)] = 0.2109 \text{ in.}^2$$
$$T = (17.4)(0.6328) + (29.0)(0.2109)$$
$$= 11.01 + 6.12$$
$$= 17.13 \text{k} \quad \text{(total tearing resistance)}$$

which is greater than the tension force in the chord.

10.4 DESIGN OF A LIGHT STEEL TRUSS WITH WELDED JOINTS

The most widely used jointing technique for steel trusses of small to medium size is welding. Let us consider the use of a truss with welded joints for the structure as shown in Figure 10.13 in the preceding section. One possibility is to use the same type of members and to simply weld, rather than bolt, the members to the gusset plates. If this is done, the bottom chord joint at midspan would appear as shown in

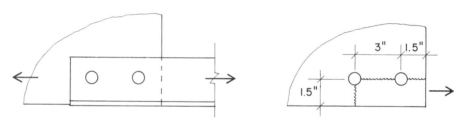

FIGURE 10.22 Tearing of the gusset plate.

DESIGN OF A LIGHT STEEL TRUSS WITH WELDED JOINTS   239

Figure 10.23. Use of welding results in the following different considerations for the choice of the truss members.

1. Leg width is not limited by the need for accommodating a bolt, so there is thus no minimum leg width and the limiting radius of gyration or other properties will establish the minimum size required.
2. Thickness of members will be based on the size of welds used. With the fillet welds as shown in Figure 10.23, the maximum size of weld is limited to the dimension of the angle thickness. Due to the form of the rolled edge, some designers prefer to limit the weld to something less—commonly $\frac{1}{8}$ in.—than the angle thickness. However, the present codes generally permit a fillet of the size of the full angle thickness.
3. When members are very thick, a minimum size of fillet weld is also required, although this is not critical in this example.

For a truss of this size most designers would prefer to use a minimum weld size of $\frac{1}{4}$ in. If this is done we make the following determinations. From Table 10.8,

$$\text{capacity of the } \tfrac{1}{4} \text{ in. weld} = 3.2 \text{ k/in. of length}$$
$$4(\text{nominal size}) = 4(\tfrac{1}{4}) = 1.0 \text{ in.} \quad (\text{minimum length})$$

With the welds placed on the ends of the members as shown in Figure 10.25, there will be a minimum of four welds on the end of each pair of angles. The minimum connection, therefore, will have the following capacity.

$$L = 4(1.0) = 4.0 \text{ in.} \quad (\text{total weld length})$$
$$T = (4.0)(3.2) = 12.8 \text{k} \quad (\text{total load capacity})$$

Since this is larger than the highest design force given in Table 10.10, the minimum end weld is sufficient for all members. However, the development of some

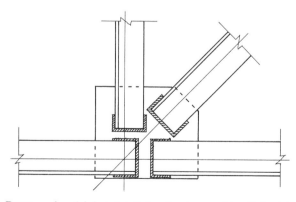

FIGURE 10.23 Bottom chord joint detail for the truss with all double-angle members, gusset plates, and welded connections.

240 DESIGN OF STEEL TRUSSES

joints may indicate the need for some additional welds to assure stability of the assemblage.

Since the minimum member for the bolted truss had a thickness of $\frac{3}{16}$ in., whereas a thickness of $\frac{1}{4}$ in. is required for welding, it is necessary to establish a new minimum member size. From the sizes listed in Table B.9, we thus consider using the following double angle.

$$2 \times 2 \times \tfrac{1}{4}, \ r_x = 0.609 \text{ in.}, \ A = 1.88 \text{ in.}^2$$

For this choice, using only the area of the attached legs, the tension capacity becomes as follows.

$$A = 2(2 \times \tfrac{1}{4}) = 1.0 \text{ in.}^2 \quad \text{(effective area)}$$
$$T = A \times F_t = (1.0)(21.6) = 21.6\text{k} \quad \text{(allowable tension)}$$

The maximum usable lengths, based on the limits for the L/r ratios, are

maximum $L = 240r = 240(0.609) = 146$ in. (for a tension member)

Maximum $L = 200r = 200(0.609) = 121.8$ in. (for a compression member)

We may thus consider the use of the minimum member for the following members: 3, 4, 5, 6, and 7. (See Figure 10.19.) For the longest of these, the limiting compression load is as follows.

$$L = 120 \text{ in.} \quad \text{(unbraced length)}$$

Then,

$$\frac{L}{r} = \frac{120}{0.609} = 197, \ F_a = 3.85 \text{ k/in.}^2 \quad \text{(from Table 10.2)}$$

$$C = \text{gross } A \times F_a = (1.88)(3.85) = 7.24\text{k} \quad \text{(allowable compression)}$$

which is adequate for the five designated members.

For the other truss members, the sizes were based on either the need for a larger r value or on combined actions of bending and compression. The sizes would thus be the same as those used for the bolted truss, except for the need for a minimum thickness of $\frac{1}{4}$ in.

Another possibility for the all welded truss is to use tee sections for the top and bottom chords. The double angle vertical and diagonal members may thus be welded directly to the webs of the tees, eliminating the need for gusset plates. Thus the midspan bottom chord joint would be formed as shown in Figure 10.24. If required, the splice for the bottom chord could be formed as shown. However, it is possible that the tee could be obtained in a single piece of sufficient length to eliminate the need for the splice.

Where the two tee chords meet at the ends of the truss, they may be directly welded to each other with a butt weld, as shown at (*a*) in Figure 10.25. This detail

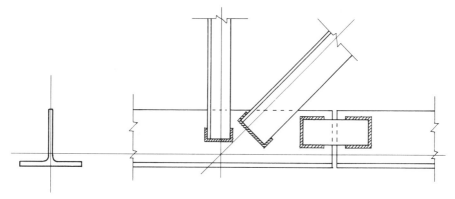

FIGURE 10.24 Bottom chord joint detail for the truss with tee chords, double-angle web members, and welded connections without gusset plates.

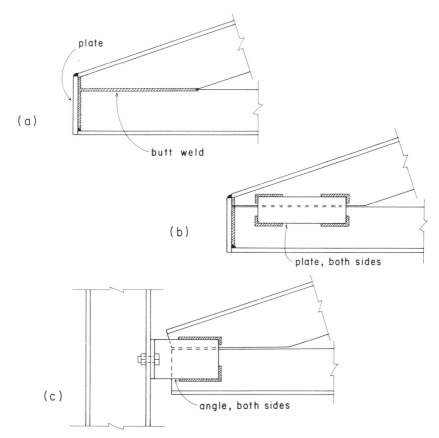

FIGURE 10.25 Alternative details for the end support (heel) joint of the truss with tee chords.

242 DESIGN OF STEEL TRUSSES

would be used if the end support is achieved by direct bearing of the bottom chord. Other possibilities for the joint, for other support conditions, are shown at (b) and (c) in Figure 10.25.

Using the criteria established for the bolted truss, we thus consider the design of member 1 as follows.

Member 1

$$\text{axial force} - C = 12{,}569 \text{ lb}, \ T = 3419 \text{ lb}$$

$$\text{length} = 126.5 \text{ in.}, \text{ bending moment} = 3312 \text{ lb-ft}$$

Try: WT6 × 15, $A = 4.40 \text{ in.}^2$, $S_x = 2.75 \text{ in.}^3$, $R_x = 1.75 \text{ in.}$, $Q_s = 0.891$

$$f_a = \frac{C}{A} = \frac{12.569}{4.40} = 2.86 \text{ k/in.}^2$$

$$\frac{L}{r} = \frac{126.5}{1.75} = 72, \ F_a = 16.22 \text{ k/in.}^2, \ F'_e = 28.81 \text{ k/in.}^2$$

$$f_b = \frac{M}{S} = \frac{(3.312)(12)}{2.75} = 14.45 \text{ k/in.}^2$$

$$\frac{f_a}{Q_s F_a} = \frac{2.86}{0.891(16.22)} = 0.198$$

Because f_a/F_a is greater than 0.15, we use the following formula for the combined action.

$$\frac{f_a}{F_a} + \frac{f_b}{(1 - f_a/F'_e)F_b} = 0.198 + \frac{14.45}{(1 - 2.86/28.81)(21.6)}$$

$$= 0.198 + 0.743$$

$$= 0.941$$

This indicates that the member is adequate. Note that we use the Q_s reduction factor for the allowable axial compression but not for the bending, since the critical combined action occurs at the midlength point where the bending stress in the tee stem is tension.

In addition to its basic structural functions, the following concerns usually affect the choice of the tee chord.

1. Since the flange will be used for direct attachment of the deck, it must be of an adequate width for this purpose. If there is to be any attachment of lateral bracing or other secondary framing to the chord flange, this may also affect the need for a particular width.
2. The stem depth must be adequate for the attachment of the truss web mem-

bers. The need for a particular dimension will be established by the size of the web members and the length of welds required for their attachment. The limits for this are most easily determined by some large-scale studies of the actual joint layouts.

3. The stem thickness is primarily limited by the need to accommodate the fillet welds on both sides, just as with the gusset plates. For small fillet welds, this limit is usually the same dimension as the nominal size of the fillet weld. Another possible requirement may exist if the end joints are of the type shown at (b) or (c) in Figure 10.25. In this case the stems of both the top and bottom should be of approximately the same thickness.

Member 3

$$\text{axial force} - C = 3.338k, \ T = 11.925k$$

$$\text{length} - 10 \text{ ft for } r_x, \ 20 \text{ ft for } r_y$$

(With lateral bracing only at midspan.)

Try: WT5 × 11, $A = 3.24$ in.^{2}, $r_y = 1.33$ in., $Q_s = 0.999$ (negligible)

Compression:

$$\frac{L}{r} = \frac{240}{1.33} = 180, \ F_a = 4.61 \text{ k/in.}^2$$

$$\text{Allowable force} = (A)(F_a) = (3.24)(4.61) = 14.94k$$

Tension (assuming the effective area to be only that of the stem):

$$A = \text{(tee depth)(stem thickness)}$$

$$= (5)(0.240) = 1.20 \text{ in.}^2$$

$$\text{allowable force} = (A)(21.6) = 25.92k$$

The member, therefore, is quite adequate. Even if the stem area is reduced by bolt holes, it should be adequate for the truss design tension forces. As with the top chord, however, there may be other considerations for the choice of the member.

In general the web members would be the same double angles as designed for the welded truss with gusset plates. As with the chords, there may be some considerations for joint detailing, attachment of bracing, and so on.

10.5 DESIGN OF A STEEL TRUSS WITH TUBULAR MEMBERS

When trusses are exposed to view, a popular form of steel truss is one with members of round steel pipe or rectangular tubes. With the joints achieved by welding the members directly to each other, a very clean-lined structure is produced. Let us consider the possibility of such a structure for the truss in the preceding examples using members of rectangular steel tubes. If the truss web members are fastened to the sides of the chords, as shown in Figure 10.26a, the following concerns affect the size of the tubes and the thickness of their walls.

Width of Tubes. The width of the web members should be slightly less than that of the chords in order to assure that the end of the web member fits onto the flat portion of the chord. The average value for the corner radius is twice the wall thickness, and the limit for the web member is thus

$$\text{maximum } w_1 = w_2 - 2(r) \quad \text{or} \quad w_1 = w_2 - 4(t_2)$$

Thickness of Tubes. The thickness of the web member should be approximately the same as the nominal size of the fillet weld. The following approximate limits are recommended.

$$\text{minimum } t_1 = \tfrac{3}{4}(f) \text{ and maximum } t_1 \; 1.5(f)$$

For the chord, the wall thickness should be adequate to resist warping or burn-

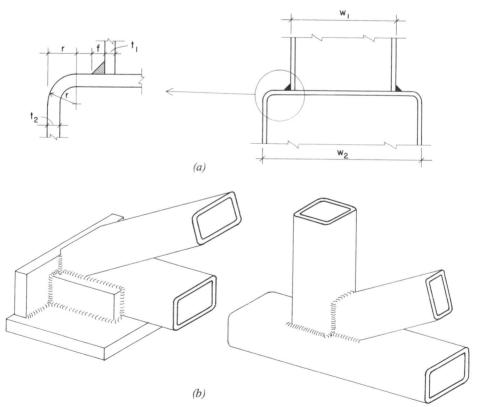

FIGURE 10.26 Joints in trusses with steel tubular members: (*a*) layout considerations for the fillet welds. (*b*) General forms of joints.

DESIGN OF A STEEL TRUSS WITH TUBULAR MEMBERS   245

through by the welds and excessive deformation by the forces from the webs. The following limits are recommended.

$$\text{minimum } w_1 = w_2 - 8(t_2)$$

$$\text{minimum } t_2 = t_1 \text{ or } f, \text{ whichever is greater}$$

With these limitations in mind, let us consider the choice of tubular members for the truss using $\frac{1}{4}$ in. fillet welds and the joint details shown in Figure 10.26b.

Member 1

$$\text{axial force} - C = 12.569 \text{ k}, \quad T = 3.419 \text{ k}$$
$$M = 3.312 \text{ k-ft} \quad \text{(bending)}$$
$$L_x = 126.5 \text{ in.} \quad \text{(length)}$$
$$(L_y = 0 \text{ due to bracing by deck.})$$

Try: $4 \times 4 \times \frac{1}{4}$, $A = 3.59$ in.^{2}, $S = 4.11$ in.^{3}, $r = 1.51$ in.

$$f_a = \frac{C}{A} = \frac{12.569}{3.59} = 3.50 \text{ k/in.}^2$$

$$\frac{L}{r} = \frac{126.5}{1.51} = 84,$$

$$F_a = 14.09 \text{ k/in.}^2,$$

$$F'_e = 21.16 \text{ k/in.}^2$$

$$f_b = \frac{M}{S} = \frac{(3.312)(12)}{4.11} = 9.67 \text{ k/in.}^2$$

$$\frac{f_a}{F_a} + \frac{f_b}{(1 - f_a/F'_e)F_b} = \frac{3.50}{14.09} + \frac{9.67}{(1 - 3.50/21.16)(21.6)}$$

$$= 0.235 + 0.536$$

$$= 0.771$$

Member 3

$$\text{axial force} - C = 3.338\text{k}, \quad T = 11.925\text{k}$$
$$\text{length} - 10 \text{ ft for } r_x, 20 \text{ ft for } r_y$$

(With lateral bracing only at midspan.)

Try: $4 \times 3 \times \frac{1}{4}$, $A = 3.09$ in.^{2}, $r_x = 1.45$ in., $r_y = 1.15$ in.

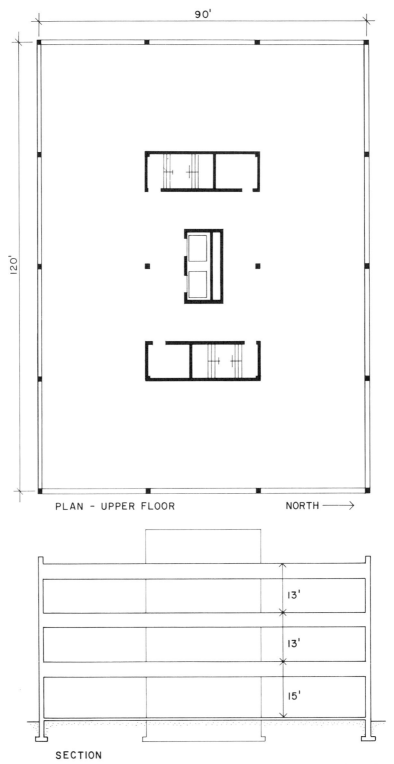

FIGURE 10.27 Upper-level floor plan and general building section for the design example building.

(Note: While the usual reference is x for the horizontal axis and y for the vertical axis, the tube will be used in a flat position, as shown in Figure 10.27. Thus the appropriate r value for the 20 ft unbraced length is actually r_x in this case.)

$$f_a = \frac{C}{A} = \frac{3.338}{3.09} = 1.080 \text{ k/in.}^2 \quad \text{(actual stress)}$$

$$\frac{L}{r} = \frac{240}{1.45} = 166, \; F_a = 5.42 \text{ k/in.}^2 \quad \text{(allowable stress)}$$

Although these chord members are stronger than required, they are chosen for their matching widths, for a width sufficient to provide for the web member attachment, and for a thickness to accommodate the $\frac{1}{4}$ in. welds.

Based on the previous recommendations, the choice of web members is limited by the following measurements.

1. A maximum width of 3 in.
2. A minimum width of 2 in.
3. A minimum thickness of $\frac{3}{16}$ in. for the $\frac{1}{4}$ in. welds.

On the basis of the member lengths, the design forces, and the considerations of the joint fabrication, the choices for members are as shown in Table 10.11. Data for members is taken from Tables B.4 and B.5.

10.6 TRUSSED BRACING FOR STEEL STRUCTURES

The use of trussed bracing for lateral forces in general is discussed in Section 8.7. The following example illustrates the design of a trussed bracing system for a low-rise building with a steel frame structure.

Figure 10.27 shows a general upper-level plan and a building section for a three-story office building. The building uses steel columns and steel beam framing systems for the roof and upper floors. A partial framing plan for one of the upper floors is shown in Figure 10.28. The extra columns placed at the locations of walls in the building core are used for the development of a trussed bracing system. The general form of this bracing system is shown in Figure 10.29. It consists of vertically can-

TABLE 10.11 Tubular Elements for the Truss Web Members

		Maximum Usable Length (in.)		
Tube Size	Minimum r (in.)	For Tension $L = 240(r)$	For Compression $L = 200(r)$	Use for Truss Members
$2 \times 2 \times \frac{3}{16}$	0.726	174	145	6,9
$2\frac{1}{2} \times 2\frac{1}{2} \times \frac{3}{16}$	0.930	223	186	8
$3 \times 2 \times \frac{3}{16}$	0.771	185	154	5,7

248 DESIGN OF STEEL TRUSSES

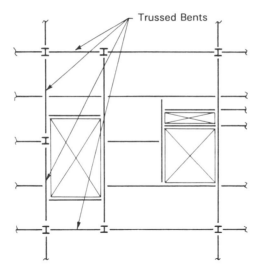

FIGURE 10.28 Partial framing plan for the steel structure.

tilevered, three-story-high, X-braced bents, placed at each corner of the building plan core.

The system shown in Figure 10.29 provides four bents for resistance to wind in each of the two major axis directions for the building. This is the general basis for design for the symmetrical building plan, so we will illustrate the design for wind in one direction only for this example. The system as constituted also provides reasonably for torsional effects, which would be of greater significance for a building with an asymmetrical plan or for seismic loading conditions.

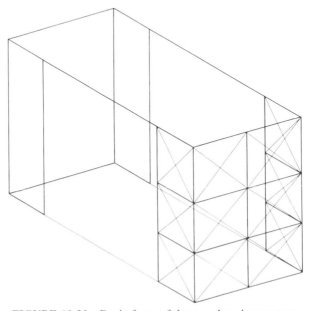

FIGURE 10.29 Basic form of the core bracing system.

The following investigation also illustrates the general use of wind load criteria as presented in current building codes.

The 1991 edition of the *Uniform Building Code* (Ref. 1) provides for the use of the projected profile method for wind using a pressure on vertical surfaces defined as

$$p = C_e C_q q_s I$$

where C_e is a combined factor including concerns for the height above grade, exposure conditions, and gust effects. From UBC Table 23-G, assuming exposure condition B,

C_e = 0.7 for surfaces from 0 to 20 ft above grade

= 0.8 from 20 to 40 ft above grade

= 1.0 from 40 to 60 ft above grade

and C_q is the pressure coefficient, which the UBC defines as follows:

C_q = 1.3 for surfaces up to 40 ft above grade

= 1.4 from 40 ft up

The symbol q_s stands for the wind stagnation pressure as related to wind speed and measured at the standard height above ground of 10 m (approximately 30 ft). For the wind speed of 80 mph assumed earlier, the UBC yields a value for q_s of 17 psf.

Table 10.12 summarizes the forgoing data for the determination of the wind pressures at the various height zones on Building Three. For investigation of the wind effects on the lateral bracing system, the wind pressures on the exterior wall are translated into edge loadings for the horizontal roof and floor diaphragms, as shown in Figure 10.30. Note that we have rounded off the wind pressures from Table 10.12 for use in Figure 10.30.

The accumulated forces noted as H_1, H_2, and H_3 in Figure 10.30 are shown applied to one of the vertical trussed bents in Figure 10.31a. For the east-west bents, the loads will be as shown in Figure 10.31b. These loads are determined by multiplying the edge loadings for the diaphragms as shown in Figure 10.30 by the 92-ft

TABLE 10.12 Design Wind Pressures for the Example Building

Height Above Average Level of Adjoining Ground (ft)	C_e	C_q	Pressure[a] p (psf)
0–20	0.7	1.3	15.47
20–40	0.8	1.3	17.68
40–60	1.0	1.4	23.80

[a] Horizontally directed pressure on vertical projected area: $p = C_e \times C_q \times 17$ psf.

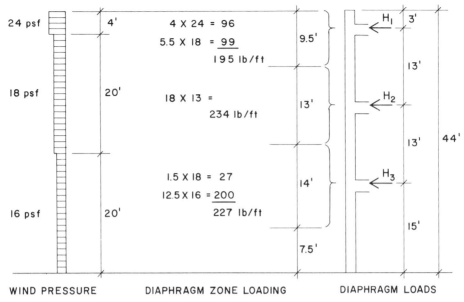

FIGURE 10.30 Wind load development for the core framing investigation.

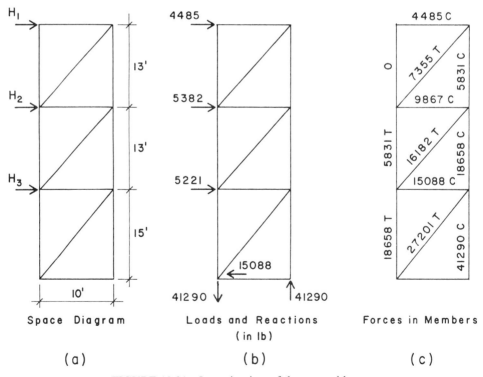

FIGURE 10.31 Investigation of the trussed bent.

250

overall width of the building on the east and west sides. For a single bent, this total force is divided by four. Thus for H_1 the bent load is determined as

$$H_1 = 195 \times 92 \div 4 = 4485 \text{ lb}$$

Analyzed as a truss, ignoring the compression diagonals, the resulting internal forces in the bent are as shown in Figure 10.31c. The forces in the diagonals may be used to design tension members, using the usual one-third increase in allowable stress. The forces in the vertical columns may be added to the gravity loads and checked for possible critical conditions for the columns previously designed for gravity load only. The anchorage force in tension at the bottom of the trusses must be investigated in combination with the dead load for consideration of the need of real anchorage to the foundation. This could become a major foundation design problem if a significant net uplift force exists.

The horizontal forces must be added to the beams in the core framing and an investigation made for the combined bending and axial compression. This can be critical, since beams are ordinarily quite weak on their minor axes (the y-axis), and it may be practical to add some additional horizontal framing members to reduce the lateral unbraced length of some of these beams.

Design of the diagonal members and of their connections to the frame must be developed with consideration of the form of the frame members and the general

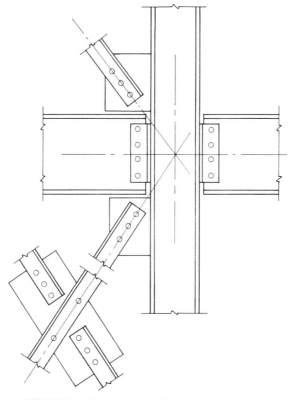

FIGURE 10.32 Joint detail for the trussed bent.

form of the wall construction that encloses the steel bents. Figure 10.32 shows some possible details for the diagonals for the bent analyzed in Figure 10.31. A consideration to be made in the choice of the diagonal members is the necessity for the two diagonals to pass each other in the midheight of a bent level. If the most common truss members—double angles—are used, it will be necessary to use a joint at this crossing, and the added details for the bent are increased considerably.

An alternative to the double angles is to use either single angles or channel shapes. These may cross each other with their flat sides back-to-back without any connection between the diagonals. However, this involves some degree of eccentricity in the member loadings and the connections, so they should be designed conservatively. It should also be carefully noted that the use of single members results in single-shear loading on the bolted connections.

11

SPECIAL PROBLEMS

Any basic form of structure has some special problems that must be considered by designers. This chapter deals with some of these problems that are not present in all situations, but which designers should be aware of, in the design of trusses.

11.1 DEFLECTION OF TRUSSES

When used in situations where they are most capable of being utilized to the best of their potential, trusses will seldom experience critical deflections. In general trusses possess great stiffness in proportion to their mass. When the deflection of a truss is significant, it is usually the result of one of two causes. The first of these is the ratio of the truss span to the depth. This ratio is ordinarily quite low when compared to the normal ratio for a beam, but when it becomes as high as that for a beam, considerable deflection may be expected. The second principal cause of truss deflections is excessive deformation in the truss joints. A particular problem is that experienced with trusses that are fabricated with bolted joints. Since the bolt holes must be somewhat larger than the bolts to facilitate the assembly, considerable slippage is accumulated when the joints are loaded. This is a reason for favoring joints made with welds, split ring connectors, or high-strength bolts, the latter functioning in friction resistance.

For most trusses deflection is essentially due to the lengthening and shortening of the members caused by the interior forces of tension and compression. Figure 11.1 shows a simple W truss in which the original, unloaded position is shown as a solid line and the deflected shape as a dotted line. In this example, with the left end held horizontally, there are two deflections of concern. The first is the vertical movement, measured as the sag of the bottom chord or as the drop of the peak of the truss. The second is the horizontal movement that occurs at the right end, where horizontal restraint is not provided.

A relatively simple procedure for determining the deflected shape of a truss due

254 SPECIAL PROBLEMS

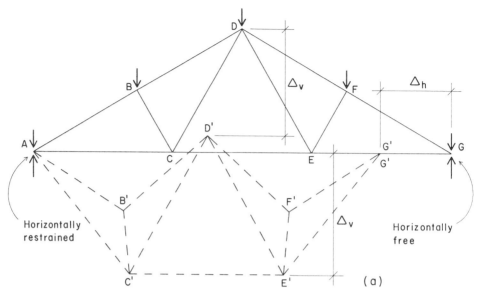

FIGURE 11.1 Exaggerated deformation of a W truss under vertical loading; left end horizontally restrained and right end on a roller.

to the length change of the members is to plot the deformations graphically. This consists simply of constructing the individual triangles of the truss with the sides equal to the deformed lengths. The procedure for this is illustrated in Figure 11.2. We begin with one truss joint as a fixed reference. In this example it is logical to use the joint at the left support, since it truly remains fixed in location, both vertically and

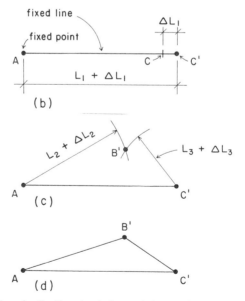

FIGURE 11.2 Procedure for finding the deformed shape of a single truss unit; in the form of length changes for the members.

horizontally. We then assume that one of the sides of this first triangle remains in its angular position. Although this is probably not true, the result can be adjusted for when the construction is completed. The deformed location of joint C (shown as C',) is now found simply as a point along the line of the fixed member by determining the length change of the member due to the internal force. For practical purposes this deformation is exaggerated by simply multiplying it by some factor. With the same factor used for all the deformations, the end produced deflections are simply divided by the same factor to find the true dimensions.

The other truss joint B' that defines the first deformed triangle is now found by using the deformed lengths of the other two members, as shown in the figure. With this triangle ($AB'C'$) constructed, we next use the known locations of joints B' and C' as a reference for the construction of the next triangle, $B'C'D'$. This procedure continues until we have produced the completed figure shown in dotted line in Figure 11.3. We then simply superimpose the original figure of the truss on the deflected figure by matching the location of joint A in both figures and by aligning the line between joints A and G' with the original position of the bottom chord. The resulting figure will be that shown in Figure 11.1.

If the construction of the deformed truss is performed only to determine the values of specific deflections, it is not necessary to rotate the original truss form, as shown in Figure 11.1. In the construction in Figure 11.3, a line has been drawn from the fixed point A to the deformed location of point G, noted as point G' on the

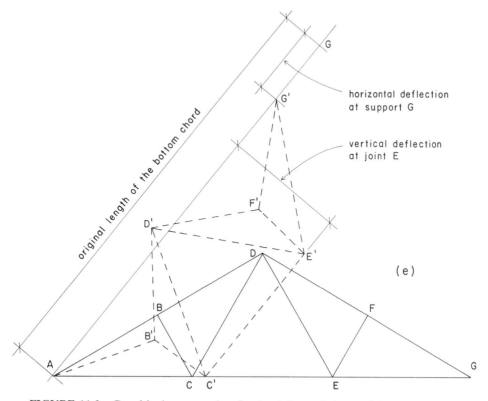

FIGURE 11.3 Graphical construction for the deformed shape of the loaded truss.

figure. As shown, the horizontal deflection at the right support and the vertical deflection of the truss joints may be determined with reference to this line.

The procedure for determining the deformed lengths of the individual members of the truss is shown in Table 11.1. The actual changes in the lengths of the members are found as follows.

1. Determine the magnitudes of the internal forces, the original lengths of the members, the gross cross-sectional area of the members, and the modulus of elasticity of the members.
2. Find the true change in the length of the members as follows.

$$\Delta L = \frac{(F)(L)}{(A)(E)}$$

Where: ΔL is the actual change in length; shortening when shortening when the member is in compression and lengthening when it is in tension.
F is the axial force in the member.
A is the gross cross-sectional area of the member.
E is the modulus of elasticity of the material.
[Note: Be sure that A and L are in the same units (in.) and that F and E are in the same units (lb/in.^{2} or k/in.^{2}).]

3. Multiply the actual length changes by some factor in order to produce dimensions that are between 5% and 10% of the original lengths. This is necessary for reasonable construction of the graphic figure. When the construction is completed, the deflections found using the figure are simply divided by this factor to find their true values.

Adjustments for True Conditions

In real situations there are always effects that produce some modification of the deflections caused by the axial forces in the members. The following effects are the most common.

Deformation of the Joints. These can be of considerable magnitude, especially with bolted joints not utilizing high-strength bolts for steel trusses or split ring connec-

TABLE 11.1 Example of the Form of Length Change of the Truss Members

Truss Member	Axial Force		Area of Cross Section (in.^{2})	Length (in.)	Length Change (in.)	
	Type	Magnitude (lb)			True[a]	for Plot[b]
1	Tension	20,000	1.6	100	0.0431	10.78
2	Compression	26,000	2.8	87	0.0279	6.96

[a] ΔL = (force)(length)/(area)(29,000,000 lb/in.^{2}) (actual change)
[b] 250 (ΔL) (for the graphic plot)

tors for wood trusses. Nailed joints will also experience considerable deformation owing to the bending of the nails and bearing stress in the wood.

Continuous Chord Members. Observation of the deflected truss form in Figure 11.1 will indicate that the top and bottom chords cannot take this form, unless they are discontinuous at joints B, C, E, and F. If the chords are continuous through any of these joints, there will be considerable reduction in the truss deflection owing to the bending resistance of the chords. If the chords are actually designed for bending, as the result of a directly applied loading, they will have considerable bending stiffness and will substantially reduce the net deflection of the truss.

Rigidity of the Joints. The typical simple analysis for the internal forces in the truss assumes a pure "pin" function of the truss joints. Except for a joint employing a single bolt, where the bolt is only moderately tightened, this is never the true condition. To the extent that the end connections of the members and the general stiffness of the joint facilitate transfer of moments between the members, the truss will actually function as a rigid frame in resisting deformations. If the joints are quite rigid (as with the all welded truss) and the members are relatively short and stiff, this effect will greatly reduce the truss deflections. The extreme case is the so-called Vierendeel truss, which is without triangulation and functions entirely on the basis of the joint stiffnesses and the shear and bending resistance of the members.

Because of these possible effects, the true deflection of trusses is quite complex and can only be determined by using considerable judgment on the results of any calculations. To be honest, the only truly reliable method is full-scale test loading of the actual structure.

11.2 INDETERMINATE TRUSSES

The situations that produce indeterminacy in trusses are discussed and illustrated in Section 6.13. The general problems of analysis and design of indeterminate trusses are beyond the scope of this book. For a thorough presentation of the theories and procedures for analysis of such structures the reader is referred to *Structural Analysis* (Ref. 7) or to any good text on advanced structural analysis.

For some situations it is possible to make some reasonable simplifying assumptions that will permit an approximate analysis using static equilibrium conditions alone. A common situation is that which occurs when a truss is supported by two columns with no provision for movement at either support. Thus, under horizontal loading, the total horizontal force must be distributed in some manner to the two supports. This is often, in effect, an indeterminate situation. However, on the basis of the details of the construction and the relative stiffnesses of the columns, it is frequently possible to make a reasonable guess as to the proportion of load carried by each column. When the columns are the same and the construction otherwise symmetrical, it is common to assume the two horizontal reactions to be equal.

Another situation in which a true indeterminate condition is reduced to a statically determinate one by simplifying assumptions is that for the X-braced structure, as discussed in Sections 6.9 and 8.7.

11.3 SECONDARY STRESSES

Secondary stresses are those induced in the truss members by effects other than the axial forces determined from the analysis of the pure pin-jointed truss actions. The principal causes of such stresses are essentially the same effects described in Section 11.1 as modifying factors for truss deflections: joint deformation, joint rigidity, member stiffness and continuous members.

In the pin-joint analysis it is assumed that all of the members meet as concurrent, axial force-carrying elements at a single point, the truss joint. If joint deformations result in the misalignment of some members, the forces applied at the ends of the members may become other than axial; that is, they may not be aligned with the centroidal axes of the members. If this is the case, bending moments will be developed as the internal forces are eccentric from the axes of the members. Details for the truss joints should be carefully developed to assure that the members are properly aligned and that the deformations that occur do not produce twisting or other unbalancing conditions in the joints.

When truss joints are quite rigid, or members are continuous through the joints, bending will be induced in the members as the truss performs partially as a rigid frame with moment-resistive joints. In addition to the bending stresses thus produced, there will also be some modification of the axial forces. The degree to which this occurs will depend on the relative stiffness of the members and the relative rigidity of the joints. When the members are quite short and stout and the joints are all welded, secondary stress effects may be considerable. When the truss members are relatively slender and the joints are capable of only moderate moment transfer to the ends of the members, secondary stresses are usually quite minor. For ordinary, light trusses for buildings, the latter is most often the case. In general, some investigation of secondary stress effects should be made when

1. Joints are quite rigid, owing to all welding or to use of large gusset plates and many fasteners in the ends of the members.
2. Members are quite stiff, as indicated by the approximate limits L/r less than 50, or I/L greater than 0.5.

11.4 MOVEMENT AT SUPPORTS

The behavior of trusses requires some allowance for change in the length of the chords. This implies that the truss supports can facilitate some overall change in the length of the truss, that is, some change in the actual distance between the support points. If movements are small in actual dimension and there is some possibility of nondestructive deformation in the truss-to-support connections, it may not be necessary to make any special provision for the movements. If, however, the actual dimension of movements are large and both the support structure and the connections to it virtually unyielding, problems will occur unless actual provision is effectively made to facilitate the movements.

A wide range in temperature can produce considerable length change in long structures. Natural shrinkage or great fluctuation in the relative humidity conditions can also cause significant length change in wood structures. These effects should be considered in terms of movements at the supports as well as length change.

Let us consider some examples.

1. Short span wood truss, supported by a wood frame, and totally enclosed within the finished building. In this case there is probably no need for any special provision. Movements will be dimensionally small and the supports sufficiently flexible.
2. Long span steel truss, supported by masonry piers. In this case the movements will be considerable and the supports essentially unyielding. Special provi-

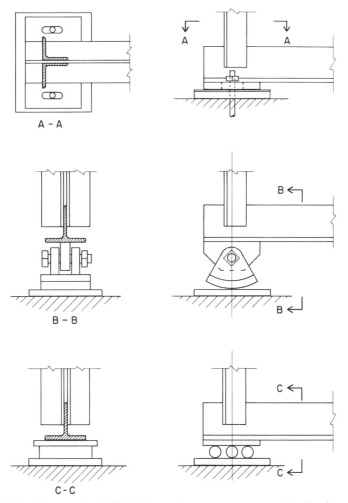

FIGURE 11.4 Support details for light steel trusses to accommodate horizontal movement at the joint. A-A shows a plan of the end with a sliding plate joint, with a slip plate between the anchored support plate and the plate welded to the truss end, and slotted holes in the truss plate. B-B shows a section through the truss end with a view of the end with a pinned rocker. C-C shows a roller end using three hardened-steel round rods for rollers.

sion must be made at one or both supports for some actual dimension of movement.

3. Truss erected during cold weather; later subjected to warm weather or to the warmer conditions maintained in the enclosed building. In this case, even though provisions for movement due to loading stresses may be unnecessary, the length changes due to thermal change should be considered.

Figure 11.4 shows several details that may be used where movement must be provided for at the truss supports. The need for these is very much a matter of judgment and must be considered in terms of the full development of the building construction.

A technique that is sometimes used to reduce the need for provision for movement at supports is to leave the support connections in a stable but untightened condition until after the building construction is essentially completed. This allows the truss deformations resulting from the dead load to accumulate during construction, so that the critical effects are limited to the deformations caused by the live loads. Where the dead load is a major part of the total design load, this is often quite effective.

12

SPECIAL TRUSS STRUCTURES

12.1 SPACE FRAMES

Except for a single post or flagpole, structures are essentially three-dimensional. They may be formed with elements that are linear or planar, but the whole structure must exist in three dimensions in space. Buildings are spatial and the structures that form them are necessarily spatial in form.

Use of planar trusses is common, although they are almost always elements in some ordered spatial arrangement, such as that shown for the simple roof structure in Figure 8.2. In many situations it is common to use bracing for planar trusses that itself constitutes a cross-bracing truss system. This is a simple case of using trussing in a spatial system, for the inherent three-dimensional stability it provides.

However, there are times when trussing is used in a more truly three-dimensional structural form. One such structure is the trussed tower, which is discussed in Section 12.3. The slender tower is in one way a linear form overall, but when fully developed with trussing, it becomes a true three-dimensional truss structure; deriving both its stability and structural strength from the three-dimensional arrangement of the truss members.

The more common example used to describe three-dimensional trussing is the two-way spanning truss structure, called—for want of a better name—a *space frame*. The remaining discussion in this section treats this form of structure.

Geometry of Trussed Space Frames

An ordinary truss system consisting of a set of parallel, planar trusses, may be turned into a two-way spanning system by connecting the trusses with cross-trussing between the parallel trusses. Thus the system shown in Figure 8.12a may actually constitute a two-way spanning system, depending on the development of joints and supports for the system. However, this system lacks stability in one plane: the

262 SPECIAL TRUSS STRUCTURES

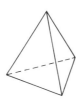
single tetrahedron

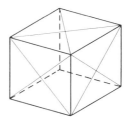
the trussed box

FIGURE 12.1 Three-dimensional trussing. The six-member, four-joint tetrahedron is the basic unit for stabilization of spatial trusses, corresponding to the triangle for a planar truss. Three-dimensional stabilization of the orthogonal (rectangular) truss system is achieved by formation of tetrahedrons with diagonal members on the sides of a cube.

horizontal. If viewed in plan, the two sets of vertical, planar trusses form rectangles, which do not have the inherent triangulation for stability in the horizontal plane. This could be rectified by adding horizontal trussing as shown in Fig. 8.20e, and indeed it sometimes is.

Whereas the planar triangle is the basic unit of planar trusses, the tetrahedron (four-sided solid) is the basic unit of the spatial truss (see Figure 12.1). If a three-dimensional system is developed with three orthoginal planes (x-y-z-coordinate system), the system basically defines rectangles in each plane and cubical forms in three dimensions. Triangulation of each of the three orthoginal planes actually produces sets of terehedrae in space, as shown in Figure 12.1.

An advantage of the use of the mutually perpendicular vertical, planar truss system is its ease in forming squared corners at the edge of the system and in relation to plan layouts beneath it. An example of this system is that used for the roof structure of the dining hall at the Air Force Academy in Colorado Springs, Colorado. Vertical trusses are used to from a square plan with a clear interior span of 266 ft. The roof

FIGURE 12.2 This large, square roof has intersecting, vertical-planar steel trusses supported on four sides by columns, The roof spans 266 ft and cantilevers 21 ft at the sides. Assembly of the trusses was performed at ground level, and the entire roof structure was lifted into position by jacks positioned on the tops of the columns. Air Force Academy dining hall, Colorado Springs, Colorado. Architects and Engineers: Skidmore, Owings, and Merrill, Chicago.

FIGURE 12.3 Air Force Academy dining hall. View of the finished building. The truss is not exposed to view, but its basic structural modules are visible in the forming of the roof fascia, the soffit of the overhang, and the details of the exterior walls.

edge is cleanly formed by the basic truss system and the exterior walls meet the bottom of the truss at natural locations of the truss chords. Although not in view on the building exterior, the truss system is visible in the modules used for the roof fascia and the exterior walls (see Figures 12.2 and 12.3).

A somewhat more pure form for the spatial truss is one that derives from the basic spatial triangulation of the tetrahedron. If the edges of the tetrahedron (or truss members) are all equal in length, the solid form described is not orthoginal; that is, it does not describe the usual x,y,z system of mutually perpendicular planes (see Figure 12.4). If used for a two-way spanning truss system, a flat structure in the horizontal plane may be developed, but it will not inherently develop rectangles in plan. Its natural plan form will be in multiples of triangles, diamonds, and hexagons.

For small size trusses, the fully triangulated system offers a possible economy in that it can be fully formed with all members of the same length. If a single type and size member and a single, simple joint can be used for most of the truss system, a significant cost savings can usually be achieved. The great multiplicity of members and joints in the space truss system makes this a critical consideration.

The pure triangulated space system offers itself to the formation of truly three-dimensional structures in overall form. That is, not just essentially flat, two-way spanning systems. Pyramidal forms of simple or of complex, crystalline-like form can be developed. Figure 12.5 shows some views of a structure of this nature.

A compromise form somewhere between the pure triangular one and the orthoginal system is one described as an offset grid. This consists in its basic form of two horizontal planes, each constituted as square, rectangular grids. However, the upper-plane grid (top chord plan) has its grid intersections located over the centers of the grids in the lower plane (bottom chord plan). The truss web members are arranged to connect the top grid intersections with the bottom grid intersections. The typical joint therefore consists of the meeting of four chord members and four web diagonal members. The web members describe vertical planes in diagonal plan directions, but do not form vertical planes with the chords.

264 SPECIAL TRUSS STRUCTURES

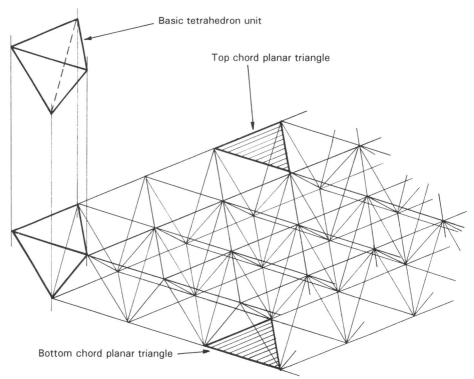

FIGURE 12.4 The equilateral tetrahedron, with all edges (truss member lengths) equal. This is the basic building unit of the fully triangulated spatial truss system.

An advantage of the offset grid is that it permits development of square plan layouts (usually more often preferred by designers), while retaining some features of the triangulated system (members the same length; joints of the same form). Some details of offset grid systems are shown in other illustrations in this section. An example system is illustrated in Section 12.2.

While truss geometry is a critical issue for the truss structure and the economies of its construction, it must relate reasonably to the purpose of the structure—often the forming of a roof for a building. For less constricting design situations—such as the forming of a canopy, or even simply a structural sculpture or theme structure—the space frame may be more dramatically developed. Figures 12.6 and 12.7 show the use of such structures.

Spatial trussing can also be used to produce just about any basic form of structural unit. Linear truss members, consisting of multiples of a three-dimensional truss unit, may form columns, trussed mullions, freestanding towers, or beams. Individual surfaces of large multiplaned structures may be developed with trussing (see Figure 12.5). Arched surfaces formed as vaults (cylindrical shapes) or domes (hemispherical shapes) may be developed with trussed surfaces. Figure 12.8 shows a dome structure developed with a double-layer trussed system using Buckminster Fuller's geodesic geometry.

In many situations trussing offers a practical, economically efficient structural solution. However, it also offers qualities of extreme lightness and a visual openness

SPACE FRAMES 265

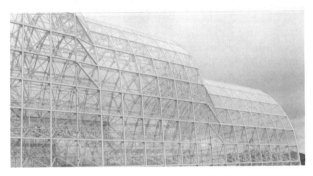

FIGURE 12.5 Views of the building complex for the Biosphere research facility in the Arizona desert. Pyramidal building forms are produced with a modular truss system designed and produced by Peter Pearce.

that may of themselves be major design considerations. Light transmission, for example, is a major factor for the structures shown in Figures 12.5 and 12.8; not just as an aesthetic consideration, but for truly functional purposes.

Span and Support Considerations

In planning of structures that have spatial trussing, if optimal utilization of the two-way action is to be realized, attention must be given to the nature of the supports.

266 SPECIAL TRUSS STRUCTURES

FIGURE 12.6 Small-scale truss system with thin metal tube members and bent sheet metal joint elements, using a fully triangulated spatial geometry. Vertical and horizontal extensions produce a fully three-dimensional structure. The wandering form exploits the freedom of the entirely sculptural structure and considerably softens the space defined by surrounding construction with precise, rectilinear concrete and masonry surfaces. Permanent strings of lighting provide an additional feature in twilight and nighttime light conditions.

The locations of supports will define not only the size of spans to be achieved, but many aspects of the structural behavior of the two-way truss system.

Figure 12.9 shows various possible support systems for a single, square panel of a two-way spanning system. In Figure 12.9*a* support is provided by four columns, placed at the outside corners. This results in a maximum span condition for the interior portion of the system. It also results in a very high shear condition in the single quadrant of the corner and requires the edges to act as one-way spanning supports for the two-way system. As a result, the edge chords will be very heavy and the

FIGURE 12.7 Use of an offset grid with bent tubular members for an entrance canopy structure. Doubletree Hotel, Salt Lake City, Utah.

FIGURE 12.8 Geodesic dome formed with a double-layer truss of aluminum tubular members. Plastic glazing is suspended in a frame beneath the dome structure. Original construction for the Climatron at the Missouri Botanical Gardens in St. Louis. Architects: Murphy and Mackey. (Photos courtesy of Rohm and Haas Company, Philadelphia.) Upper photo shows the exposed truss structure. Lower photo shows a closeup of the typical construction elements.

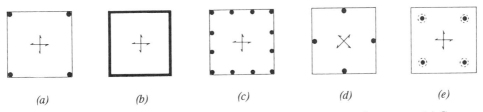

FIGURE 12.9 Alternative supports for a single unit, two-way spanning system. (*a*) Corner columns. (*b*) Perimeter bearing walls. (*c*) Closely-spaced, perimeter columns. (*d*) Columns at centers of the sides; corners cantilevered. (*e*) Corner columns with four-side overhang.

267

corner web members will be heavily loaded for transfer of the vertical force to the columns.

Figures 12.9b and c show supports that eliminate the edge spanning and corner shear by providing either bearing walls or closely spaced perimeter columns. The trade-off for the building is a lower cost for the truss, but higher costs for the support system and its foundations. It is also a more restrictive architectural planning solution.

An interesting possibility is that shown in Figure 12.9d, consisting of four columns placed in the centers of the sides. While this requires a considerable edge structure to achieve the corner cantilevers, it actually reduces the span of the interior system and further reduces its maximum moment by the overhang effect of the cantilever corners. The clear span of the building interior remains the same and the high shear at the four columns is the same as in Figure 12.9a.

For an ideal column shear condition, the solution shown in Figure 12.9 places the columns inside the edges. This provides a wider edge spanning strip as well as a total perimeter of the truss system at each column with shear divided between more truss members. It does reduce the clear interior span to something less than the full width of the truss. If the exterior wall is at the roof edge, the interior columns may be intrusive in the plan. However, if a roof overhang is desired, the wall may be placed at the column lines and the structure relates well to the architectural plan. This is the solution illustrated in Figure 12.3 and in the example design in Section 12.2.

Single units of a two-way spanning system should describe a square as closely as possible. Even though the assemblage has a potential for two-way spanning, if the support arrangement provides oblong units, the structural action may be essentially one-way in function. Figure 12.10 shows three forms for a span unit, with sides in ratios of 1:1, 1:1.5, and 1:2. If the system is otherwise fully symmetrical, the square unit will share the spanning effort equally in each direction.

If the ratio of sides gets as high as 1:1.5, the shorter span becomes much stiffer for deflection and will attract as much as 75% of the total load. This ratio is usually the maximum one for any practical consideration of two-way action.

If the ratio of the sides gets as high as 1:2, scarcely any load will be carried in the long direction, except for that near the ends of the plan unit, adjacent to the short sides. Two-way action relates generally to the development of two-way curvature (domed or dish-like form), and it should be clear that the long, narrow unit will bend mostly in single, arch-like form.

Two-way systems—especially those with multiple spans—are often supported on

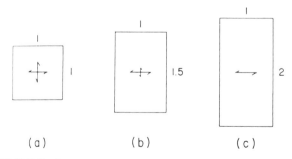

FIGURE 12.10 Variations of spans in two-way systems.

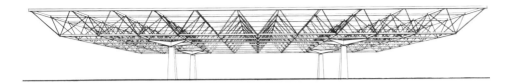

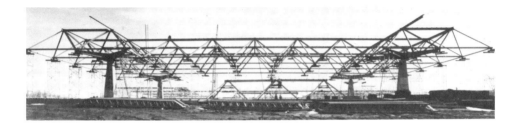

FIGURE 12.11 Large two-way spanning truss system supported on only four columns. Columns have a four-fingered, cantilevered top unit that picks up four truss node points at the bottom chord plane. Overall size of the roof is 216 ft by 297 ft. Pekin High School gym, Pekin, Illinois. Architects: Foley, Hackler, Thompson and Lee, Peoria, Illinois. Structural Engineers: The Engineers Collaborative, Chicago.

columns. This generally results in a high shear condition at the supporting columns, and various means are used to relieve the interior force concentrations in the spanning structure. A common solution for the two-way spanning concrete flat slab is one with the effective perimeter size of the column extended by an enlarged top (called a *column capital*) and the strength of the slab increased locally by a thickened portion at the column (called a *drop panel*). These elements have their analogous counterparts in systems using two-way trussing.

Figure 12.11 shows the structure for a large offset grid truss, supported by only four columns. On the top of each column is a cross-shaped member that extends to pick up four truss joints, whereas the single column would ordinarily be able to support only a single joint. In this case, the shear in the truss is shared by four times as many web members as it would be without the column-top device. Another application of this type is shown in Figure 12.12.

In some truss systems it is possible to develop something analogous to the drop panel in the concrete slab. Thus an additional layer of trussing, or a single dropped-down unit, may be provided to achieve the local strengthening within the truss system itself.

Joints and System Assembly

Many of the issues to be considered in developing joints for spatial truss systems are essentially the same as those discussed elsewhere in this book for planar truss sys-

270 SPECIAL TRUSS STRUCTURES

FIGURE 12.12 Offset grid truss used for a moderate span length, with extended support units on the column tops. Main terminal roof structure, Mitchell Field Airport, Milwaukee, Wisconsin.

tems. Primary considerations are for materials, truss member shape, arrangement of members at a joint, and magnitude of loads. Selection of jointing methods (welding, bolting, nailing, etc.) and the use of intermediate devices (gusset plates, nodal units, etc.) must first deal with these same concerns.

For the spatial system there are two additional concerns of some significance.

The first is that most joints are of a three-dimensional character, relating to the specific geometry of the system. This generally calls for a somewhat more complex joint development than the simple alternatives often possible for planar trusses. A second concern is for the typically large number of joints, which requires that some relatively simple, economically feasible joint construction be used.

Proprietary truss systems often derive basically from some special, clever jointing system that accommodates the necessary variations of jointing for the truss system; is low in cost when mass-produced, and can be reasonably quickly and easily assembled. Although a particular joint system may relate specifically to a single truss member shape, the member selection is usually not as critical a design problem as the joint development.

Most spatial trusses are of steel, and we are thus mostly concerned here with methods for assemblage of steel frameworks. The primary jointing methods are therefore:

By welding of members directly to each other, or of all members to a primary joint element (gusset, node, etc.).

By bolting, in this case most likely to a joint element of some form.

By direct connection with threaded, snap-in, or other attachments.

As with other truss systems, it is often possible and desired to fabricate large units of the total structure in a fabricating shop and transport them to the building site. Site erection then consists typically of bolting the fabricated units together and to their supports. Working out of the field joints and of the shapes and sizes of shop-fabricated units is a major part of the system design for this situation. General selection of the jointing methods may well relate critically to the erection problems.

The procedure just described is more often the case with relatively large-scale trusses of individual design. Proprietary systems mostly consist of individual members with joints that are individually field-assembled.

Joints may also be developed to facilitate other functions, besides that of the truss assemblage. Figure 12.13 shows a jointing system used for an offset grid system. Joints are achieved by bolting of steel plate elements; some occurring as attachments to the member ends, and others as bent plate node units. An additional function required here is the support of an ordinary timber frame roof and plank deck. A typical support unit is developed on top of the top chord joints to achieve the support of the wood roof system.

An additional consideration in the structure shown in Figure 12.13 is the facilitation of slope for the roof surface. In this case, this is done by simply varying the height of the top chord support element, permitting the truss system to be built dead flat in the horizontal plane, which is a really simple, pragmatic solution for an otherwise quite sophisticated system.

Joints may also need to accommodate thermal expansion, seismic separation, or some specific controlled structural action, such as pinned joint response, to avoid transfer of bending moments. Support of suspended elements is also often a requirement. Elements supported may relate to the module of the truss as defined by the joints or the member spacing. As shown in Figure 12.14, supported elements may even relate to the basic form of the truss.

A special joint is one that occurs at a support for the truss system. Besides the

272 SPECIAL TRUSS STRUCTURES

FIGURE 12.13 Offset grid truss used for a roof system for a shopping center concourse area. The truss system provides support for an ordinary timber frame and plank deck roof and a series of skylights. Members are steel pipe, and joints are achieved with steel plates and bolts.

usual requirement for direct compression bearing, there may be need for tension uplift resistance or lateral force for wind or seismic actions. It may also be necessary to achieve some real pin joint response to avoid transfer of bending to column tops as the truss deflects. Lengthening and shortening of chords due to thermal expansion or live load stress changes may present the same problems as those discussed for planar trusses. Of course, the nature of the supporting structure must be considered, whether it is a steel column or a pier of concrete or masonry.

FIGURE 12.14 Truss system shown in Fig. 12.12. In view here are supported elements including signs, lighting, a sprinkler system, loudspeakers, and decorative sculptures.

12.2 DESIGN EXAMPLE: SPATIAL TRUSS

This section presents some possibilities for the development of the roof structure for a medium-size sports arena, possibly big enough for a swim stadium or a basketball court (see Figure 12.15). Options for the structure here are related strongly to the desired building form. Functional planning requirements derive from the specific activities to be housed and from the seating, internal traffic, overhead clearance, and exit and entrance arrangements.

In spite of all of the requirements, there is usually some room for consideration of a range of alternatives for the general building plan and overall form. Choice of the truss system shown here relates to a commitment to a square plan and a flat roof profile. Other choices for the structure may permit more flexibility in the building form or also limit it. Selection of a dome, for example, would pretty much require a round plan.

In addition to the long-span structure in this case, there is also a major problem in developing the 42-ft-high curtain wall. Braced laterally only at the top and bottom, this is a 42-ft span structure sustaining wind pressure as a loading. With even modest wind conditions, producing pressures in the 20-psf range, this requires some major vertical mullion structure to span the 42 ft.

In this example the fascia of the roof trusses, the soffit of the overhang, and the curtain wall are all developed with products ordinarily available for curtain wall construction. The vertical span of the tall wall is developed by a series of custom-designed trusses that brace the major vertical mullions. These vertical mullions are

274 SPECIAL TRUSS STRUCTURES

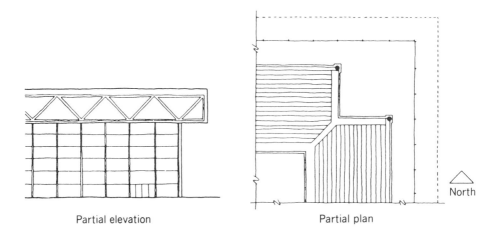

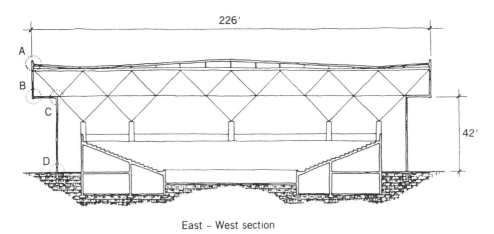

FIGURE 12.15 General form of the building for the design example. The large offset grid truss is supported by edge columns.

themselves custom developed to relate to the standard units of the basic wall system, something that may well be done by the supplier of the basic curtain wall system.

General Building Planning Issues

As an exposed structure, the truss system is a major visual element of the building interior. The truss members define a pattern that is orderly and pervading. There will, however, most likely be many additional items overhead, within and possibly beneath the trusses. These may include:

Leaders and other elements of the roof drainage system

Ducts and registers for the HVAC system.

A general lighting system

Signs, scoreboards, etc.

Elements of an audio system

Catwalks for access to the various equipment

To preserve some design order, these should be related to the truss geometry and detailing, if possible. However, some amount of independence is to be expected, especially with items installed or modified after the building is completed.

Structural Alternatives

The 168-ft span used here is definitely in the class of long-span structures, but not so great as to severely limit the options. A flat-spanning beam system is definitely out, but a one-way or two-way truss system is a feasible choice for a flat span. Most other structural options generally involve some form other than a flat profile; this includes domes, arches, shells, folded plates, suspended cables, cable-stayed systems, and pneumatic systems.

The structure shown in Figure 12.15 uses a two-way spanning steel truss system of the form described in Section 12.1 as an offset grid. The basic planning of this type of structure requires the development of a module relating to the frequency of nodal points (joints) in the truss system. Supports for the truss must be provided at nodal points, and any concentrated loads should preferably be applied at nodal points. While the nodal point module relates basically to the formation of the truss system, its ramifications in terms of supports and loads typically extend it to other aspects of the building planning. At the extreme, this extension of the module may be used throughout the building—as has largely been done in this example.

The basic module here is 3.5 ft, or 42 in. Multiples and fractions of this basic dimension (X) are used throughout the building, in two and three dimensions. The truss nodal module is actually 8X or 28 ft. The height of the exterior wall, from ground to the underside of the truss is 12X, or 42 ft, and so on.

This is not exactly an ordinary building, although the need for such a building for various purposes has created many examples. There is an inevitable necessity for some innovation here, unless an exact duplicate of some previous example is used—not the usual case. There is nothing particularly unique about the construction shown here, but it cannot really be called common or standard. We have not pursued the customized aspects of this construction, as it is not the basic purpose of this book.

Even in "unique" buildings, however, it is common to use as many standard products as possible. Thus the roof structure on top of the truss and the curtain wall system for the exterior walls use off-the-shelf products (see Figures 12.16 through 12.19.) The supporting columns and general seating structure may also be of conventional construction.

Selection of the Truss System

The truss system shown here might be produced from various available proprietary systems, manufactured by different producers. It is generally advisable to pursue the

276 SPECIAL TRUSS STRUCTURES

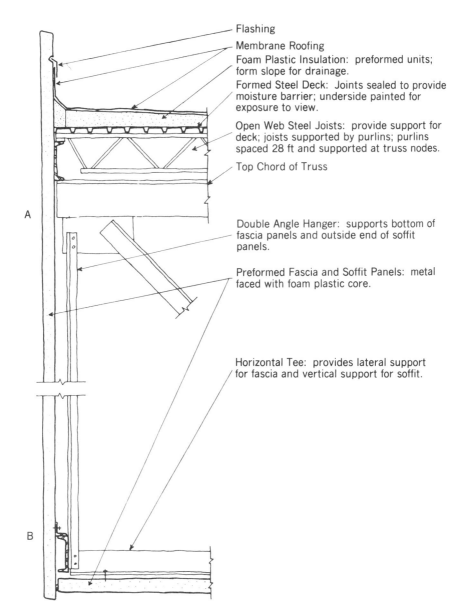

FIGURE 12.16 General construction of the roof and edge. Metal panels attached directly to the trusses form the surfaces of the facia and soffit. The general roof surface is formed with open-web steel joists and a formed sheet steel deck (see also Figure 12.19).

availability of these when beginning the design of such a structure. A completely custom designed structure of this kind requires enormous investments of design time for basic development and planning of the structural form, development of nodal joint construction, and reliable investigation of structural behavior of the highly indeterminate structure. Development of assemblage and erection processes are themselves a major design problem. And, of course, all of the design time and effort has to be paid for.

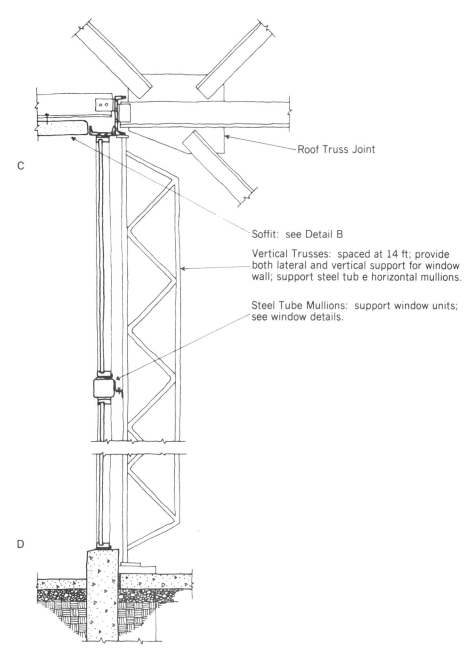

FIGURE 12.17 Formation of the tall glazed walls, using steel trusses for the vertically spanning wall structure. Windows are supported by horizontal steel tubing attached to the vertical trusses.

278 SPECIAL TRUSS STRUCTURES

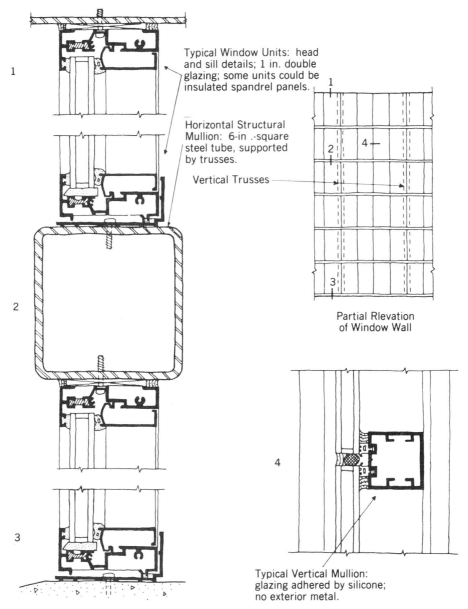

FIGURE 12.18 Details for the walls.

If the project deserves this effort, the time for the design work is available, and the budget can absorb the cost, the end result might justify the expenditures. But if what is really desired and needed can be obtained with available products and systems, a lot of time and money can most likely be saved.

Aside from the considerations of the plan form, planning module, and development of the supports, there is the basic issue of the particular form and general nature of the truss structure. The square plan and general biaxial plan symmetry seem to indicate the logic for a two-way spanning system here. This is indeed what is

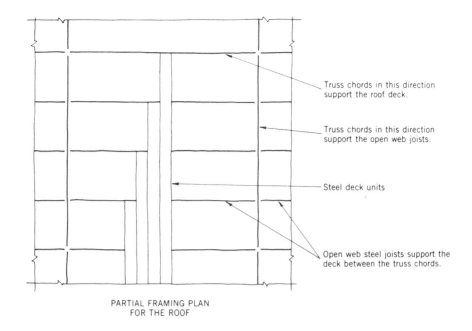
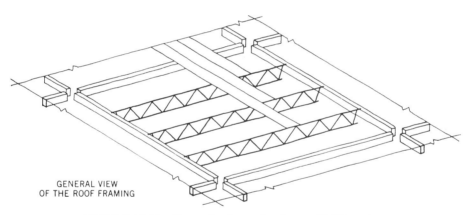

FIGURE 12.19 General scheme for the roof infill system.

shown in Figure 12.15. The particular system form here is called an offset grid, which describes the relationship between the layouts of the top and bottom chords of the truss system. The squares of the top chords are offset from those of the bottom chords so that the top chord nodules (joints) lie over the center of the bottom squares. As a result, there are no vertical web members and generally no vertical planar sets in the system (see Figure 12.20).

Development of the Offset Grid System

The drawings in Figure 12.15 indicate the use of three columns on each side of the structure, providing a total of 12 supports for the truss system. The tops of the columns are dropped below the spanning truss to permit the use of a pyramidal module of four struts between the top of the column and the bottom chords of the

280 SPECIAL TRUSS STRUCTURES

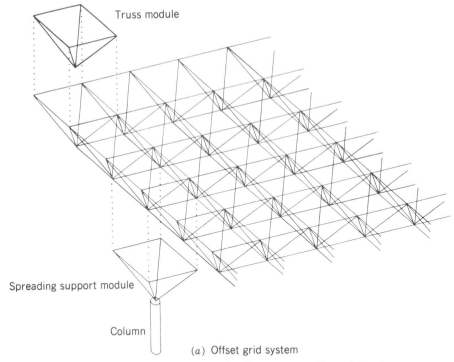

FIGURE 12.20 General scheme for the two-way offset grid system.

truss system. This considerably reduces the maximum shear required by the truss interior members, with the entire gravity load being shared by 12 times 4 equals 48 truss members. If the total design gravity load is approximately 100 psf (0.1 kips/ft^2), the load in a single diagonal column strut is thus approximately

$$C = (226)^2(0.1)/48 = 106 \text{ kips}$$

and since each strut picks up a truss node with four interior diagonals, the maximum internal force in the interior diagonals is

$$C = 106/4 = 26 \text{ kips}$$

If steel pipe are selected for the 28-ft long members, possible choices would be

6 in. standard for the truss member (37 kip capacity)
8 in. extra strong for the strut (137 kip capacity)

The relatively closely spaced edge columns plus the struts constitute almost a continuous edge support for the truss system, with only a minor edge cantilever. The spanning task is thus essentially that of a simple beam span in two directions. For an approximation, we may consider the span in each direction to carry half the load. Thus, taking half the clear span width of 168 ft as a middle strip, the total load for design of the "simple beam" is

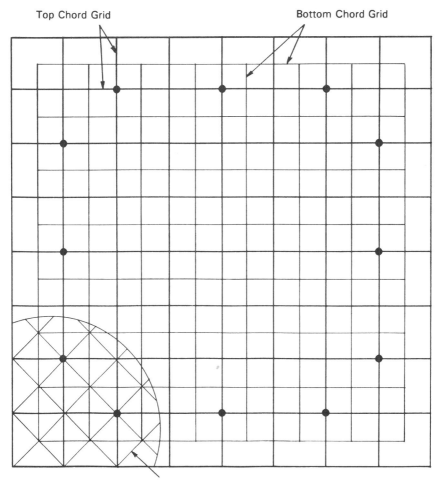

Plan of Truss Web Members

(b) Plan Layout of the Offset Grid System

FIGURE 12.20 (*Continued*)

$$W = \text{(span width)(span length)}(0.1 \text{ kips/ft}^2)$$
$$= (84)(168)(0.1) = 1411 \text{ kips}$$

and the simple beam moment at midspan is

$$M = WL/8 = (1411)(168)/8 = 29631 \text{ k-ft}$$

Sharing this with three top chord members in the middle strip, and assuming a center-to-center chord depth to be approximately 19 ft, the force in a single chord is

$$C = 29631/(3)(19) = 520 \text{ kips}$$

282 SPECIAL TRUSS STRUCTURES

If the compression chord is fully unbraced for its 28-ft length, this is beyond the capacity of a pipe (at least from the AISC Tables), but could be achieved with a W-shape (W12×136, with F_y of 50 ksi) or a pair of thick angles (8 × 8 × 1 in., with F_y = 50 ksi).

For the bottom chords the critical problem will be development of joints at nodes or splices. It is unlikely that chord members will be more than a single module (28 ft) long, so each joint must be fully developed with welds or bolts.

These are not record sizes for large steel structures, but it does seem to indicate some strong efforts for reduction of the design loads (lightest possible general roof construction). It also probably means some variation of sizes will be used in the truss with some minimal members used for low stress situations.

This is a highly indeterminate structure, although its symmetry and the availability of computer-aided procedures for investigation make its design accessible to most professional structural designers.

Development of the Vertical Planar Two-Way System

A second possibility for the truss form is shown in Figure 12.21, consisting of perpendicular, intersecting sets of vertical planar trusses. In this system the top chord grid squares are directly above the bottom chord grid squares.

Figure 12.21b shows a layout for this system in which the vertical truss planes are offset from the columns. The principal structural reason for this is to permit the use of the spread unit at the column, similar to the one used for the preceding example.

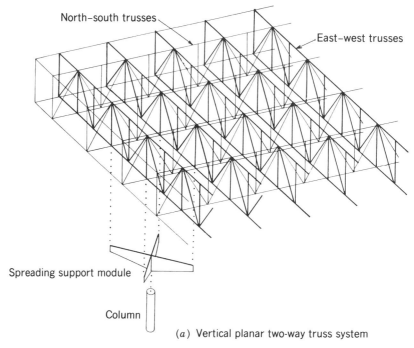

FIGURE 12.21 General scheme for the two-way system with vertical planar trusses.

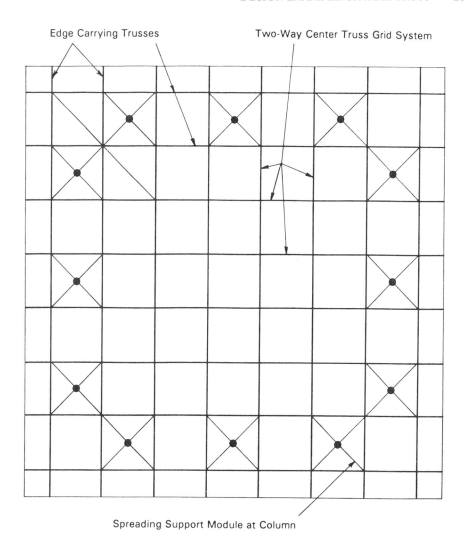

(b) Plan Layout of the Vertical Planar Two-Way System

FIGURE 12.21 (*Continued*)

This unit does not relate to the basic truss system form as it did in the preceding example, but could take various shapes to fulfill its task. As shown here, it could literally be quite similar to the one for the offset grid structure.

Approximation of the chord forces for this two-way spanning system could be made in a manner similar to that for the preceding example. An advantage here is the possibility for use of interior vertical members to reduce the lateral support problem for the chords. This is possible with the addition of some members in the offset grid system, but not so neatly achieved. Using the same force approximations as determined for the chords in the preceding example, but with unsupported lengths of only 14 ft, some considerably smaller members will be obtained.

For both the two-way systems, a major consideration is the planning for the erection. Major considerations are what size and shape of unit can be assembled in the

284 SPECIAL TRUSS STRUCTURES

shop and transported to the site and what temporary support must be provided. Working this out may well relate to the design of the truss jointing details.

Development of the One-Way Spanning System

Figure 12.22 shows the form of a system that uses a set of one-way spanning, planar trusses to achieve a system with a general appearance very similar to that of the system in the preceding example. In fact, the general truss form is identical, the difference consisting of the manner in which trusses are individually formed and joints are achieved.

In this example, the span is achieved by the set of trusses spanning in one direction, while the cross-trussing is used only for spanning between the carrying trusses

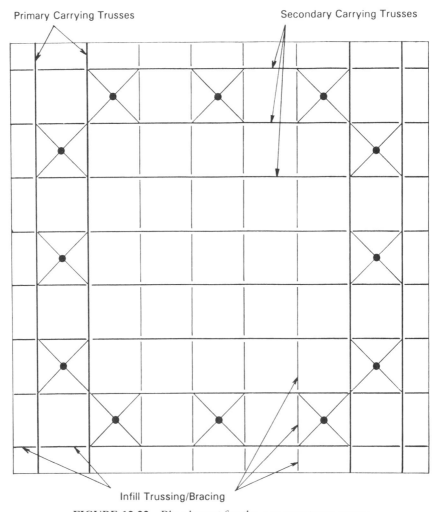

FIGURE 12.22 Plan layout for the one-way truss system.

and providing lateral bracing for the system. The cross-trussing also cantilevers to develop the facia and soffit on two sides of the building.

This system lends itself to relatively simple design procedures, as the main trusses are simple, planar, determinate trusses. Although we will not do so, we could fully illustrate their investigation and design with the simple procedures developed in this book. This is maybe not a compelling reason for choosing this scheme, but is food for thought when one considers the complexity of investigations of highly indeterminate systems.

The 28-ft on center carrying trusses will be slightly heavier than the trusses in the preceding scheme, since their share of the load is slightly more in the one-way structure. However, this is compensated for by the minor structural demands for the cross-trussing system. The designer would have to make a philosophical decision about how far to go to make the structure appear to be symmetrical in the otherwise

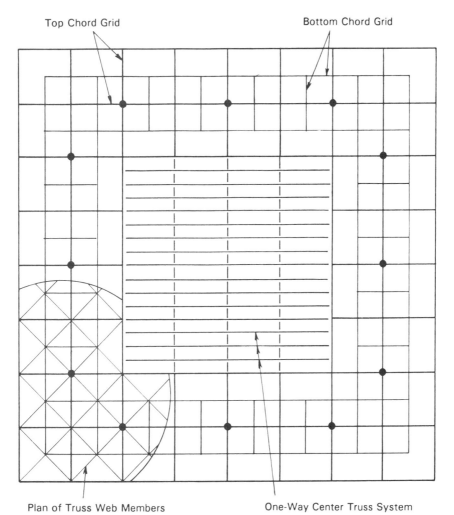

FIGURE 12.23 Plan layout for the structure with a perimeter offset grid and a center one-way infill.

FIGURE 12.24 Truss system for a large square space, using one-way trusses and infill (as shown in Figure 12.22). The supported roof consists of square pyramidal skylights. Hyatt Regency Hotel, Chicago, Illinois. The upper photo shows a general view of the system looking through the north side glazed wall. The middle photo gives a detail view of the main carrying trusses, which are formed with W-shape members. The lower photo is an upward-looking view of the roof skylights. Despite the heaviness of the trusses as seen in the detail view, the fully skylit roof produces a feeling of lightness for the viewed structure.

biaxially symmetrical building. In reality, however, if no effort is made, most nonprofessionals will probably never notice the lack of symmetry, unless someone points it out. This is most likely true of most of the subtlety we design into our buildings; it is lost on all but our fellow professionals.

A principal potential advantage for this scheme is its simplified assemblage and erection. Single carrying trusses may be erected in one piece with very little temporary support necessary. Once two that straddle a column are in place, the development of the cross-trussing can begin, serving both temporary and permanent bracing functions. This may be the single critical deciding factor for favoring this scheme. Discuss it with steel fabricators and erectors and you will become educated in many realities.

One More Possibility

The drawings in Figure 12.23 show a system consisting of a variation on the offset grid system. In this example, the structure at the four sides is achieved with a system of the same form as in that scheme. However, this structure is limited to achieving a maximum span of 56 ft between the edge columns. It defines the building edges and a square opening with 112-ft-long sides.

To fill the center space, the drawings show the use of one-way spanning trusses on 7-ft centers. These may consist of manufactured steel trusses, available from various suppliers for this span and load situation. Bridging for these trusses plus a one-way formed sheet steel deck complete the system for the roof structure.

The perimeter offset system here could be almost identical to that developed in the first scheme. However, the lack of biaxial symmetry causes the trusses and supports on two sides to be slightly more heavily loaded.

Figure 12.24 shows some views of a one-way spanning system used for the roof of a large open space at the lobby level of a hotel. The trusses support a glazed system consisting of pyramidal skylights. An additional consideration was the desire for an open view on the street side, which faces north and has among other things, a view of the John Hancock building on the horizon. Although the space is essentially square, the one-way spanning system was chosen for various reasons, including the ability to achieve the very light mullion system for the north-facing wall. The main spanning trusses are quite heavy, with large W shapes for members, but the system is actually quite light-appearing due to the fully glazed roof above.

12.3 THE DELTA TRUSS

A popular truss structure for various applications is the so-called delta truss. In its typical form this truss consists of three parallel chords with web members in three planes connecting the chords. as shown in Figure 12.25. The truss derives its name from the shape of the truss cross section. which resembles the Greek capital letter delta (Δ).

A primary advantage of the delta truss is its ability to function as a self-stabilizing structure. This property and the basic functioning of the truss can be demonstrated as follows. If the truss is used as a spanning structure, as shown in Figure 12.26, the beam functions required result in forces in the truss members as shown in the illus-

288 SPECIAL TRUSS STRUCTURES

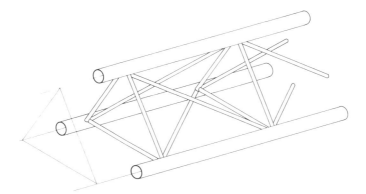

FIGURE 12.25 Delta truss formed with steel pipes.

tration. The web members of the truss serve a dual function, taking the normal forces from the truss action while also providing lateral bracing for the chords.

When a delta truss is used as a spanning roof or floor structure, the bending axis of the truss cross section is preselected. The chords are thus designed specifically for their particular functions, as tension or compression members. For the structure shown in Figure 12.26, the design would be as follows.

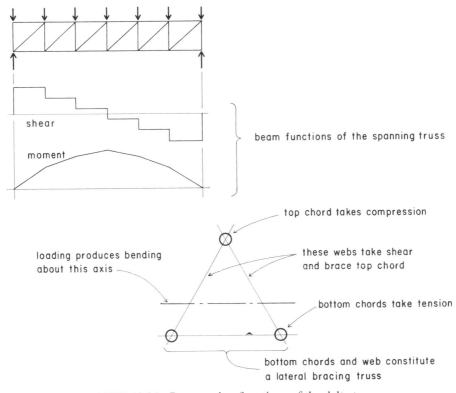

FIGURE 12.26 Beam action functions of the delta truss.

1. The single top chord is designed for compression due to the bending moment.
2. The two bottom chords share the tension due to the bending moment.
3. The two webs that connect the top and bottom chords share the shear force due to the beam action. As shown in Figure 12.27, the forces in the web members will be somewhat higher than the shear forces, due to the slope of the webs.
4. The bottom web is used only to produce a lateral bracing truss, working in conjunction with the two bottom chords. The top chord is braced against this truss through the two sloping webs.

Because of its self-stabilizing character, the delta truss is often used where separate lateral bracing is not desired or not possible for some reason. One such situation is where a single truss is used as a freestanding structure.

Figure 12.28 shows the use of a set of delta trusses for the support of a roof in an airport terminal. The trusses are partially exposed, and the exposed columns are developed to relate to both the trusses and the rest of the exposed building structure. In this case the truss cross section is reversed from that shown in Figure 12.25, with two top chords and a single bottom chord, relating to the development of the roof and ceiling constructions.

The delta shape cross section is also useful for other purposes, such as columns or simply very long compression members in large truss structures. Examples of these uses are discussed in the next two sections.

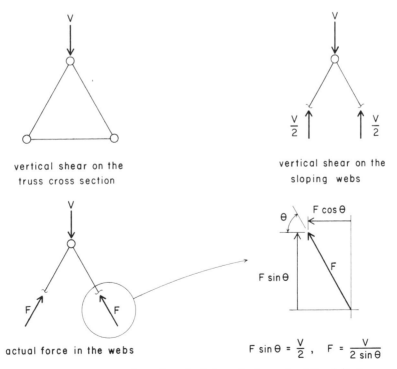

FIGURE 12.27 Resolution of vertical shear in the web of the delta truss.

290 SPECIAL TRUSS STRUCTURES

FIGURE 12.28 Delta trusses of welded steel pipe form the structure for the roof of this space. Inverted from the position as shown in Figure 12.25, the bottom chords here are composed of two pipes, while the trusses are spaced at their top to allow for a skylight between the trusses. United Airlines terminal, Ohare International Airport, Chicago; Helmut Jahn, architect.

12.4 TRUSSED COLUMNS AND TOWERS

Trussed columns usually consist of multiple separated parallel elements laced together with triangulated webs. The delta truss, shown in Figure 12.25, is one such structure, and it is widely used in its most popular form with chords of round steel pipe and webs of pipe or solid bars. Another popular form is the column with four chords arranged in a rectangular cross section. The rectangular column is often made with four single angles, as shown in Figure 12.29, but it can also be made with chords of round pipe or rectangular tubing.

There are two investigations that must be made for a trussed column. With the internal compression force determined for an individual chord element, the single member must be analyzed for column action using a laterally unbraced length equal to the distance between web members—L_1 as shown in Figure 12.29. For the column action of the whole trussed structure the lateral unbraced length will be the total column height, and the r value to be used must be determined for the whole truss cross section.

The procedure for determining the r value for the truss cross section is illustrated in Figure 12.30 for both a delta and rectangular configuration. The moment of inertia of the cross section is approximated by the parallel axis method and the r value is simply determined from its basic definition: $r = (I/A)^{1/2}$.

A special form of trussed column is the guyed, or outrigger, column shown in Figure 12.31. This consists of a single column that is laterally braced by a series of struts and ties. For buckling in a single direction, the tie and struts in that direction form a truss with the column in that direction. Since a very small restraining force is required to keep a slender column from buckling—ordinarily a maximum of 3% of the compression in the column—the required force in the compression strut is correspondingly small. However, the tension force in the ties will be considerably larger, and the struts should be of sufficient length to reduce the tie forces. The tension in the ties will produce a corresponding compression in the column, which should be deducted from the compression capacity of the column to determine its allowable loading.

When the guyed column is freestanding, it must be braced in all directions, which is usually achieved by using three or four ties. In this case each strut actually consists of a set of three or four elements—one for each tie. When the column is braced otherwise on one axis, as in the case of a column occurring in a wall plane with other

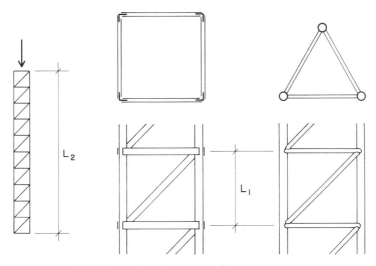

FIGURE 12.29 Examples of trussed columns.

292 SPECIAL TRUSS STRUCTURES

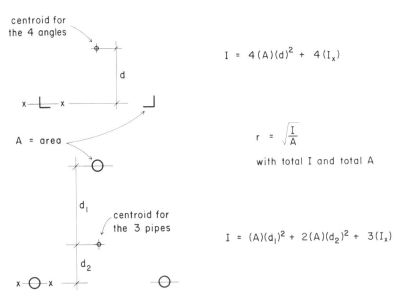

FIGURE 12.30 Procedure for determining moment of inertia (I) and radius of gyration (r) for the cross section of the trussed column.

framing elements, the strut and tie system may be reduced to that required for buckling in the direction perpendicular to the wall plane.

The trussed column form may be used for individual freestanding columns or may be used for compression members in various situations. In a very large spanning truss the individual truss members can become exceedingly long and may themselves be developed as individual trussed structures. There is thus a progression of development of members as the scale changes, with a range from the single

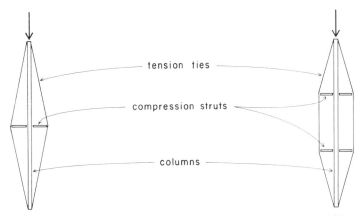

FIGURE 12.31 Guyed columns, tension-braced for lateral stability.

solid bar in open-web joists to the large trussed elements used for the chords of long-span bridges or very tall towers.

A familiar example of the cable-guyed compression structure is the large crane (see Figure 12.32), which uses very long, trussed members for the construction of the mast and boom elements.

Tall towers may be developed in one of two basic ways: as freestanding, or as guyed structures. The guyed structure, with cable bracing at intermittent locations in its height, is a familiar one for large broadcasting transmission towers. The tower itself in this case is limited to primarily compression functions and in most cases consists of a trussed form.

The freestanding tower combines the structural functions of a vertical column (compression member) and a cantilever beam. A critical concern becomes the stability in resistance to horizontal wind or seismic forces, which produce bending in the tower and an overturning (toppling) effect at the base. The simplest solution to the overturning problem is generally to spread the base out to a sufficient width, producing a profile such as the familiar one of the Eiffel Tower in Paris. A common example of this form for a trussed tower is that used for the many electrical power transmission line structures that traverse the landscape of the United States (see Figure 12.33).

FIGURE 12.32 A familiar example of the large trussed compression member: the structure for an erection crane.

294 SPECIAL TRUSS STRUCTURES

FIGURE 12.33 A highly familiar trussed tower form.

12.5 TRUSSED MULLIONS

When glazed walls exceed a single story in height (10 to 15 ft or so), the achieving of the vertical span becomes a major structural task. The usual critical loading is that of wind, exerted as either inward or outward (suction) pressure on the wall surface. This has a direct analogy to the spanning of a horizontal structure for a roof or floor surface under uniformly distributed loading. And, as for any spanning structure, some major bending-resistive elements must be used for achieving the span.

If vertical elements of the general building structure exist at the plane of the wall, they may perform the major vertical spanning task. Thus this window wall function may be developed by building perimeter columns or the outer edges of interior walls that abut the exterior wall. Thus the spanning by the wall construction may be reduced to that of a horizontal one with a span defined by the spacing of the building structural elements in plan.

There are situations, however, where a major vertical span for a glazed wall must be achieved by elements (possibly vertical mullions) of the wall construction, independently of the rest of the vertical structure of the building. An example of this is the exterior wall for the building illustrated in Section 12.2, where a 42-ft-high wall exists, braced only at the ground plane and at the underside of an overhead spanning structure. The solution illustrated in the example consists of a series of light, delta-form trusses in vertical position, to achieve the 42-ft span.

Other than its general structural efficiency in this case, the trussed mullion offers the advantage of a visibly lightened form, with its open, lacy appearance, versus a heavy solid one for a beam required for the same span. Figure 12.34 shows the use of similar trussed mullions for a glazed wall in an airport terminal.

FIGURE 12.34 Large trussed mullions used for a glazed wall similar to that shown for the design example in Figure 12.17. United Airlines terminal, Los Angeles International Airport.

296 SPECIAL TRUSS STRUCTURES

Another possibility for a light-appearing structure for this situation is one using a guyed structure, such as those shown in Figure 12.31. Steel cables are even less massive than pipes or other compression-capable members, and the overall effect would be to have a real minimum of visible interference for viewing in or out through the glazed surface. This is about as invisible a structure as can be imagined, other than that using structural glass mullions.

12.6 TRUSSED ARCHES AND BENTS

As has been discussed in many other places in this book, trussing may be used to achieve many structures, other than the basic one of a horizontal spanning element than emulates beam functions. Trussed arches of either wood or steel (or combinations thereof) have been used for many buildings throughout the past three centuries. In recent times, large trussed arches, vaults, and domes have been used for many of the longest-spanning roof structures in every era. These examples include many of the large train sheds and exhibition buildings of the nineteenth century and recent structures such as the Houston Astrodome.

Examples of several historic structures are illustrated in Chapters 9 and 10. Although the details of the construction and the materials have changed, these structural forms are generally still valid ones for contemporary construction.

The trussed bent is also a popular form, occurring in many forms This was widely used for industrial buildings in the late nineteenth and early twentieth centuries. In recent times welded steel frames and glued-laminated wood frames have replaced many of these structures, but the trussed solutions remain as alternatives, being popular with architects because of their lightweight appearance when used as exposed structures.

Three-Hinged Structures

Figure 12.35 shows a special structure in which two elements are connected to each other and to their supports with pinned joints. Without the internal pin—between the two elements—this structure would be indeterminate. However, the existence of

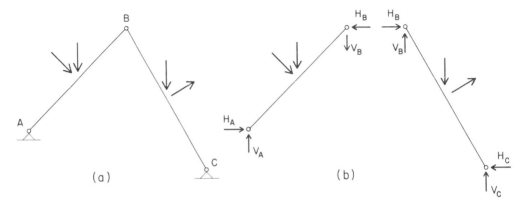

FIGURE 12.35 Action of a three-hinged structure of gabled form.

the internal pin makes the structure statically determinate. A general solution for such a structure is made by considering the individual free body diagrams of the two elements, as shown in the illustration. Thus, considering the left half,

$$\text{for } \sum M_A = 0, \; P(H_B) + Q(V_B) = R$$

and considering the right half,

$$\text{for } \sum M_C = 0, \; S(H_B) + T(V_B) = U$$

in which P, Q, R, S, T, and U are established by the equilibrium equations using the actual loads and dimensions of the structure.

These two equations may be solved simultaneously to yield the values for H_B and V_B. Then consideration of the vertical and horizontal equilibrium for the two halves will yield the values for H_A, V_A, H_C, and V_C.

The basic stability of such a structure is independent of the form of the two elements. Common forms are the gable form shown in Figure 12.35, and the arch and bent, shown in Figure 12.36. All three of these basic forms may be built with elements of solid cross section or with elements consisting of trussed structures. Except for the determination of the reactions, the analysis and design of a three-hinged gable truss essentially is not different from that for a pair of ordinary spanning trusses. The form of the analysis for a simple trussed three-hinged arch is shown in Figure 12.37. The Maxwell diagram is shown for the left half of the structure only, under gravity loading on the structure.

Trussed bents may be formed in a number of ways. When formed without internal pins, they are generally statically indeterminate and their proper analysis is beyond the scope of this book. If an internal pin is used, as shown in Figure 12.36, the bent reactions become statically determinate and the individual truss halves may be designed by the procedures developed in this book. Figure 12.38 illustrates the form of the graphic solution for a simple three-hinged trussed bent with gravity and wind left loadings.

The investigation for member forces using the Maxwell diagram is explained for a three-hinged trussed bent in Section 15.4 (see Figures 15.34 and 15.35).

Drawings for two structures built for the 1893 Columbian Exposition in Chicago, using three-hinged arches, are shown in Figure 2.1.

FIGURE 12.36 The most common forms for three-hinged structures.

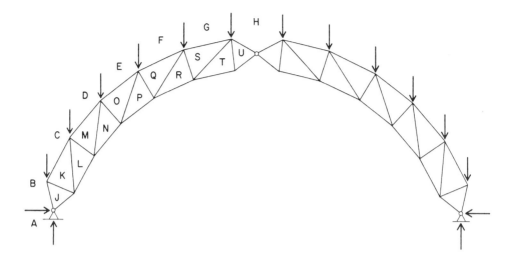

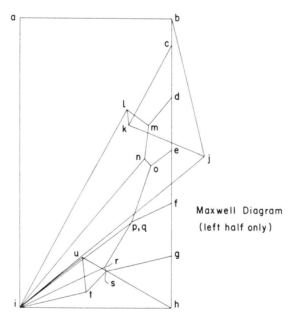

FIGURE 12.37 Graphical analysis for member forces due to gravity loads in a three-hinged trussed arch.

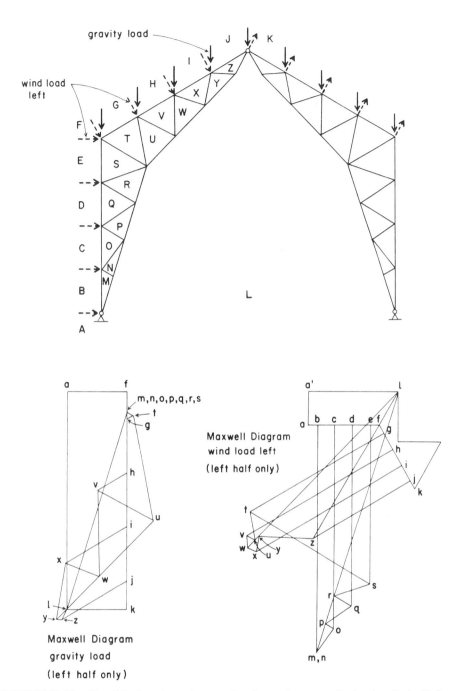

FIGURE 12.38 Graphical analyses for member forces due to gravity load and wind left on a three-hinged trussed bent.

PART FOUR

CONSTRUCTION PLANNING AND DESIGN

Design of trusses, and of structural systems using trusses, requires more than the average considerations for the construction process. This process begins with the manufacture of basic elements for truss members and connecting devices. The putting together of trusses also typically extends from some factory (called the *shop*) where partial assemblage of the finished truss is achieved. Depending on its size, a single truss may be almost entirely fabricated in the shop. However, some work must be done at the building site (called the *field*), usually consisting at a minimum of attachment to supports and installation of bracing for the trusses. Very large trusses are often fabricated in segments, limited in overall size by considerations for transportation to, and erection at, the site. For the large truss, or for extensive systems such as space frames, the amount of construction assemblage work at the site may be considerable. The exact nature of all of this construction will usually be worked out mostly by the manufacturers, shop fabricators, and erection contractors. However, anticipating it, making some allowances for it, and having some appreciation for its potentialities and limitations are problems for the truss designers. The discussions in this part points out some design considerations that may affect choices for the materials, form, details, and general use of trusses.

13

DETAILS FOR CONSTRUCTION

The confrontation between designer and builder occurs mostly in the development of the details for the construction. The builder is the one who must do the work, but the designer must try to work out the problems and tell the builder how to do the job. Of course, the more the designer knows what the builder's job is like, the better for all. This education for the designer is typically built up over many years of friendly feedback from patient and tolerant builders. This chapter presents some of the major concerns for the designer in terms of basic construction details that must be considered in the design development process.

13.1 JOINT DETAILS

For an individual planar truss, or for the truss system in general for a space frame, a major aspect of the design is the development of the construction details for the truss joints. This is principally an economic problem that stems from the multiplicity of joints. The cost of a single joint—repeated many times by type—must be reasonable, or the multiplication quickly leads to unreasonable overall cost for the truss system.

The development of truss joints relates to several considerations, including the following principal ones.

1. The overall size of the truss and the resulting implications of the size of members and the magnitudes of force in the members.
2. The material, size, and shape of the members.
3. The nature of force (tension or compression) in the members.
4. The geometry of the joint, in terms of the arrangements of the members.
5. The nature of the assemblage, whether done in the shop or the field, or possibly both for some joints or systems.
6. Any needs for special considerations, such as demountability, control joint

actions (sliding, etc.), or attachments for other construction (suspended ceiling, supported roof, etc.).

In ancient times, wood trusses had joints achieved by many means. Compression was often transferred by simple bearing: one member against the other. Careful fitting of joints was required to minimize the movements in the joints once the truss was erected and loaded. Tension connections in wood trusses were a problem, and this was a limiting factor for the use of wood trusses, as well as the motivation for development of various truss forms that emphasized the role of compression or arranged tension members in selected positions for jointing. The first combination trusses were timber trusses with iron or steel tension members, which eliminated the need for the tension connection of a wood member.

Timber truss construction was altered when it became feasible to use steel connecting plates with through bolts. Now joints are sometimes developed with welded steel plate assemblages that develop connection for all the members at a joint; transferring both tension and compression through bolts that hold the plates and members together.

Bolted connections—in both wood and steel—have the problem of slippage that is inherent due to the oversizing of the bolt holes. The answer in steel is the use of high-strength bolts that squeeze the joint so hard, the members do not slip at service load levels; the answer in wood is the use of shear-developing elements: split-rings or shear plates mostly.

In the early stages of a truss design, basic decisions must be made about the materials, member forms, and jointing methods. The truss arrangement itself may be affected by these choices. However, the investigation for forces and design of members and joints must be carried to a fairly complete level before the basic judgment can be made as to whether the first choices for member type and jointing method are feasible. Members of a particular type (pipe, W-shape, double angle, etc.) may not be available in the sizes necessary for the forces required. Or the jointing method selected may not be reasonable; too many bolts may be required, or bolting may not be feasible for the small members appropriate for very low forces.

Welding is used increasingly for joint assemblage, especially for any assembly that can be achieved in the shop. Use of the computer to work out the complex geometry for the cutting of the ends of members has made the fully welded joint for members of various cross section possible. The double-angles and tees that facilitate bolted joints are less favored when fully welded joints are used. When trusses are exposed to view, the truss with steel pipe or tube members and fully welded joints yields a very trim-looking structure.

There are many standards and industry common practices that determine features of joints and the use of particular fastening devices. Many of these are discussed in Chapters 9 and 10. For special situations, it may be necessary to work out a unique jointing method and details. However, a full awareness of the standard methods should precede any such speculative design work.

13.2 CONTROL JOINTS IN TRUSSED STRUCTURES

One of the difficult problems of joint design is that of joints that must provide some controlled behavior. This typically means the allowance of some movement and is associated with problems of thermal expansion, seismic separation, or movements at supports due to truss deformation.

Control usually means limited freedom. That is, the provision of some form of support or attachment, while permitting some specific movement by not resisting it. A typical case of this is the truss support that provides resistance to vertical force (the truss reaction), but allows horizontal movement. Some means for dealing with this are discussed in Section 11.4. Some other cases of control joints are the following:

1. *Shear resisting, rotation free.* This is the classic *pinned joint.* Resistance is given to direct force in any direction, but rotation-causing moments are not transferred through the joint. This is the simplifying condition assumed for truss joints in most cases, although some very rigid jointing accompanied by short, stiff members will cause a lot of bending in the members, as discussed under the topic of *secondary stresses* in Section 11.3 and the Vierendeel truss in Section 5.5. A reasonably true pinned joint is required for the controlled behavior of the three-hinged arch, and some very special details may be necessary to make it happen.
2. *Fully-free movement.* This essentially means no joint, which may be the necessary case for a thermal expansion joint or a seismic separation joint. This cannot usually occur within a spanning truss, but may happen between trusses in a continuing series. It can become quite complex for space frames that are extensive in two directions. In addition to the basic nature of the joint, there is the necessity to determine the exact dimension of separation required; otherwise, although no attachment is made, the abutting structures may collide.

Control of the structure is usually only one part of the problem of control. The rest of the construction that is attached to or supported by the structure must also be handled for the effects of the controlled movements. The location for or particular details for a structural control joint may be optimal for the behavior of the structure, but may present real problems for the general elements of the construction. Maintaining seals for the building enclosure, effects on alignments of doors or elevators, and effects on piping are a few of the problems that must often be considered. A special problem of control is that of deliberate details that control the structural behavior. For truss construction it is generally desired that the ultimate failure limit for the structure not be failure in the joints. This leads to some design code restrictions for minimum requirements for joints in various situations. In general, it is best to have joints that can develop the strength of the members, whether or not the truss investigation shows them to need such resistance.

13.3 JOINTING FOR ASSEMBLAGE AND ERECTION

All buildings must be assembled from many separate pieces. For trussed structures, the problem becomes twofold. First is the consideration of a single truss, not a problem for relatively small trusses, but a major one when the truss is very large—say, 100 ft or longer. Since most assemblage must be done in the shop for economic reasons, some means must be developed in the jointing of the truss to facilitate one or more field joints.

Depending on the truss pattern, the location of a field joint may or may not be apparent. It may very well involve not a single joint, but several joints and members. Breaking up a single truss into pieces for field erection also creates the problem of how the large pieces are actually erected, usually requiring some considerable shoring and temporary bracing for stability of the partially erected structure.

When steel trusses were mostly either riveted or bolted, and field riveting was possible, the field joint was usually not much different from a shop joint. Gusset plates were simply left protruding from member ends with their holes awaiting the field bolts or rivets. Now that welding is almost exclusively used in the shop, the field joint—which is typically bolted—becomes a special detail. And it is possible that the truss pattern itself may be altered to make the field joints feasible.

The second major consideration is for the development of the complete trussed structure. This typically consists of a number of trusses plus some secondary framing system, and in some cases a separate bracing system for the main trusses. Once assembled, the system may be expected to have a whole-system stability, but during erection a lot of temporary bracing may be required for safety of the construction crews. Facilitating the secondary systems, the bracing, any temporary bracing, and other construction attachments must also be considered in the development of the truss details.

Systems consisting of separate, one-way spanning elements are usually much easier to erect than two-way spanning (space frame) systems. Individual elements in the one-way system are independently functional as structures, but the two-way system really does not work until the whole system is in place. Furthermore, a single planar element, no matter how large, is easier to assemble in the shop, transport to the site, and erect, than a truly three-dimensional assemblage. Where to make field joints in a large two-way system is a real problem for designers and builders. Detailing for both shop and field joints and developing a plan for the sequence of erection of the parts for the space frame are also challenging problems.

Most commercially successful manufactured truss systems for space frames are developed around some reasonably clever means for dealing with the jointing problems: economics, ease of field assemblage of both single joints and the whole system, and good use of shop fabrication capabilities. This is also true generally for all manufactured truss systems, but especially for the space frames.

13.4 DEVELOPMENT OF THE GENERAL CONSTRUCTION

The building structural designer is mostly concerned about the truss as a part of the overall development of the building structural system, or of the building construction as a whole. The truss itself must be achieved, but this is not enough. The building doesn't really *need* a truss, it needs a roof; and whatever it takes to create the roof must be considered as an extension from the basic structural unit of the truss.

Modular dimensions, which are usually a part of the truss design, must relate to logical ones derived from the rest of the construction or from building planning. Support for major elements of the construction is best provided at joints of the truss (*panel points*, as they are sometimes called), which is directly related to the modular dimensions of secondary framing.

As with any structure, movements of the structure must be considered for their effects on the rest of the construction. Flat roofs are never truly flat, but must provide for some drainage with minimum slope angles. For long-span trusses a major concern is *ponding*, which occurs when a truss deflects (L/360 on a 200-ft span is a lot of deflection), causing a loss of some drainage angle and possible collection of water in the center of the truss span. The weight of the collected water causes more deflection, which causes more water collection, and so on—a classic roof collapse story.

Deflections may also cause effects on interior construction, both that supported above, or standing immediately beneath a truss. If the bottom of a long-span truss provides stability for the tops of walls beneath it, some provision must be made for accommodating the truss live load deflection. This may well be the source for real criteria for the limitation of deflections.

The idea of allowing horizontal movement at some truss supports to alleviate internal forces or horizontal pushing of supports may look good for structural design, but the effects on the whole construction must be considered. If a truss end is allowed to slip on top of an exterior wall pier, what happens to the roof surface above? And to the supported ceiling below? And to any piping for roof drains or sprinklers? And to an interior wall above or beneath the truss, parallel to the truss, and intersecting the exterior wall?

When a structural system is designed by specialists, and the rest of the construction by other people, these are issues that must be cooperatively dealt with. The basic idea is that the truss is a part of the building and must be fully integrated into the whole construction. A *good* truss design cannot deal only with the structural tasks.

13.5 THE REVEALED STRUCTURE

For purposes of both design and construction, some concern must be given to the situation for a building structure regarding its condition as a viewed object. For simplification, three general cases may be considered.

1. The fully concealed structure.
2. The fully revealed structure.
3. The partly revealed structure.

Considerations for trusses for these situations include general considerations for any structure and some special ones for the particular nature of the truss. Some general considerations are the following:

Fire protection. Structures of wood or steel, with relatively thin elements, are highly vulnerable to fire. Except for the heavy timber structure, no real fire resistance can be expected for the fully exposed structure of wood or steel. Even the timber truss, however, may be vulnerable, if its joints are achieved with revealed steel elements.

Deterioration. Structures exposed to the weather are subject to effects of water, freezing, and extreme ranges of temperature, as well as special problems for acid rain, salt water, or wind-driven sand.

Viewed form. The viewed structure is often quite prominent in the viewed field, and its overall form and closely viewed details are of concern as architectural design considerations. Building form in general may be strongly derived from the structure, and the viewed structure is thus the strongest element of the general architectural expression.

Modular units and dimensions. If the structure has a repeating unit of particular form and size, it will be hard to ignore this in development of other modular units in the building planning, such as windows, lighting, ceilings, and room sizes.

For the truss in particular, the strongest concerns are the following:

General shape. This includes the overall truss profile (gabled, flat, arched, etc.) and the pattern of arrangement of members.

Materials and member shapes. Wood or steel; single-piece or multiple-piece members; complex shapes (W, double angle, multiple-piece wood) or simple shapes (pipe, tube, single timber).

Truss joints. Unobtrusive (directly welded steel pipe) or highly articulated (bolted joints with gusset plates).

Let us consider the three general cases for the degree of exposure as regards these various concerns.

The Concealed Structure

If out of sight, the structure is of practically no concern as a viewed object. This permits use of the most pragmatic and economical means for construction in general. Style, grace, neatness, symmetry, and finishes are of no real concern. The steel truss may be coated with really ugly sprayed-on fiber insulation; the wood structure may be executed with stained, scratched, nicked, indelibly stamped wood pieces; and the joints may take any form that works best for purely functional purposes.

Deterioration is generally not a problem, although steel may rust and wood may rot, and the condition may remain undetected if the structure is concealed.

While fire resistance of the structure itself may be enhanced by containment by more resistive materials, the creation of the concealed spaces in ceilings and walls are a real problem for fire fighters. The interstitial spaces that we find so handy for architectural design—allowing for clean interior spaces—are really treacherous for fire fighters when fires spread into the trapped spaces. Ask the local fire fighters how they feel about an overhead structure that is in view, versus one that is concealed by a suspended ceiling.

The Revealed Structure

The revealed structure becomes a major architectural feature and can be a dominant aspect of the building. Figure 13.1 shows a building, which virtually is a truss for one of its major parts. In this case, the building also has other parts, but the viewed truss—exposed both on the interior and exterior—is a dominant architectural feature. Obviously, the form, modular unit, and dimensions for the truss had to work for the structural task, but also for the building planning. And, the relatively clean details of the fully welded joints and single-piece members are consistent with the rest of the building construction as viewed.

THE REVEALED STRUCTURE   309

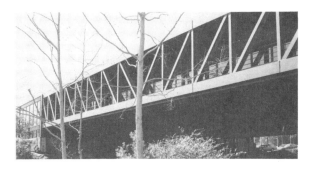

FIGURE 13.1 For site development considerations, this building was made to span across a ravine on its hillside location. For this portion of the building, the structure consists of a full story-high truss that is exposed to view on both the exterior and interior. Art Center, Pasadena California; architects: Craig Ellwood and James Tyler; structural engineer: Norman Epstein.

Figure 13.2 shows another building with a highly revealed truss structure; this time viewed mostly only on the inside of the building. Design of the saw-tooth roof form was obviously strongly influenced by the desire for use of daylight illumination of the interior, but also relates to the repeating bays of the truss system. In this example, the truss is somewhat less neatly developed—with relatively ordinary

310 DETAILS FOR CONSTRUCTION

FIGURE 13.2 The structure for the roof of this one-story library building consists of a series of trussed bents that form the saw-tooth roof configuration and also develop resistance to lateral forces. Trusses use conventional double-angle members with gusset plate joints and welded fastening. The structure is not much in view on the exterior, but is fully revealed as part of a high-tech interior design. Thousand Oaks Library, Thousand Oaks, California; architects and engineers: Albert C. Martin, Los Angeles.

double-angle members and gusset plate joints. However, the general interior development works into an overall "high-tech" style, with all other functional elements exposed: lighting, drain piping, sprinklers, HVAC elements, and so on.

The Partly Revealed Structure

In some cases the building structure may not actually be fully in view, and yet its existence may be revealed to various degrees. An example of this is the structure shown in Figures 12.2 and 12.3. The two-way spanning truss system is fully encased by finish materials, and yet its repeating module and dimensions are revealed on the interior by the details of the ceiling and on the exterior by the form of the fascia and soffit.

THE REVEALED STRUCTURE 311

Another example is the John Hancock building in Chicago, shown in the frontispiece photo and here in Figure 13.3. Again, the building columns and spandrel beams, and the diagonal truss members, are fully encased by finish materials. But the form of the whole exterior structure is clearly revealed. Indeed, as the close view shows, the existence of the gusset plate joints is even expressed.

Whether fully concealed, or fully or partly revealed, any structural system will present some important coordination concerns for the general building design development. The more the structure is actually viewed, however, the more constrained become both the structure and the general building design and planning. And when the actual structure is fully in view, considerable portions of the structural design may be strongly affected. Add exterior exposure to the weather, and a lot must be dealt with besides structural behaviors in designing the structure.

FIGURE 13.3 This giant building dominates the landscape of the north side of Chicago just as the Sears Tower dominates the south side. The close views show the details of the jacket construction that encloses, and yet fully reveals the general shape of, the exterior structure. John Hancock Building, Chicago; architects and engineers: Skidmore, Owings, and Merrill, Chicago.

FIGURE 13.3 (*Continued*)

14

GENERAL CONSTRUCTION CONCERNS

The making of trusses themselves is a task related to the materials, size, and problems of fabrication for the particular type of truss. Using trusses in a building presents the problem of creating the trussed structure in relation to the general construction. This issue has been mentioned frequently elsewhere in this book, since consideration for it must be made in many stages of design for any truss and for the general structural system of which it is a part. This chapter treats some of the general concerns for the building construction in a more direct way, as a setting for the use of trusses.

14.1 ERECTION PROBLEMS

For any building that uses large structural components, a major construction planning concern is for the erection of the large elements. Because of the problems of transporting large elements to most sites, there are problems in putting very large individual elements (trusses, arches, domes, etc.) together and in place. Some of these concerns are as follows:

1. *Moving large erection equipment around the site.* Large cranes must be used before site construction, or even some ground-level building construction, is in place.
2. *Provision of temporary shoring.* If a large truss must be fabricated in several pieces, these must be held up in place temporarily until the last connection is complete, before the truss can do its intended job, including holding itself up.
3. *Provision for the making of field joints (at the site).* If really necessary, welding can

be used for this, but bolting is much preferred for ease and reliability under field conditions.

4. *Provision of temporary bracing for the partially erected structure.* Most structural systems are self-supporting and self-bracing when they are completed. After all, *they* are supposed to support the rest of the building. However, many structures are not self-sufficient until the last part is in place. For example, the roof purlins, or even the roof deck, may be part of the bracing for a truss. Before the necessary construction is completed, some temporary bracing may be required.

5. *Protection from the elements.* Structures that are eventually to be totally covered or enclosed within the construction may need some temporary protection after erection and before the enclosing construction is complete. Protection from rusting is one concern. Another may be the appearance of the structure, if it is exposed to view after the construction is complete.

6. *Avoidance of unanticipated loading.* During construction many large elements of the construction or large packaged quantities of materials need to be installed or temporarily located near the work area. There is a temptation to use the truss that is just standing there for temporary support. A special problem is lifting of elements by hanging from the truss. If this is not done by using truss joints as pickup points, the chord members may be damaged. Unintended lateral loadings may also be involved. The truss may look strong, but it is highly oriented to its design loading conditions, and probably quite weak for other loads.

These are mostly problems for the builders, and this book is basically about design of trusses. However, the more the designer anticipates these problems, the less possibility that they will develop as problems at all. Some design considerations that may respond to the list of issues above are the following:

1. If the building construction is such that considerable construction must be in place before the trusses are erected, the system should be studied for the possibility of using the fewest number of major trusses as a starting point for a system that can be subsequently self supporting. This in-place structure may then be used to help with erection of the rest of the system.

2. The design considerations relating to lifting as described for point 1 may also apply to the reduction of need for temporary shoring. Temporary support for other trusses may be provided by those in place.

3. Large trusses and space frame systems should be very carefully studied for the incorporation of field joints. Best is a jointing system that easily converts from welding to bolting, or one that works equally well for either shop or field connection. At the least, the truss should yield to being divided into parts with the splice involving the least number of members and joints.

4. A lot of temporary bracing may be eliminated if the permanent bracing system is easily installed during erection and is all part of the basic structural erection process. Other parts of the construction used for bracing may be installed by separate contractors and not reliable as to timing. If temporary bracing must

be in place a long time, it may well get in the way for other construction work—presenting the temptation to remove it while it is still really needed.
5. Trusses that are exposed to view and are expected to have a reasonably attractive finish after the construction is complete require special consideration. This is generally the case for exposed structures of any kind. The structure is ordinarily considered to be rough construction and out of sight upon completion of the construction. If it is not, special workmanship may be required and some protection during the construction activity may be necessary. The design specifications, notes on the design drawings, and possibly some considerations for design details should reflect this concern. Don't expect the structural erection contractor and his crew to treat the structure like a fine piece of cabinet work.
6. There really should not be anything such as an *unanticipated load.* You must *expect* that people will hang things from the bottom chords of a truss during construction, and after construction, during the life of the building, for that matter. Excessive abuse cannot be provided for, but some consideration can be given.

14.2 ACCOMMODATION OF OTHER CONSTRUCTION

The general idea of integrating the structure into the whole building has been raised in many places in this book. Some specific issues that relate to the construction needs for other parts of the building are the following:

Roofs

Roofs must drain and the drainage water must run off or be collected. The profile of the top of a roof truss must relate to this and to the type of roofing that is appropriate to the slopes developed. For a flat roof, the minor slope must be developed by the truss itself or by some built-up construction.

Roofing materials relate to slope angles, and it may not be possible to use a single type of roofing for a roof surface of many different slopes. The extreme case is an arch or dome that goes from dead flat at its top to a considerable angle at its edge. Roofing is expensive, and the need for expensive roofing may well work against the feasibility of a truss system.

Roof drains must be accommodated in the general roof construction, but drainage piping may well need to run through the truss system. It should be noted that this piping needs to drain well, preferably at least $\frac{1}{2}$ in. per ft. For a long horizontal run through many trusses this is no small problem.

HVAC Ducts

Large ducts fit more easily into the spaces between trusses, but allowance for crossbracing must be made. If ducts run through the truss—a selling point for the open trusswork in some cases—the exact sizes of the largest ducts plus their encasing insulation must be known early in the design process. Choices for the truss pattern, truss depth, modular dimensions, and even sizes of members may be affected. A problem in this is that the true sizes of the ducts are probably not really known in the design process early enough to be used with any reliability. It is advisable to be conservative in such allowances, if it involves critical sizing of trusses.

Other Piping

Piping for fire sprinklers or for HVAC elements may also run through the truss system. This is pressurized piping and technically does not need drainage, but is usually installed to provide it. The same comment made for roof drain piping applies here. A long run through many trusses will change its vertical position considerably and can hopefully do so without having to cut out any truss web members.

Supported Ceilings

Two cases are involved here: direct attachment to bottom chords and suspension of a secondary frame. For direct attachment the ceiling system must be closely related in its modular dimensions to that for the truss system. Support might be provided by the cross-bracing in some cases, avoiding direct loading of the truss chords. Otherwise, the chords must of course be designed for the combined loading condition of tension plus bending, which requires special form for the chords and usually much heavier chord members.

The suspended ceiling will hopefully be suspended from truss joints. This provides a modular spaced situation for the supports, but the suspended frame may be able to make a transition to a separate module dimension for the ceiling itself.

In any event, the dead load of the ceiling—and possibly a live load, if the ceiling frame can support anything on its top—must be considered in the truss investigation. A light ceiling construction may be planned, but it is probably smart to allow for some heavier future ceiling; ceilings are easily replaceable for modifications of interior building spaces.

HVAC Equipment

Ductwork, registers (for delivery), and exhaust air grilles are common elements contained in overhead construction. More serious for loading are large fans, zone reconditioners, or unit heaters, which may represent significant concentrated loads. Rooftop equipment, of course, may be minor in size or of considerable bulk. This is a major area of concern, and information about it should be pursued, although it is not always easy to get hard facts in time for early design work.

Also of concern are needs for piping, drains, and penetrations of the roof for piping or ducts. Supports for heavy equipment may be carried directly by a secondary framing to avoid bad placements on the roof structure, but the total load on the trusses must be considered in early investigations.

Floor Construction

Floors must generally be dead flat, so it is assumed that most floor trusses will be flat chorded (also called parallel chorded.). The depth of such a truss becomes a critical decision; in this case, both for the truss and the floor construction. For fire protection, it is likely that a floor truss system must be encased by a ceiling below it.

For the general building design the critical planning dimension is from the top of the finished floor surface above to the underside of the finished ceiling below. Within this space, the truss itself will occupy an out-to-out depth dimension somewhat smaller. And smaller than that will be the working depth for the truss, which is from the center of top chord joints to the center of bottom chord joints. The latter is what is usually meant by the truss depth for investigation, and establishing it requires considerations for the exact form of both the truss and the general constructions.

A generous dimension of 4 ft for a floor/ceiling space may thus yield a much smaller working truss depth. Adding to the problem, it is usually desired to restrict floor structure deflections considerably, to avoid actual sag and also bouncing. This works to restrict span/depth ratios for floor trusses to ones well below those used for floor beams or slabs. Since floor trusses are not likely to be used unless the spans are at the upper limit for beams, this issue should be handled carefully in early design work, so that problems do not come up later with the building planners.

14.3 DEAD LOADS

Dead load consists of the weight of the materials of which the building is constructed such as walls, partitions, columns, framing, floors, roofs, and ceilings. In the design of a beam, the dead load must include an allowance for the weight of the beam itself. Table 14.1, which lists the weights of many construction materials, may be used in the computation of dead loads. Dead loads are due to gravity and they result in downward vertical forces.

Dead load is generally a permanent load, once the building construction is completed, unless frequent remodeling or rearrangement of the construction occurs. Because of this permanent, long-time, character, the dead load requires certain considerations in design, such as the following:

1. It is always included in design loading combinations, except for investigations of singular effects, such as deflections due to only live load.
2. Its long-time character has some special effects causing sag and requiring reduction of design stresses in wood structures, producing creep effects in concrete structures, and so on.
3. It contributes some unique responses, such as the stabilizing effects that resist uplift and overturn due to wind forces.

TABLE 14.1 Weights of Building Construction

	lb/ft^2	kN/m^2
Roofs		
3-ply ready roofing (roll, composition)	1	0.05
3-ply felt and gravel	5.5	0.26
5-ply felt and gravel	6.5	0.31
SHINGLES		
Wood	2	0.10
Asphalt	2–3	0.10–0.15
Clay tile	9–12	0.43–0.58
Concrete tile	8–12	0.38–0.58
Slate, $\frac{1}{4}$ in.	10	0.48
Fiber glass	2–3	0.10–0.15
Aluminum	1	0.05
Steel	2	0.10
INSULATION		
Fiber glass batts	0.5	0.025
Rigid foam plastic	1.5	0.075
Foamed concrete, mineral aggregate	2.5/in.	0.0047/mm
WOOD RAFTERS		
2 × 6 at 24 in.	1.0	0.05
2 × 8 at 24 in.	1.4	0.07
2 × 10 at 24 in.	1.7	0.08
2 × 12 at 24 in.	2.1	0.10
STEEL DECK, PAINTED		
22 ga	1.6	0.08
20 ga	2.0	0.10
18 ga	2.6	0.13
SKYLIGHT		
Glass with steel frame	6–10	0.29–0.48
Plastic with aluminum frame	3–6	0.15–0.29
PLYWOOD OR SOFTWOOD BOARD SHEATHING	3.0/in.	0.0057/mm
Ceilings		
Suspended steel channels	1	0.05
LATH		
Steel mesh	0.5	0.025
Gypsum board, $\frac{1}{2}$ in.	2	0.10
FIBER TILE	1	0.05
DRY WALL, GYPSUM BOARD, $\frac{1}{2}$ IN.	2.5	0.12
PLASTER		
Gypsum, acoustic	5	0.24
Cement	8.5	0.41
SUSPENDED LIGHTING AND AIR DISTRIBUTION SYSTEMS, AVERAGE	3	0.15

TABLE 14.1 (*Continued*)

	lb/ft²	kN/m²
Floors		
Hardwood, $\frac{1}{2}$ in.	2.5	0.12
Vinyl tile, $\frac{1}{8}$ in.	1.5	0.07
Asphalt mastic	12/in.	0.023/mm
CERAMIC TILE		
$\frac{3}{4}$ in.	10	0.48
Thin set	5	0.24
Fiberboard underlay, $\frac{5}{8}$ in.	3	0.15
Carpet and pad, average	3	0.15
Timber deck	2.5/in.	0.0047/mm
Steel deck, stone concrete fill, average	35–40	1.68–1.92
Concrete deck, stone aggregate	12.5/in.	0.024/mm
WOOD JOISTS		
2 × 8 at 16 in.	2.1	0.10
2 × 10 at 16 in.	2.6	0.13
2 × 12 at 16 in.	3.2	0.16
Lightweight concrete fill	8.0/in.	0.015/mm
Walls		
2 × 4 studs at 16 in., average	2	0.10
Steel studs at 16 in., average	4	0.20
Lath, plaster; *see* Ceilings		
Gypsum dry wall, $\frac{5}{8}$ in.	2.5	0.12
Stucco, $\frac{7}{8}$ in. on wire and paper or felt	10	0.48
WINDOWS, AVERAGE, GLAZING + FRAME		
Small pane, single glazing, wood or metal frame	5	0.24
Large pane, single glazing, wood or metal frame	8	0.38
Increase for double glazing	2–3	0.10–0.15
Curtain walls, manufactured units	10–15	0.48–0.72
BRICK VENEER		
4 in., mortar joints	40	1.92
$\frac{1}{2}$ in., mastic	10	0.48
CONCRETE BLOCK		
Lightweight, unreinforced— 4 in.	20	0.96
6 in.	25	1.20
8 in.	30	1.44
Heavy, reinforced, grouted— 6 in.	45	2.15
8 in.	60	2.87
12 in.	85	4.07

14.4 BUILDING CODE REQUIREMENTS

Structural design of buildings is most directly controlled by building codes, which are the general basis for the granting of building permits—the legal permission required for construction. Building codes (and the permit-granting process) are administered by some unit of government: city, county, or state. Most building codes, however, are based on some model code, of which there are three widely used in the United States:

1. *The Uniform Building Code* (Ref. 1), which is widely used in the West, as it has the most complete data for seismic design.
2. *The BOCA Basic National Building Code*, used widely in the East and Midwest.
3. *The Standard Building Code*, used in the Southeast.

These model codes are more similar than different, and are in turn largely derived from the same basic data and standard reference sources, including many industry standards. In the several model codes and many city, county, and state codes, however, there are some items that reflect particular regional concerns.

With respect to control of structures, all codes have materials (all essentially the same) that relate to the following issues:

1. *Minimum Required Live Loads.* All codes have tables similar to those shown in Tables 14.2 and 14.3, which are reproduced from the *Uniform Building Code.*
2. *Wind Loads.* These are highly regional in character with respect to concern for local windstorm conditions. Model codes provide data with variability on the basis of geographic zones.
3. *Seismic (Earthquake) Effects.* These are also regional with predominant concerns in the western states. This data, including recommended investigations, is subject to quite frequent modification, as the area of study responds to ongoing research and experience.
4. *Load Duration.* Loads or design stresses are often modified on the basis of the time span of the load, varying from the life of the structure for dead load to a fraction of a second for a wind gust or a single major seismic shock. Safety factors are frequently adjusted on this basis. Some applications are illustrated in the work in the design examples in this part.
5. *Load Combinations.* These were formerly mostly left to the discretion of designers, but are now quite commonly stipulated in codes, mostly because of the increasing use of ultimate strength design and the use of factored loads.
6. *Design Data for Types of Structures.* These deal with basic materials (wood, steel, concrete, masonry, etc.), specific structures (towers, balconies, pole structures, etc.), and special problems (foundations, retaining walls, stairs, etc.) Industry-wide standards and common practices are generally recognized, but local codes may reflect particular local experience or attitudes. Minimal structural safety is the general basis, and some specified limits may result in questionably adequate performances (bouncy floors, cracked plaster, etc.)
7. *Fire Resistance.* For the structure, there are two basic concerns, both of which produce limits for the construction. The first concern is for structural collapse or significant structural loss. The second concern is for containment of the fire to control its spread. These concerns produce limits on the choice of materials

TABLE 14.2 Minimum Roof Live Loads

	Minimum Uniformly Distributed Load					
	(lb/ft²)			(kN/m²)		
	Tributary Loaded Area for Structural Member					
	(ft²)			(m²)		
Roof Slope Conditions	0–200	201–600	Over 600	0–18.6	18.7–55.7	Over 55.7
1. Flat or rise less than 4 in./ft (1:3). Arch or dome with rise less than $\frac{1}{8}$ span.	20	16	12	0.96	0.77	0.575
2. Rise 4 in./ft (1:3) to less than 12 in./ft (1:1). Arch or dome with rise $\frac{1}{8}$ of span to less than $\frac{3}{8}$ of span.	16	14	12	0.77	0.67	0.575
3. Rise 12 in./ft (1:1) or greater. Arch or dome with rise $\frac{3}{8}$ of span or greater.	12	12	12	0.575	0.575	0.575
4. Awnings, except cloth covered.	5	5	5	0.24	0.24	0.24
5. Greenhouses, lath houses, and agricultural buildings.	10	10	10	0.48	0.48	0.48

Source: Adapted from the *Uniform Building Code*, (Ref. 1), with permission of the publishers, International Conference of Building Officials.

TABLE 14.3 Minimum Floor Loads

Use or Occupancy		Uniform Load		Concentrated Load	
Description	Description	(psf)	(kN/m^2)	(lb)	(kN)
Armories		150	7.2		
Assembly areas and auditoriums and balconies therewith	Fixed seating areas	50	2.4		
	Movable seating and other areas	100	4.8		
	Stages and enclosed platforms	125	6.0		
Cornices, marquees, and residential balconies		60	2.9		
Exit facilities		100	4.8		
Garages	General storage, repair	100	4.8	*	
	Private pleasure car	50	2.4	*	
Hospitals	Wards and rooms	40	1.9	1000	4.5
Libraries	Reading rooms	60	2.9	1000	4.5
	Stack rooms	125	6.0	1500	6.7
Manufacturing	Light	75	3.6	2000	9.0
	Heavy	125	6.0	3000	13.3
Offices		50	2.4	2000	9.0
Printing plants	Press rooms	150	7.2	2500	11.1
	Composing rooms	100	4.8	2000	9.0
Residential		40	1.9		
Rest rooms		**			
Reviewing stands, grandstands, and bleachers		100	4.8		
Roof decks (occupied)	Same as area served				
Schools	Classrooms	40	1.9	1000	4.5
Sidewalks and driveways	Public access	250	12.0	*	
Storage	Light	125	6.0		
	Heavy	250	12.0		
Stores	Retail	75	3.6	2000	9.0
	Wholesale	100	4.8	3000	13.3

Source: Adapted from the *Uniform Building Code* (Ref. 1), with permission of the publishers, International Conference of Building Officials.

*Wheel loads related to size of vehicles that have access to the area.
**Same as the area served or minimum of 50 psf.

(e.g., combustible or noncombustible) and some details of the construction (cover on reinforcement in concrete, fire insulation for steel beams, etc.)

The work in the design examples in this part is based largely on criteria from the *Uniform Building Code* (Ref. 1). The choice of this model code reflects only the fact of the degree of familiarity of the author with specific codes in terms of his recent experience.

Live Loads

Live loads technically include all the nonpermanent loadings that can occur, in addition to the dead loads. However, the term as commonly used usually refers only to the vertical gravity loadings on roof and floor surfaces. These loads occur in combination with the dead loads, but are generally random in character and must be dealt with as potential contributors to various loading combinations.

Roof Loads

In addition to the dead loads they support, roofs are designed for a uniformly distributed live load that includes snow accumulation and the general loadings that occur during construction and maintenance of the roof. Snow loads are based on local snowfalls and are specified by local building codes.

Table 14.2 gives the minimum roof live load requirements specified by the 1991 edition of the *Uniform Building Code*. Note the adjustments for roof slope and for the total area of roof surface supported by a structural element. The latter accounts for the increase in probability of the lack of total surface loading as the size of the surface area increases.

Roof surfaces must also be designed for wind pressure, for which the magnitude and manner of application are specified by local building codes based on local wind histories. For very light roof construction, a critical problem is sometimes that of the upward (suction) effect of the wind, which may exceed the dead load and result in a net upward lifting force.

Although the term *flat roof* is often used, there is generally no such thing; all roofs must be designed for some water drainage. The minimum required pitch is usually $\frac{1}{2}$ in./ft, or a slope of approximately 1:50. With roof surfaces that are this close to flat, a potential problem is that of *ponding*, a phenomenon in which the weight of water on the surface causes deflection of the supporting structure, which in turn allows for more water accumulation (in a pond), causing more deflection, and so on, resulting in an accelerated collapse condition.

Floor Loads

The live load on a floor represents the probable effects created by the occupancy. It includes the weights of human occupants, furniture, equipment, stored materials, and so on. All building codes provide minimum live loads to be used in the design of buildings for various occupancies. Since there is a lack of uniformity among different codes in specifying live loads, the local code should always be used. Table 14.3 contains values for floor live loads as given by the 1991 edition of the *Uniform Building Code*.

Although expressed as uniform loads, code-required values are usually established large enough to account for ordinary concentrations that occur. For offices, parking garages, and some other occupancies, codes often require the consideration of a specified concentrated load as well as the distributed loading. Where buildings are to contain heavy machinery, stored materials, or other contents of unusual weight, these must be provided for individually in the design of the structure.

When structural framing members support large areas, most codes allow some reduction in the total live load to be used for design. These reductions, in the case of roof loads, are incorporated into the data in Table 14.2. The following is the method given in the 1991 edition of the *Uniform Building Code* for determining the reduction permitted for beams, trusses, or columns that support large floor areas.

Except for floors in places of assembly (theaters, etc.), and except for live loads greater than 100 psf [4.79 kN/m^2], the design live load on a member may be reduced in accordance with the formula

$$R = 0.08 (A - 150)$$
$$[R = 0.86 (A - 14)]$$

The reduction shall not exceed 40% for horizontal members or for vertical members receiving load from one level only, 60% for other vertical members, nor R as determined by the formula

$$R = 23.1 \left(1 + \frac{D}{L}\right)$$

In these formulas

R = reduction in percent,
A = area of floor supported by a member,
D = unit dead load/sq ft of supported area,
L = unit live load/sq ft of supported area.

In office buildings and certain other building types, partitions may not be permanently fixed in location but may be erected or moved from one position to another in accordance with the requirements of the occupants. In order to provide for this flexibility, it is customary to require an allowance of 15–20 psf [0.72–0.96 kN/m^2] which is usually added to other dead loads.

Lateral Loads

As used in building design, the term *lateral load* is usually applied to the effects of wind and earthquakes, as they induce horizontal forces on stationary structures. From experience and research, design criteria and methods in this area are continuously refined, with recommended practices being presented through the various model building codes, such as the *Uniform Building Code* (UBC) (Ref. 1).

Space limitations do not permit a complete discussion of the topic of lateral loads and design for their resistance. The following discussion summarizes some of the criteria for design in the latest edition of the UBC. Examples of application of these criteria are given in the chapters that follow containing examples of building structural design. For a more extensive discussion the reader is referred to *Simplified Building Design for Wind and Earthquake Forces* (Ref. 13).

Wind. Where wind is a major local problem, local codes are usually more extensive with regard to design requirements for wind. However, many codes still contain relatively simple criteria for wind design. One of the most up-to-date and complex standards for wind design is contained in the *American National Standard Minimum Design Loads for Buildings and Other Structures*, ANSI A58.1-1982, published by the American National Standards Institute in 1982.

Complete design for wind effects on buildings includes a large number of both architectural and structural concerns. The following is a discussion of some of the requirements for wind as taken from the 1991 edition of the UBC (Ref. 1), which is in general conformance with the material presented in the ANSI Standard just mentioned.

Basic Wind Speed. This is the maximum wind speed (or velocity) to be used for specific locations. It is based on recorded wind histories and adjusted for some statistical likelihood of occurrence. For the continental United States the wind speeds are taken from UBC, Figure No. 4. As a reference point, the speeds are those recorded at the standard measuring position of 10 m (approximately 33 ft) above the ground surface.

Exposure. This refers to the conditions of the terrain surrounding the building site. The ANSI Standard describes four conditions (A, B, C, and D), although the UBC uses only two (B and C). Condition C refers to sites surrounded for a distance of one-half mile or more by flat, open terrain. Condition B has buildings, forests, or ground-surface irregularities 20 ft or more in height covering at least 20% of the area for a distance of 1 mile or more around the site.

Wind Stagnation Pressure (q_s). This is the basic reference equivalent static pressure based on the critical local wind speed. It is given in UBC Table No. 23-F and is based on the following formula as given in the ANSI Standard:

$$q_s = 0.00256 V^2$$

Example: For a wind speed of 100 mph

$$q_s = 0.00256 V^2 = 0.00256(100)^2$$
$$= 25.6 \text{ psf } [1.23 \text{ kPa}]$$

which is rounded off to 26 psf in the UBC table.

Design Wind Pressure. This is the equivalent static pressure to be applied normal to the exterior surfaces of the building and is determined from the formula

$$p = C_e C_q q_s I$$

(UBC Formula 11-1, Section 2311)

in which p = design wind pressure in psf,

C_e = combined height, exposure, and gust factor coefficient as given in UBC Table No. 23-G,

C_q = pressure coefficient for the structure or portion of structure under consideration as given in UBC Table No. 23-H,

q_s = wind stagnation pressure at 30 ft given in UBC Table No. 23-F,

I = importance factor.

The importance factor is 1.15 for facilities considered to be essential for public health and safety (such as hospitals and government buildings) and buildings with 300 or more occupants. For all other buildings the factor is 1.0.

The design wind pressure may be positive (inward) or negative (outward, suction) on any given surface. Both the sign and the value for the pressure are given in the UBC table. Individual building surfaces, or parts thereof, must be designed for these pressures.

Design Methods. Two methods are described in the Code for the application of the design wind pressures in the design of structures. For design of individual elements particular values are given in UBC Table 23-H for the C_q coefficient to be used in determining p. For the primary bracing system the C_q values and their use is to be as follows:

Method 1 (Normal Force Method). In this method wind pressures are assumed to act simultaneously normal to all exterior surfaces. This method is required to be used for gabled rigid frames and may be used for any structure.

Method 2 (Projected Area Method). In this method the total wind effect on the building is considered to be a combination of a single inward (positive) horizontal pressure acting on a vertical surface consisting of the projected building profile and an outward (negative, upward) pressure acting on the full projected area of the building in plan. This method may be used for any structure less than 200 ft in height, except for gabled rigid frames. This is the method generally employed by building codes in the past.

Uplift. Uplift may occur as a general effect, involving the entire roof or even the whole building. It may also occur as a local phenomenon such as that generated by the overturning moment on a single shear wall. In general, use of either design method will account for uplift concerns.

Overturning Moment. Most codes require that the ratio of the dead load resisting moment (called the restoring moment, stabilizing moment, etc.) to the overturning moment be 1.5 or greater. When this is not the case, uplift effects must be resisted by anchorage capable of developing the excess overturning moment. Overturning may be a critical problem for the whole building, as in the case of relatively tall and slender tower structures. For buildings braced by individual shear walls, trussed bents, and rigid-frame bents, overturning is investigated for the individual bracing units. Method , is usually used for this investigation, except for very tall buildings and gabled rigid frames.

Drift. Drift refers to the horizontal deflection of the structure due to lateral loads. Code criteria for drift are usually limited to requirements for the drift of a single story (horizontal movement of one level with respect to the next above or below). The UBC does not provide limits for wind drift. Other standards give various recommendations, a common one being a limit of story drift to 0.005 times the story height (which is the UBC limit for seismic drift). For masonry structures wind drift is sometimes limited to 0.0025 times the story height. As in other situations involving structural deformations, effects on the building construction must be considered; thus the detailing of curtain walls or interior partitions may affect limits on drift.

Combined Loads. Although wind effects are investigated as isolated phenomena, the actions of the structure must be considered simultaneously with other phenomena. The requirements for load combinations are given by most codes, although common sense will indicate the critical combinations in most cases. With the increasing use of load factors the combinations are further modified by applying different factors for the various types of loading, thus permitting individual control based on the reliability of data and investigation procedures and the relative significance to safety of the different load sources and effects. Required load combinations are described in Section 2303 of the UBC.

Special Problems. The general design criteria given in most codes are applicable to ordinary buildings. More thorough investigation is recommended (and sometimes required) for special circumstances such as the following:

Tall Buildings. These are critical with regard to their height dimension as well as the overall size and number of occupants inferred. Local wind speeds and unusual wind phenomena at upper elevations must be considered.

Flexible Structures. These may be affected in a variety of ways, including vibration or flutter as well as the simple magnitude of movements.

Unusual Shapes. Open structures, structures with large overhangs or other projections, and any building with a complex shape should be carefully studied for the special wind effects that may occur. Wind-tunnel testing may be advised or even required by some codes.

Earthquakes. During an earthquake a building is shaken up and down and back and forth. The back-and-forth (horizontal) movements are typically more violent and tend to produce major unstabilizing effects on buildings; thus structural design for earthquakes is mostly done in terms of considerations for horizontal (called lateral) forces. The lateral forces are actually generated by the weight of the building—or, more specifically, by the mass of the building that represents both an inertial resistance to movement and the source for kinetic energy once the building is actually in motion. In the simplified procedures of the equivalent static force method, the building structure is considered to be loaded by a set of horizontal forces consisting of some fraction of the building weight. An analogy would be to visualize the building as being rotated vertically 90 to form a cantilever beam, with the ground as the fixed end and with a load consisting of the building weight.

In general, design for the horizontal force effects of earthquakes is quite similar to design for the horizontal force effects of wind. Indeed, the same basic types of lateral bracing (shear walls, trussed bents, rigid frames, etc.) are used to resist both force effects. There are indeed come significant differences, but in the main a system of bracing that is developed for wind bracing will most likely serve reasonably well for earthquake resistance as well.

Because of its considerably more complex criteria and procedures, we have chosen not to illustrate the design for earthquake effects in the examples in this part. Nevertheless, the development of elements and systems for the lateral bracing of the buildings in the design examples here is quite applicable in general to situations where earthquakes are a predominant concern. For structural investigation, the principal difference is in the determination of the loads and their distribution in the building. Another major difference is in the true dynamic effects, critical wind force being usually represented by a single, major, one-direction punch from a gust, while earthquakes represent rapid back-and-forth, reversing-direction actions. However, once the dynamic effects are translated into equivalent static forces, design concerns for the bracing systems are very similar, involving considerations for shear, overturning, horizontal sliding, and so on.

For a detailed explanation of earthquake effects and illustrations of the investigation by the equivalent static force method the reader is referred to *Simplified Building Design for Wind and Earthquake Forces* (Ref. 13).

PART FIVE

USE OF GRAPHICAL METHODS

Use of graphical methods for analysis of force actions offers a unique experience, most notably for persons already extensively experienced in graphical work in general. For this reason, the graphical investigation of force actions was highly developed in past years by various teachers and writers who worked with architects. A particular favored presentation was that of the investigation of planar trusses. Much of the presentation in Chapter 15 of this part is adapted from work produced by various authors whose work appeared in the early part of the twentieth century. A principal reference for this work is *Simplified Design of Roof Trusses* (Ref. 12.).

Practicing engineers have long since pretty much abandoned the general use of graphics, and the age of the computer has almost entirely eliminated the need for people to use mathematics at all. However, the involvement experience and actual practical use of graphical methods is still a valuable learning opportunity. Some possibilities for use of these antique graphical methods in serious design applications are discussed in Chapter 16.

15

GRAPHICAL METHODS FOR INVESTIGATION OF FORCES

This chapter develops the use of graphical methods for investigation of the actions of planar trussed structures. Actions investigated include the reaction forces at supports, the internal forces in truss members, and the deflection (general deformed shape) of the truss frame.

15.1 FUNDAMENTALS OF FORCE INVESTIGATION

The basic nature of forces and the use of graphical representation of force vectors is presented in Chapter 6. The material presented here builds on that work, and the reader should refer to Chapter 6 for the following basic presentations:

Properties of forces, vector representation of forces, equilibrium of force systems—Section 6.1.
Composition and resolution of forces, components of a force, classification of force systems—Section 6.2.
The force polygon, notation for concentric force systems—Sections 6.3 and 6.5.
Basic graphical analysis for internal forces in planar trusses; use of space diagrams, Maxwell diagrams, and separated joint diagrams—Sections 6.5 and 6.6.

The graphical work in Chapter 6 deals only with concentric force systems; thus there is no need to treat moments that occur in a general planar force system (nonconcurrent). The following section presents a graphical method for dealing with equilibrium of moments, which permits investigation for reaction forces.

15.2 THE FUNICULAR POLYGON

The *funicular polygon* is a device that permits the determination of equilibrium for a planar force system that is not necessarily concurrent. The following demonstration develops the basic concept, and succeeding illustrations show applications to various investigations. A principal use of the funicular polygon is in the investigation for reactions, which is developed in Section 15.3.

Consider the system of forces, *AB*, *BC*, and *CA* shown in Figure 15.1a. If these three forces constitute a system in equilibrium, both the force polygon and funicular polygon will close. Let us apply the test.

The force polygon *abc*, Figure 15.1b, is first drawn. This polygon closes. Thus far it appears that the system may be in equilibrium. If the force polygon had not closed there would be no need to continue further; the system would not be in equilibrium.

Now let us construct the *funicular polygon*. To do this we begin by selecting any point in connection with the force polygon. This point is called the *pole* and is marked *o*, Figure 15.1b. From this point draw the lines *oa*, *ob*, and *oc*. These lines are known as *rays* and represent the stresses in an imaginary frame connecting the three forces *AB*, *BC*, and *CA* shown in Figure 15.1a. Our next step is to construct this imaginary frame known as the *funicular or equilibrium polygon*.

It is seen in Figure 15.1b that the force *ab* is held in equilibrium by the rays *oa* and *ob*. Therefore, in Figure 15.1a select any point on the line of action of the force *AB* and draw lines *oa* and *ob* parallel respectively to rays *oa* and *ob*. These lines, called strings, are continued until they intersect the adjacent forces. In the force polygon, Figure 15.1b, force *bc* is held in equilibrium by rays *ob* and *oc*, hence in the space diagram, Figure 15.1a, *BC*, *ob*, and *oc* must have a point in common. Consequently, from the point where *ob* intersects force *BC*, draw the string *oc* parallel to ray *oc*. We now observe in Figure 15.1a that strings *oa* and *oc* intersect on the line of action of force *CA*, that is to say, *the funicular polygon closes*, and, since the force polygon also closes, the given system is in equilibrium.

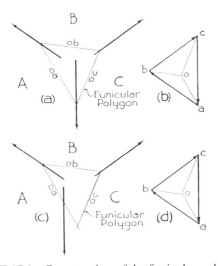

FIGURE 15.1 Construction of the funicular polygon.

If a different point had been selected as a pole, a different funicular polygon would have resulted. However, the last two rays drawn would have intersected on the remaining force, showing that the funicular polygon closed and proving that the system is in equilibrium. If we continue the lines of action of the forces *AB*, *BC*, and *CA*, in Figure 15.1a, we notice that they intersect at a common point so that actually the system is concurrent.

Again consider the three forces *AB*, *BC*, and *CA* in Figure 15.1c. Let us see if this system is in equilibrium. By drawing the force polygon, *abc*, Figure 15.1d, we find that it closes. Next select a pole and draw the rays and strings of the funicular polygon as previously described, taking the forces in the same sequence. It is seen, in this instance, that *oa* and *oc*, the sides of the funicular polygon, do not intersect on the force *CA*. *The funicular polygon does not close*, hence this system is not in equilibrium.

Notice that if the lines of action of the forces *AB*, *BC*, and *CA* are extended in Figure 15.1c, they do not intersect at a common point. If we were required to alter the system to produce equilibrium, we could move the force *CA* to the right so that its line of action passes through the intersection of strings *oa* and *oc*. We would then have a system of forces whose force and funicular polygons close and hence would be in equilibrium. But observe! If *CA* is moved so that its line of action passes through the intersection of strings *oa* and *oc* we alter the system so that forces *AB*, *BC*, and *CA* constitute a concurrent system. This illustrates a principle well worth remembering, namely, *three non-parallel forces in equilibrium must meet in a point*.

Funicular Polygon Used as a Test

Figure 15.2a consists of a system of three forces *AB*, *BC*, and *CA*; let it be required to find whether or not they are in equilibrium. In accordance with the principle stated previously, we know that equilibrium exists provided both the force polygon and funicular polygon close. Let us construct these polygons. First draw the force polygon, Figure 15.2b. We observe that it closes. If the system had been concurrent it would not be necessary to proceed further, for the closing of the force polygon would have assured us that the forces were in equilibrium. But the system is not concurrent; therefore we must test it further by constructing a funicular polygon.

To do this, select any point *o* in the force polygon and draw the rays *oa*, *ob*, and *oc*. In the force polygon we find that force *ca* is held in equilibrium by *oa* and *oc*; therefore draw from point *r*, any point in the line of action of *CA*, lines parallel to *oa* and *oc*. *AB* is intersected by line *oa* at *s*, and from this point draw a line *ob* parallel to

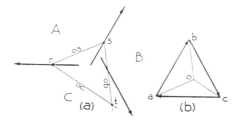

FIGURE 15.2 Use of a funicular polygon as a test for equilibrium.

334 GRAPHICAL METHODS FOR INVESTIGATION OF FORCES

the ray *ob*. In this problem we find that *oc* and *ob* intersect at point *t* which is *not* on the line of action of the force *BC*. "The polygon does not close," and therefore the system is not in equilibrium. Note that, if the force *BC* is moved to a parallel position whereby its line of action passes through point *t*, the altered system would be concurrent and hence in equilibrium.

Actually the given system in this problem is equivalent to a mechanical couple. This is apparent if we consider the forces in pairs. The resultant of *CA* and *AB* is *CB*, found by the force polygon. But the resultant of two concurrent forces passes through the point their lines of action have in common. The resultant of *CA* and *AB* and the remaining force *BC* constitute a mechanical couple, as they are two equal parallel forces acting in opposite directions and having different lines of action.

A system of forces whose force polygon closes, but whose funicular polygon does not close, constitutes a mechanical couple.

Finding the Resultant by Use of the Funicular Polygon

Let it be required to find the resultant of the system of four forces, *AB*, *BC*, *CD*, and *DE* shown in Figure 15.3*a*. The system is nonconcurrent, and the resultant might be found by use of the parallelogram of forces, taking the forces in pairs. This method, however, would be unnecessarily long, and the most direct procedure is to construct the funicular polygon. To do this, draw the force polygon, Figure 15.3*b*. This gives us the resultant *ae* in magnitude and direction but does not tell us its position with respect to the system. The pole is selected (any point such as *o*) and the rays are drawn, *oa*, *ob*, *oc*, *od*, and *oe*. Next select some point on the line of action of one of the forces, as point *r* on force *AB*, and draw lines parallel to the rays *oa* and *ob*. Where the line *ob* intersects *BC* draw the line *oc* parallel to the ray *oc*. In a similar manner continue to draw the sides of the funicular polygon until finally we have the two sides *oa* and *oe* which intersect at point *s*. Since, in the force polygon we find that *ae*, the resultant, is held in equilibrium by *oa* and *oe*, the point of intersection, *s*, determines the point through which the resultant will pass, thus determining its line of action.

Another example of finding the resultant of a system of forces by constructing the funicular polygon is shown in Figure 15.4. The given system consists of P_1, P_2, P_3, P_4, and P_5, Figure 15.4*a*. Drawing the force polygon, Figure 15.4*b*, shows us that the resultant is a downward force marked *R*. In this example the rays have been identified as 1, 2, 3, 4, 5, and 6. Since the rays 1 and 6 hold *R* in equilibrium, the sides of the funicular polygon corresponding to these rays intersect at a point through which the resultant will pass and thus determines the line of action of the resultant.

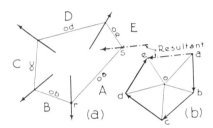

FIGURE 15.3 Finding the location of a resultant with a funicular polygon.

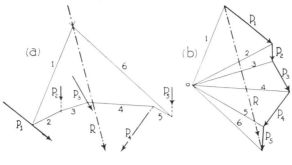

FIGURE 15.4

Resultant of Parallel Forces

A method of finding the resultant of parallel forces was explained previously. A more convenient solution of finding the resultant is by use of the funicular polygon. Consider, for example, the parallel forces *AB*, *BC*, *CD*, *DE*, and *EF* shown in Figure 15.5a. The force polygon is drawn, Figure 15.5b, and, since the forces are parallel and are both upward and downward, the polygon overlaps, the resultant being *AF*. Select the pole *o*, draw the rays and construct the funicular polygon connecting the forces as shown in Figure 15.5a. The intersection of the sides of the funicular polygon *oa* and *of* determines the point through which the resultant *AF* will pass, the magnitude and direction being found by the force polygon.

Reactions Found by Funicular Polygon

One of the most important uses of the funicular polygon is in finding the reactions of beams and trusses. The reactions are the upward forces under a beam that hold the downward forces in equilibrium. If a beam is symmetrically loaded, of course the reactions are equal, and each has a magnitude equal to one-half the sum of the downward forces or loads. However, if the loading is not symmetrical, the reactions are not necessarily equal and although we know their directions to be upward their magnitudes are unknown. They may be solved mathematically by the use of the principle of moments but their magnitudes may also be found by constructing the funicular polygon. Frequently the latter method is more advantageous.

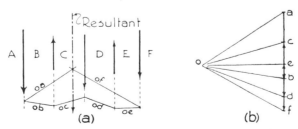

FIGURE 15.5 Finding the location of the resultant of a set of parallel forces with a funicular polygon.

336 GRAPHICAL METHODS FOR INVESTIGATION OF FORCES

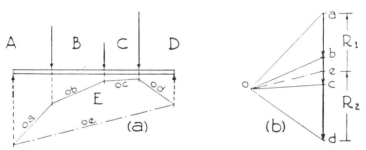

FIGURE 15.6 Reaction forces found with a funicular polygon.

Figure 15.6a represents three parallel forces *AB*, *BC*, and *CD* exerting pressure on a beam. The two upward forces *EA* and *DE*, known respectively as R_1 and R_2, are the two reactions that hold in equilibrium the three downward forces. We know the lines of action of these reactions and also their directions, but their magnitudes are unknown.

First draw as much of the force polygon as possible, Figure 15.6b. This consists of the downward forces *AB*, *BC*, and *CD*. The next force in order is *DE*, an upward force. Since its magnitude is unknown as yet, the point *e* cannot be located on the force polygon. Obviously it will be at some point on the load line, for *DE* + *EA* must equal *AB* + *BC* + *CD* as the five forces are parallel and in equilibrium, and the force polygon must close. Now select a point *o* and draw the rays *oa*, *ob*, *oc*, and *od*.

The next step is to construct the funicular polygon, taking, for instance, some point on the line of action of the force *AB* and drawing the sides *oa* and *ob* parallel to the rays *oa* and *ob*. At the point where *ob* intersects the line of action of *BC* draw *oc* parallel to the ray *oc*. Continuing in the same manner, draw *od*. Now we remember that if a system of forces is in equilibrium both the force polygon and funicular polygon must close. Therefore, from the intersection of the side *oa* and the force *EA* draw the line *oe* to the point of intersection of *od* and *DE*. This is the closing line of the funicular polygon and is called the "closing string." Having determined *oe*, we draw a parallel line from the pole to the force polygon, Figure 15.6b. This line, known as the "closing ray," determines the point *e* on the force polygon, consequently establishing the magnitudes of the reactions, *DE* and *EA*.

Figure 15.7a represents four wind loads, *AB*, *BC*, *CD*, and *DE*, which exert pressure on a truss. The truss is shown by the double lines, the web members being omitted. Let us assume the reactions at the supports to be parallel to the direction of the wind. We know then their lines of action and direction but not their magnitudes. By using the method explained in the previous problem these magnitudes are readily found.

First draw the loads *AB*, *BC*, *CD*, and *DE*, select the point *o* and draw the rays, Figure 15.7b. Next select some point on the line of action of one of the loads and draw the two sides of the funicular polygon. For instance, take a point on *BC* and draw the lines *ob* and *oc*. The funicular polygon is developed as has been explained and, knowing that the forces are in equilibrium and that the funicular polygon must close, we draw *of*, the closing string. From point *o* in Figure 15.7b we draw a line parallel to the closing string *of*, and this determines the point *f* in the force polygon. Hence the magnitudes of R_1 and R_2, the two reactions, are established.

In many problems it is convenient to know the position of the resultant of the

THE FUNICULAR POLYGON   337

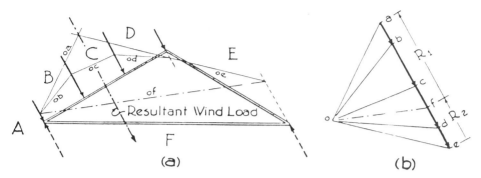

FIGURE 15.7 Wind load reactions found with a funicular polygon.

wind loads. By referring to Figure 15.7b we see that the resultant is *ae* and that it is held in equilibrium by the rays *oa* and *oe*. Therefore, the intersection of *oa* and *oe* in Figure 15.7a determines a point through which the resultant must pass.

To Find Reactions, Line of Action of One Being Unknown

The following problem frequently occurs in the solution of roof trusses. It consists of finding the reactions or supporting forces, when the line of action of one reaction is known while all that is known of the other reaction is a point in its line of action. The example given here consists of forces on a beam, but the method employed is applicable to forces on a truss as well.

Figure 15.8a represents three forces *BC*, *CD*, and *DE* acting on a beam. Let it be assumed that the right-hand reaction, *EA*, is vertical and that its magnitude is unknown. Assume also that the line of action of *AB*, the left reaction, is unknown as well as its magnitude. The points of support are, of course, points in the line of action of the two reactions. Let it be required to determine completely *EA* and *AB*, the reactions.

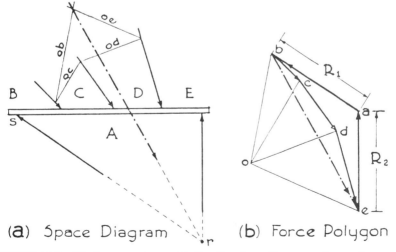

FIGURE 15.8 Finding reactions with the direction of one reaction unknown.

The problem may be simplified if we consider that we have three forces in equilibrium, the resultant of the loads and the two reactions. Here is where we employ the knowledge we have gained concerning resultants. The loads consist of the forces *BC, CD,* and *DE*. To find their resultant is a simple matter. By drawing the force polygon in Figure 15.8*b* the resultant is found to be *be*, a downward force to the right. Next select a pole, *o*, and draw the rays. By constructing the funicular polygon the sides *ob* and *oe* intersect at a point through which the resultant *BE* must pass. Now we may forget the individual forces *BC, CD,* and *DE* and consider in their stead their resultant *BE*. The system now consists of three forces *BE, EA,* and *AB*. Since by data they are in equilibrium and are not parallel, they must have a point in common. This is readily found; we merely continue the line of action of *BE* until it intersects the vertical force *EA* at *r*. This is the point the three forces have in common. We know that *AB* must pass through *s*, the point of support, and therefore by drawing a line through *r* and *s* we determine the line of action of *AB*.

Returning to the force polygon, we have already drawn the three forces *BC, CD,* and *DE*. The next force in order is *EA*. Therefore, through the point *e* draw a line parallel to *EA*; this, of course, is vertical, and the point *a* will occur somewhere on this line. The next force is *AB*, and through the point *b* draw a line parallel to *AB*. The point *a* will be somewhere on this line. Since *a* is on the line through *e* parallel to *EA* and also on the line through *b* parallel to *AB*, it must be at their point of intersection. This, then, completes the force polygon and we have completely determined the two reactions R_1 and R_2, their magnitudes being found by measuring their lengths in the force polygon.

15.3 DETERMINATION OF TRUSSES REACTIONS

Before the internal forces in a structure can be investigated, it is necessary to resolve the equilibrium of the total system of external forces. The external forces consist of the active forces (loads on the structure) and the reactive forces (resistance developed by the structure's supports, generally called simply the *reactions*. Accepting the loads as given values, the first step of investigation then becomes the determination of the reactions. This section presents demonstrations of the use of the graphical methods so far developed in solving for reactions of trusses.

Reactions for Vertical Loads, Trusses Loaded Symmetrically

Since for vertical loads the reactions are vertical, determining the reactions is a very simple matter. The system of forces is nothing more than a number of parallel forces in equilibrium. See the discussion of truss reactions in Section 16.2. If the truss has vertical loads and is symmetrically loaded, which is a very common condition, the reactions are equal and vertical and each reaction is equal in magnitude to one-half the sum of the loads.

Figure 15.9 represents the outline of a truss, the web members being omitted to avoid confusion. This truss has six equal panels, the panel loads being 2,000 lb each with the exception of the loads over the supports which, of course, are half-panel loads or 1,000 lb each. The total vertical load therefore is

$$(5 \times 2{,}000) + (2 \times 1{,}000) = 12{,}000\#$$

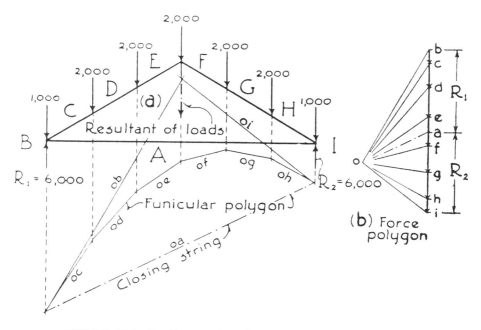

FIGURE 15.9 Finding reactions for a symmetrically loaded truss.

As the truss is loaded symmetrically each reaction is equal to 12,000 ÷ 2, or 6,000 lb, their directions being vertical.

In this problem, the forces consist of the downward forces, *BC, CD, DE, EF, FG, GH, HI* and the two reactions *IA* and *AB*. To draw the force polygon we select a convenient scale of so many pounds to the inch and, beginning with *bc*, lay off the load line *bc, cd, de, ef, fg, gh*, and *hi*, Figure 15.9b. We have seen that R_2, the right-hand reaction *IA*, is 6,000 lb, therefore, at the same scale we measure upward 6,000 lb from point *i* and thus determine point *a*. The next force in order is *AB*, the left-hand support. Since points *a* and *b* are established, this completes the force polygon of the external forces.

For the above elementary problem the solution just given is all that is necessary. The magnitudes of R_1 and R_2 are known by merely inspecting the load diagram. If, however, we wish to determine their magnitudes by graphical methods we can construct a funicular polygon as explained in Art. 26. This polygon is shown directly below the truss. The point *a* in the force polygon, which determines the magnitudes of *IA* and *AB*, shows that each reaction has a magnitude of 6,000 lb.

Note that the resultant of the loads is *bi*, shown in the force polygon, and that its magnitude is 12,000 lb. The two rays holding this force in equilibrium are *ob* and *oi*. Hence their intersection in the funicular polygon determines a point through which the resultant of the loads must pass. This is unimportant in this particular problem but very often, particularly for wind loads, it is of great convenience to find this line of action of the resultant.

Reactions for Vertical Loads, Unsymmetrical Loading

The outline of a six-panel truss with web members omitted is shown in Figure 15.10a. A suspended load, *IJ* of 6,000 lb, results in unsymmetrical loading and consequently the reactions, although vertical, will not be equal in magnitude.

340 GRAPHICAL METHODS FOR INVESTIGATION OF FORCES

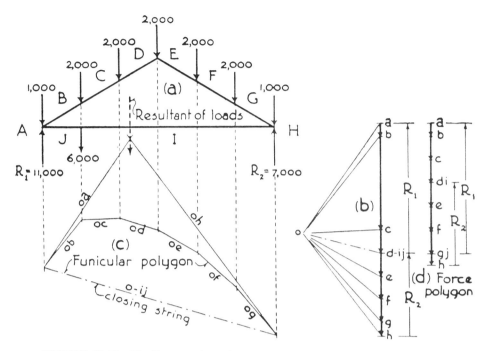

FIGURE 15.10 Finding reactions for a truss with unsymmetrical loading.

A convenient method of finding the magnitudes of R_1 and R_2 is by constructing a funicular polygon. Note that *BC* and *IJ* have the same line of action. Therefore, for the purpose of drawing the funicular polygon, we may consider the force at *BC* to be 2,000 + 6,000 or 8,000 lb. The force polygon used in constructing the funicular polygon is shown in Figure 15.10*b*. First draw *ab*, a downward force of 1,000 lb. The next force is *BC*, 8,000 lb. This is followed by *CD*, *DE*, *EF*, *FG*, each of which is 2,000 lb, and finally *GH*, a force of 1,000 lb. This completes the load line. As the 6,000 lb load is combined with *BC*, the space between the supports and below the truss may be called temporarily "*IJ*" instead of the two separate letters. This then results in calling the right reaction *H-IJ* and the left reaction *IJ-A*. The pole *o* is selected and the funicular polygon, Figure 15.10*c*, is constructed as has been explained. A line drawn from *o* parallel to the closing string determines the point *ij* on the force polygon and hence establishes the magnitudes of R_1 and R_2. The sum of all the loads is 12,000 + 6,000 or 18,000 lb and is represented by *ah* in Figure 15.10*b*. By measuring the lengths of *h-ij* and *ij-a*, in this force polygon, we find R_2 = 7,000 lb and R_1 = 11,000 lb.

Although it is not required for this particular problem, note that the intersection of lines *oa* and *oh* in the funicular polygon determines the line of action of the resultant of the loads.

We have now determined the magnitudes of the reactions, but Figure 15.10*b* cannot be used as a force polygon of external forces to be used in drawing a stress diagram. However, to construct a true force polygon is a simple matter now that all the forces are known. It is shown in Figure 15.10*d*. Begin with *ab*, a downward force of 1,000 lb, and follow with *bc*, *cd*, *de*, ef, *fg*, each 2,000 lb in magnitude and finally *gh*, 1,000 lb. These are the loads on the upper chord. Next comes *HI* an upward force of

DETERMINATION OF TRUSSED REACTIONS   341

7,000 lb found in Figure 15.10b, thus determining point i. The next force is IJ, the downward force of 6,000 lb, and finally ja the left reaction, 11,000 lb, completes the true force polygon of the external forces. Since this is a system of parallel forces, the sides of the force polygon overlap.

Reactions for Wind Load

Wind loads generally produce some combination of vertical and horizontal forces on a structure. For a horizontally spanning structure, the vertical aspect of the wind is similar to the vertical gravity loads and may be dealt with as described in the preceding examples. However, the horizontal component of the wind force requires some special considerations for determination of the reactions.

This problem has been discussed considerably elsewhere in this book, so we will not present all the considerations here. For the limited concerns of determination of the reactions, a judgment must be made as to how the horizontal force is distributed between the supports. The two simple cases are an equal distribution (half the total horizontal force assigned to each support) or having all the resistance at one support; the latter occurring when a roller or other device prevents horizontal transfer of force at one support.

The following examples illustrate the use of graphical methods for various situations of wind loading.

Wind-Load Reactions—Fixed Ends—Reactions Parallel

A triangular truss with wind loads is indicated in Figure 15.11a. *Assuming that the reactions are parallel to the direction of the wind*, let it be required to draw the force polygon of the external forces. Note that the external forces consist of the loads, AB, BC, CD, DE, EF and the two reactions FG and GA. The truss is, of course, loaded unsymmetrically and, although we assume the direction of the reactions to be parallel to the wind and know points in their lines of action, their magnitudes as yet are unknown. First, lay off the load line beginning with ab, Figure 15.11b. Completed it is ab, bc, cd, de, and ef. Now one method of determining the reactions, though not the simplest, is to draw the funicular polygon as shown at Figure 15.11c. A line drawn from point o, Figure 15.11b, parallel to the closing string determines the point g and consequently FG and GA are established. Now all that need be done to find the magnitudes of the reactions is to scale the lengths of fg and ga.

Here is a much simpler method of accomplishing the same results found above. First draw the load line ab, bc, cd, de, and ef, Figure 15.11d. The resultant of the wind loads is AF, 8,000 lb. Since the wind loads on the side of the truss are placed symmetrically it is obvious that their resultant lies at the midpoint, or on the same line of action as CD. This was verified by the funicular polygon, Figure 15.11c. The problem is now simplified for we may say that there are but three external forces, the wind-load resultant AF and the two reactions FG and GA. It can be shown that the magnitudes of the reactions are in the same proportion as the two divisions of the lower chord cut by the wind-load resultant. These divisions are marked x and y, Figure 15.11a. All that remains, therefore, is to divide the load line in the same proportion as x is to y. To accomplish this, draw any line from f, as fm, and lay off divisions fn and nm, proportional to x and y. A line drawn from n to the load line, parallel to mu, divides the load line into the desired proportion, hence establishing

342 GRAPHICAL METHODS FOR INVESTIGATION OF FORCES

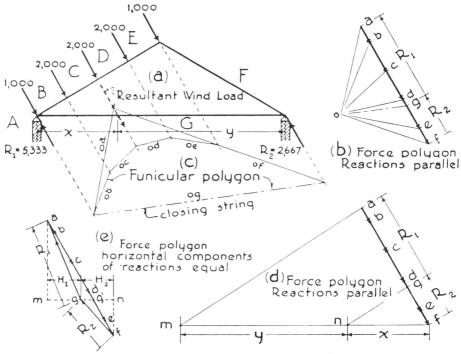

FIGURE 15.11 Finding wind load reactions.

fg and *ga*, the two reactions. In using this convenient method it is necessary to inspect the truss diagram to see which reaction is the greater so that the divisions *x* and *y* may be properly located with respect to the load line. In this particular instance it is noted that since the resultant wind load is nearer R_1 than R_2, R_1 or *GA* will have the greater magnitude and consequently *x* and *y* are located in this manner with respect to the load line.

Finding the reactions by this *proportional line method* may be accomplished very quickly indeed. Note that the results agree with those found by using the funicular polygon in Figure 15.11*b*.

Horizontal Components Equal

The assumption in the previous example was that the two reactions are parallel. The other possibility is that the horizontal components of the reactions are equal. This is quite simple.

Draw the load line *ab*, *bc*, *cd*, *de*, and *ef*, Figure 15.11*e*, and determine the reactions, assuming them to be parallel as has just been explained. This establishes the point *g'*. Now the horizontal component of the wind is simply the horizontal distance between points *a* and *f*. Therefore, draw a horizontal line through *g'* and vertical lines through points *a* and *f*. The intersections, *m* and *n*, determine the horizontal component of the wind load. Next divide *mn* into two equal parts, *mg* and *gn*, *the two equal horizontal components*. R_2 therefore will be *fg*, and R_1 will be *ga*, not necessarily parallel. Attention should be called to the fact that the vertical components of the reactions remain unchanged in this construction.

Wind-Load Reactions—Roller Under One End

The end connections for the previous trusses have been fixed, that is, restrained against horizontal action. Now let us investigate reactions for trusses having one end supported on a roller. Since, theoretically, such a support is incapable of offering horizontal resistance, the reaction at this end must be vertical and the tendency to move horizontally will be resisted by the fixed end.

Figure 15.12a indicates in outline a six-panel truss with wind loads on the left and a roller at the right support, the left support being fixed. By data we know the direction, lines of action and magnitude of the wind loads. We also know the line of action of R_2, but all that is known of R_1 is a point in its line of action, namely, the support. Let it be required to draw a force polygon of the external forces.

One method of accomplishing this is by constructing a funicular polygon. First draw the load line beginning with bc and continue with cd, de, and ef, Figure 15.12b. FA, the next force, is the right reaction and this we know to be vertical. Therefore, through f extend a vertical line. The point a will lie somewhere on this line and when it is found we will have determined fa and ab, the two reactions. Next select a pole o and draw the rays. Since the only point we know on the line of action of AB is the point of support, we must begin the funicular polygon here. Therefore, through this point draw oc, the side of the funicular polygon, and continue with the remaining sides as previously described. A line through o parallel to the closing string intersects the vertical line through f at a, Figure 15.12b, thus establishing fa, the right reaction. A line drawn from a to b determines AB, the left reaction, in both magnitude and direction.

The intersection of the sides of the funicular polygon of and ob gives us a point through which the resultant of the wind load must pass. This resultant is not used in the method just explained, but often it is convenient to know its position.

Roller Under One End—A Shorter Solution

The previous method of determining the reactions when a roller is employed required the construction of a funicular polygon. When the loads are symmetrical, as they most frequently are, the work may be simplified by the following procedure.

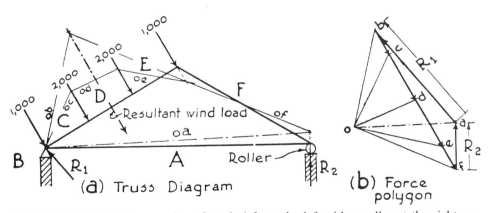

FIGURE 15.12 Finding reactions for wind from the left with a roller at the right reaction.

344 GRAPHICAL METHODS FOR INVESTIGATION OF FORCES

Again in Figure 15.13a we have a diagram representing wind loads on the left side of a truss with a roller at the right. Our problem is to draw the force polygon of the external forces. Fundamentally, the system consists of three forces, the wind load and the two reactions. By data they are in equilibrium. Since *FA* is vertical, the forces are not parallel and therefore must meet in a common point, as we know that any system of three forces, not parallel and in equilibrium, must be concurrent. To find the resultant of the wind loads we can construct a funicular polygon as shown. This would be necessary if the loading was unsymmetrical but in this instance, *as is generally the case*, we can locate its position by observation, namely at the midpoint of the loads. Therefore all that is necessary is to continue the line of action of the resultant until it meets the line of action of *FA*, at *m*. This gives us the point in common of the three external forces. Consequently a line drawn from *m* to the left-hand support determines the *direction* of *AB*.

To draw the force polygon we begin with *bc* and lay off the load line *bc, cd, de*, and *ef*, Figure 15.13b. From *f* draw a vertical line; point *a* will lie somewhere on this line. As we have determined the direction of *AB*, we now draw a line through *b* parallel to the direction of *AB* found in Figure 15.13a; the point *a* will occur on this line. Therefore, the intersection of this line and the vertical line through *f* determines the point *a* and consequently the two reactions *FA* and *AB*.

For the wind coming from the right, with a roller at the right support, a similar construction is shown for determining the reactions in Figures 15.14a. Again the resultant is found by the funicular polygon, a step not required for this loading, for obviously its line of action is at the center point of the upper chord. The important principle involved in finding reactions by this method is that three forces, not parallel and in equilibrium, must meet in a common point. By applying this principle, the direction of the unknown reaction is established, thus permitting the completion of the force polygon.

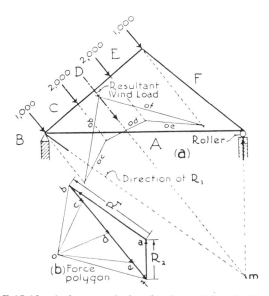

FIGURE 15.13 A shorter solution for the problem in Figure 15.12.

DETERMINATION OF TRUSSED REACTIONS   345

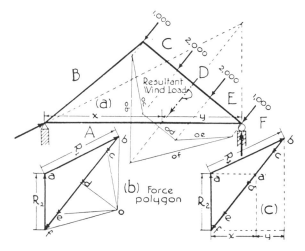

FIGURE 15.14 Reactions for wind right, roller right.

Suppose we had been given the truss shown in Figure 15.14a with the stipulation that the truss had fixed ends. By the principle of the proportional line, fa' and $a'b$ are quickly found, Figure 15.14c, thus giving us the two reactions but with the assumption that the reactions are parallel. Now if a roller is substituted at the right reaction, the supporting force can no longer be fa'; it can only be its vertical component. Consequently if we draw a horizontal line through a' to intersect a vertical line through f, their intersection gives us the desired vertical component and therefore the two reactions FA and AB are found. Note that this is the same result found in Figure 15.14b.

Resultant Reactions

To find the pressure exerted by a truss on its supports, it is necessary to find the resultant reactions. As an illustration, Figure 15.15 represents a six-panel truss sub-

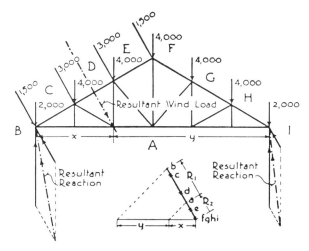

FIGURE 15.15 Find reactions for combined loads.

jected to a total vertical load of 24,000 lb and a total wind load of 9000 lb. Since the truss is symmetrically loaded *with respect to the vertical loads*, it is obvious that each reaction will be equal to one-half of the total vertical load or 12,000 lb. Their directions, of course, are vertical and the reactions are drawn to scale at the supports.

Assuming the truss to be fixed at both ends and that the reactions due to the wind loads are parallel, the wind-load reactions are found by the proportional line method to be $R_1 = 6,000$ lb and $R_2 = 3,000$ lb. These reactions likewise are drawn to scale at the supports. The true or resultant reactions, due to the vertical loads and wind loads acting simultaneously, may be readily found in direction and magnitude by completing the parallelogram of forces as shown. By scaling the resultant reactions, R_1 is found to be 17,500 lb and R_2 equals 14,500 lb. It should be noted that the wind is shown coming from the left in Figure 15.15, resulting, in this instance, in R_1 being greater in magnitude than R_2. However, if the direction of the wind changes and comes from the right, R_2 becomes the greater reaction. The customary procedure is to assume the wind to come from either the right or left and to use for each reaction whichever force is the greater magnitude.

15.4 APPLICATIONS OF GRAPHICAL METHODS

The potential variety of trussed structures is virtually endless. However, many classical forms have derived from specific uses for common forms of buildings. This section presents a large number of examples of graphical investigations of trussed structures. The presentation begins with relatively simple forms of trusses and goes on to more complex ones. The process is more fully explained for the earlier examples, but is presented mostly only with the finished graphic constructions for later examples.

The work presented here is derived from publications produced some time ago. Present design codes provide for some different loading conditions, but the general form of application of the loads for truss investigation is not basically different from that shown here.

Pratt Truss

Two commonly used roof trusses are the Pratt and Howe trusses, Figures 15.16 and 15.17. They differ in the direction of the diagonal web members. Both trusses are shown as having six panels but any number of panels may be used, depending, of course, on the length of the span. Figure 15.16a represents a Pratt truss with vertical loads. Since the diagonal members are in tension and are relatively long, this truss is better adapted to steel construction than timber.

The construction of the stress diagram presents no difficulties. The total load is 18,000 lb and, as the truss is symmetrical and is symmetrically loaded, each reaction is vertical and has a magnitude of $18,000 \times \frac{1}{2}$ or 9,000 lb.

First draw the force polygon of the external forces. These forces are all known. Draw bc, de, ef, fg, gh, and hi, the loads, Figure 15.16b. The line thus drawn is known as the *load line*. The next force in order is IA, an upward force of 9,000 lb, thus determining the point a. An upward force of 9,000 lb, ab, completes the force polygon of the external forces.

APPLICATIONS OF GRAPHICAL METHODS   347

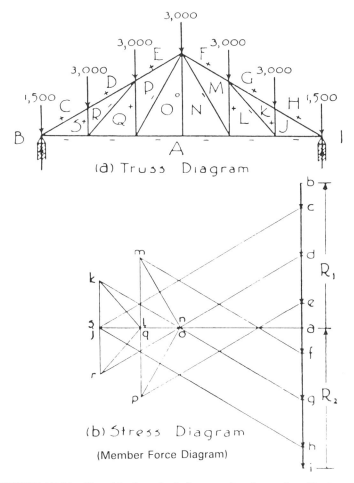

FIGURE 15.16 Graphical analysis for member forces in a Pratt truss.

The only joints at which we might begin to construct the stress diagram are *BCSA* or *HIAJ*, the joints at the supports. At each of these joints there are four concurrent forces two of which are known, two being unknown. First let us consider *BCSA*. In connection with the load line just drawn, draw a line through *c* parallel to the member *CS*. The point *s* will lie at some position on this line. The next member in order is *SA*, therefore through point *a* draw a line parallel to member *SA*. The point *s* lies somewhere on this line and since it also occurs on the line through *c* parallel to *CS*, it must be at their point of intersection.

The joint *SRQA* may not be considered next, for at this point there are four forces and as yet only one, *AS*, is known. Therefore take joint *CDRS*; here there are only two unknowns. A line through *d* parallel to *DR* and a line through *s* parallel to *RS* intersect at a point which is *r*. In a similar manner we may take the remaining joints in the truss and thus complete the stress diagram as shown in Figure 15.16b.

On inspecting the stress diagram it is noted that the letters *n* and *o* occur at the same point. This indicates that, since *on* has no length, the member *ON* has a zero

348 GRAPHICAL METHODS FOR INVESTIGATION OF FORCES

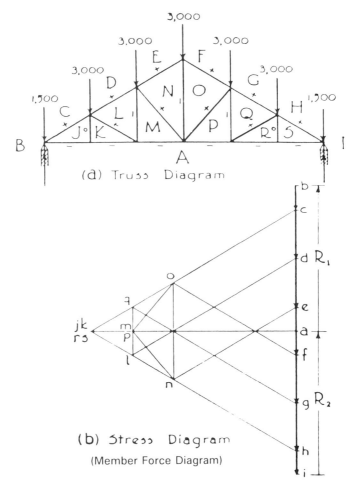

FIGURE 15.17 Graphical analysis for member forces in a Howe truss.

stress. This is called a *redundant member* and is only employed to prevent a possible sag in this portion of the lower chord. For the stress diagram just drawn, only loads on the upper chord were considered, but if a ceiling had been suspended from the lower chord, a tensile stress would have been developed in member *ON*.

A redundant member may be readily identified by inspecting the truss diagram. At the joint *ONA* three forces are in equilibrium. Since *AO* and *NA* have the same line of action, one of the forces must be the equilibrant of the other since any other single force acting at this joint would disturb the state of equilibrium. Hence the stress in member *ON* must be zero. Whereas the construction of the stress diagram readily indicates a redundant member, it is frequently convenient to identify such members by merely inspecting the truss diagram.

Howe Truss

Figure 15.17*a* represents a Howe truss having vertical loads on the upper chord only. This truss differs from the Pratt truss in the direction of the diagonal web members.

It is commonly used in timber construction and, as the vertical members are ties, rods are used since metal is more readily adapted to tensile stresses than timber. Although the lower chord is in tension, it is customary to use timber for these members to facilitate the detailing of the joints. A ceiling suspended at the lower chord likewise suggests the use of timber for these members owing to the convenience in framing.

The stress diagram for this truss is shown in Figure 15.17b. It presents no difficulties in its construction. Note that the two vertical members, *JK* and *RS* are redundant members and may be omitted if there are no suspended loads.

Cambered Trusses

When a portion of the lower chord of a truss is raised, as shown in Figure 15.18b, the truss is said to be *cambered*. The purpose is to provide greater head-room under the central part of the truss and also, possibly, to enhance its appearance. Trusses having horizontal lower chords, particularly trusses of long spans, often appear to sag, and cambering overcomes this illusion.

Figures 15.18a and b represent trusses of similar loads and spans. Their stress diagrams at c and d show how cambering increases the stresses in certain members. It is evident from Figure 15.18d that the stresses increase in magnitude with the degree of camber. The most common practice is to camber the lower chord one-sixth the height of the truss.

Crane Truss

Let it be required to determine the stresses in the Crane truss shown in Figure 15.19a. In order to draw the force polygon of the external forces, first consider of what they

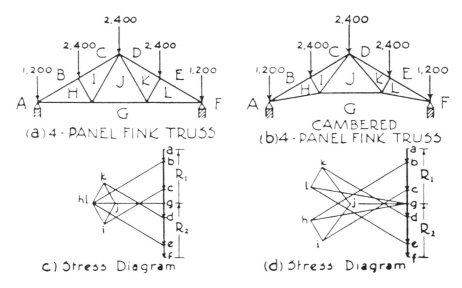

(Member Force Diagram)

FIGURE 15.18 Graphical analysis for member forces in cambered trusses.

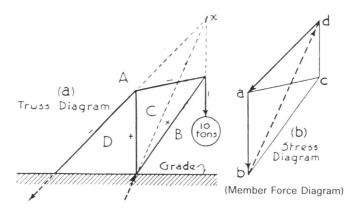

FIGURE 15.19 Graphical analysis of a crane truss.

consist. The load *AB*, 10 tons, is completely known. The resultant pressure at the ground, *BD*, is the resultant of *DC* and *CB*. Thus far we know only a point in its line of action, since *DC* and *CB* are of unknown magnitudes. The remaining external force is *DA*, the equilibrant of the stress in the member *DA*. This force is a tensile force in the earth and its line of action must coincide with the member *DA*. Since the three external forces, *AB*, *BD*, and *DA*, are in equilibrium, and are not parallel, they must have a point in common. This point is readily found by extending the lines of action of *AB* and *DA*. They intersect at point *x*. A line drawn from *x* to the point at which *DC* and *CB* meet at the surface of the ground determines the *direction* of the external force *BD*. Thus the directions of the three external forces are now established.

To draw the force polygon of these external forces draw, at a convenient scale, *ab* parallel to the force *AB*, having a length equal to 10 tons, Figure 15.19*b*. Next draw a line through *b* parallel to the external force *BD* and also a line through point *a* parallel to the external force *DA*. Their intersection establishes point *d*, thus determining the three external forces. In Figure 15.19*b*, *bd* is shown as a dotted line since it is not a stress in a member of the truss but represents the resultant pressure under the members *DC* and *CB*. To find the stresses in *BC* and *CA*, draw a line through *b* parallel to *BC* and also a line through point *a* parallel to *CA*; their intersection determines point *c*. Connecting points *c* and *d* gives us the stress in the member *CD*. Note that this line must be parallel to member *CD* and if this condition does not exist an error has occurred at some point in constructing the stress diagram.

The magnitudes of the stresses are found by measuring the lengths of the lines in the stress diagram. Note that member *CD* has a greater length than member *AC*, whereas the stress in *CD*, shown in the stress diagram, is smaller than the stress in member *AC*. Again we see that the length of a member is not an indication of its stress; the magnitude of the stresses are represented by the lengths of their respective lines in the stress diagram. Note also that the pressure of the truss on the ground is approximately twice the magnitude of the "pull" exerted by member *DA*.

Sawtooth Truss

The sawtooth roof is used principally in industrial buildings, its purpose being to aid in the distribution of daylight over floor areas. The construction consists of a glass area in the steepest portion of the roof, preferably facing the north, the remainder being composed of a suitable roofing surface. In its erection, particular care must be taken to guard against water leakage and the possibility of condensation.

One type of truss supporting a sawtooth roof is shown in Figure 15.20a. As indicated, this truss is unsymmetrical as well as the loading. To construct the stress diagram, first draw the load line *ab, bc, cd, de, ef, fg*, and *gh*. The reactions *HI* and *IA* are unknown and must be determined in order to complete the polygon of external forces. These reactions may be computed by the principle of moments, or by the construction of a funicular polygon, the latter being shown in Figure 15.20. The pole *o* is selected, the rays are drawn and the funicular polygon constructed directly beneath the truss diagram. A line drawn from *o*, parallel to the closing string *oi*, determines the point *i* in the force polygon of external forces, thus establishing the reactions *HI* and *IA*. To draw the stress diagram we may begin at either joint *ABJI* or *GHIQ*. The stress diagram is developed as has been previously described and is shown in Figure 15.20b.

The characters of the stresses are purposely omitted in the truss diagram. Before analyzing these stresses, as explained in Art. 66, try to identify them by inspection, beginning with the compressive stress *PQ*.

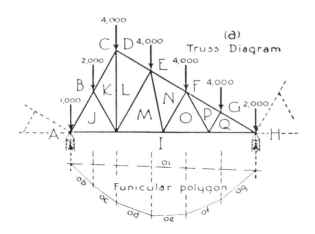

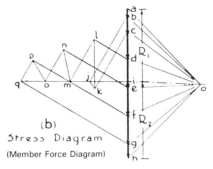

FIGURE 15.20 Graphical analysis of a saw-tooth truss.

Grandstand Truss

Another unsymmetrical truss is shown in Figure 15.21a. This is not a simple truss since the left-hand end overhangs R_1, the left reaction. The two reactions not being equal, it is necessary to determine their magnitudes in order to draw the force polygon of external forces. As the loads in this instance are placed at equal distances horizontally, their resultant, 20,000 lb, acts at $2\frac{1}{2}$ units of distance from R_2, the right reaction. Then, by the principle of moments,

$$3 \times R_1 = 20,000 \times 2.5$$
$$R_1 = 16,666\tfrac{2}{3} \text{ lb}$$

and

$$R_2 = 20,000 - 16,666\tfrac{2}{3} = 3,333\tfrac{1}{3} \text{ lb}$$

Figure 15.21 illustrates how readily the funicular polygon may be employed to obtain the same result. Note that the resultant of the loads is AG and that it acts at the midpoint between the ends of the truss. Considering then the three forces AG, GH, and HA, the funicular polygon is readily drawn, thus determining point h and consequently R_1 and R_2. Attention is called to the fact that pains must be taken to locate point h accurately, otherwise the stress diagram will not close. The complete stress diagram is shown in Figure 15.21b. Note the character of stresses in the chord members.

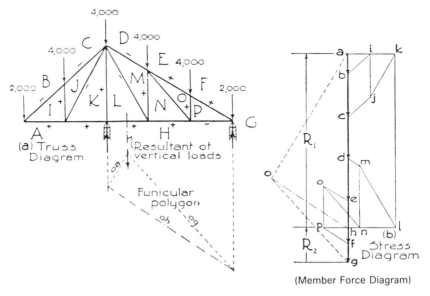

FIGURE 15.21 Graphical analysis of a grandstand truss with one end cantilevered.

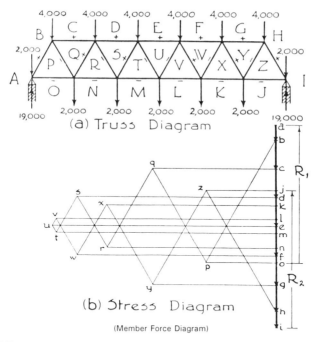

FIGURE 15.22 Graphical analysis of a Warren truss with suspended loads.

Warren Truss—Suspended Loads

Trusses with parallel chords are often used for flat roofs or for floor supports instead of plate girders. The advantage of the truss over the plate girder lies in the fact that the open spaces between web members permit the placing of pipes and ducts. For roofs, the upper chord may be inclined slightly, if desired, to provide for drainage. A depth of one-fifth or one-sixth of the span generally proves to be most economical.

A Warren truss is shown in Figure 15.22a. In addition to a load of 28,000 lb on the upper chord, there is a suspended load of 10,000 lb. As the truss is symmetrically loaded, each reaction is equal to one-half the total vertical load, or 19,000 lb. To draw the force polygon of the external forces, begin with *ab* and draw *bc, cd, de, ef, fg, gh*, and *hi*. The next force in order is *IJ*, an upward force of 19,000 lb; therefore measure upward 19,000 lb locating point *j*. This is followed by the downward forces *jk, kl, lm, mn*, and *no*. The remaining force is *OA*, 19,000 lb, thus completing the force polygon of the external forces. Beginning with joint *ABPO* or *HIJZ*, the stress diagram is readily drawn, there being no complications, Figure 15.22b.

Fan Truss—Suspended Load

A Fan truss having a load of 24,000 lb on the upper chord and a suspended load of 6,000 lb is shown in Figure 15.23a. The truss being unsymmetrically loaded, it is necessary to find the magnitudes of the reactions in order to draw the force polygon of the external forces. It is observed that the suspended load *IJ* has the same line of action as the force *CD*. Therefore, for the purpose of determining R_1 and R_2, we may assume *CD* to be 4,000 + 6,000 or 10,000 lb. Force and funicular polygons are con-

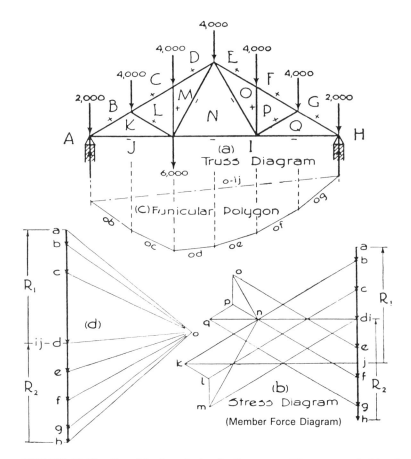

FIGURE 15.23 Graphical analysis of a fan truss with a suspended load.

structed as shown in Figures 15.23d and c, respectively, and the two reactions thus obtained. This force polygon cannot, of course, be used in drawing the stress diagram since CD and IJ act at different parts of the truss. Our objective thus far has been merely to find the magnitudes of R_1 and R_2.

If we had wished, the reactions might have been computed by the principle of moments. The loads are equally spaced horizontally and the resultant of the loads on the upper chord, 24,000 lb, acts through the center line of the truss. Then

$$6 \times R_1 = (6{,}000 \times 4) + (24{,}000 \times 3)$$

or

$$R_1 = 16{,}000 \text{ lb}$$

and

$$R_2 = (24{,}000 + 6{,}000) - 16{,}000 = 14{,}000 \text{ lb}$$

APPLICATIONS OF GRAPHICAL METHODS   355

The polygon of external forces may now be drawn beginning with the downward force *ab* and continuing with *bc*, *cd*, *de*, *ef*, *fg*, and *gh*, Figure 15.23*b*. From point *h* measure upward 14,000 lb, establishing point *i*, and next lay off *IJ* the downward force of 6,000 lb. The force *JA*, an upward force of 16,000 lb, completes the polygon. The stress diagram, as shown in Figure 15.23*b*, may be started at joint *ABKJ* or *GHIQ*. It presents no difficulties.

Fink Truss—Substitute Member

One of the most commonly used steel trusses is the Fink truss shown in Figure 15.24*a*. The truss diagram shows both vertical and wind loads, and stress diagrams for each will be constructed separately.

Consider first the vertical loads. The truss being symmetrically loaded, the reactions are each equal to one-half the total vertical load. The force polygon of external forces consists then of the loads *ab*, *bc*, *cd*, *de*, *ef*, *fg*, *gh*, *hi*, and *ij*, followed by the reactions *jk* and *ka*, Figure 15.24*b*.

To construct the stress diagram, begin with joint *ABLK* and follow successively with joints *BCML* and *LMNK*. We now find that joints *CDPONM* and *NORK* each

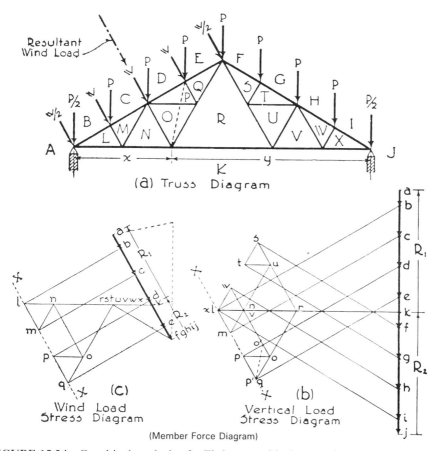

FIGURE 15.24 Graphical analysis of a Fink truss with the use of a substitute member.

contain more than two unknowns, hence we can proceed no further by the usual methods.

It is noted in the truss diagram that the horizontal member *OP* divides a quadrangle. For the time being, remove *OP* and substitute in its stead the other diagonal shown by the dotted line and call it *O'P'*. This substitution does not disturb the state of equilibrium of the truss. It enables us to draw the stress diagram for the forces about joint *CDO'NM* as there are now only two unknowns. Therefore, a line through *d* parallel to *DO'*, and a line through *n* parallel to *O'N*, determines point *o'*. As a result of the substitute member, *P'Q* becomes redundant and hence *p'* and *q* will occupy the same point. Hence a line through *e* parallel to member *E—QP'*, and a line through *o'* parallel to *QP'—O'* establishes the point *qp'*. Note particularly that we have now determined the stress in member *EQ*, for the point *q* has been established. The stress in this member has not been affected by the substitute member *P'O'*. We now remove the substitute member *P'O'* and in its place return the original horizontal member *PO*. By drawing a line through point *q* parallel to *QP*, and a line through *d* parallel to *PD* we find point *p*. Next, a line through *p* parallel to *PO*, and a line through *n* parallel to *ON* determines point *o*. By the usual methods point *r* is found, thus completing one-half of the stress diagram. The remaining portion of the diagram is drawn in a similar manner. The method of adding a substitute member is applicable regardless of the magnitudes of the loads, although in the problem just solved the truss is symmetrically loaded. This is the condition most frequently met with in practice.

Fink Truss—Another Solution

A simpler method of drawing the stress diagram for the Fink truss, *when the truss is symmetrically loaded*, is as follows: Draw the polygon of external forces and complete the stress diagram for joints *ABLK*, *BCML*, and *LMNK*. This establishes points *l*, *m*, and *n*, Figure 15.24*b*. For this type of Fink truss, symmetrically loaded and having equal divisions of the upper chord, each successive member of the upper chord, beginning with the member adjacent to the reaction, has the same rate of decrease in stress. By referring to the portion of the stress diagram just completed, note that stresses in *BL* and *CM* have been established and hence the rate of decrease in stress is determined. Members *DP* and *EQ* will have stresses decreased in the same degree. Therefore draw a line through points *l* and *m*, as *XX*. Points *p* and *q* will lie on this line at the intersections, respectively, of lines through *d* parallel to *DP* and through *e* parallel to *EQ*. The remaining portion of the stress diagram is readily completed.

Wind-Load Stress Diagram—Fixed Ends

As well as the vertical loads shown on the Fink truss in Figure 15.24*a*, loads for wind left are indicated. Assuming the truss to have fixed ends (no roller), let it be required to draw the stress diagram for the wind loads only. To draw the polygon of external forces, draw the wind loads *ab*, *bc*, *cd*, *de*, and *e—fghij*, Figure 15.24*c*. In drawing a stress diagram it is well to include all the letters shown on the truss diagram. In this instance, since there are no loads between letters *F*, *G*, *H*, *I*, and *J*, these letters will fall at one point. The load line having been drawn, the next two forces to consider are the reactions R_2 and R_1. It is observed in the truss diagram that the wind loads are symmetrical with respect to the left side of the truss, consequently the resultant of the wind loads will act at the midpoint of the upper chord as shown. If the loading had been unsymmetrical, we could have drawn a funicular polygon to find the line of action of the resultant. The extended resultant wind load divides the lower chord into the two segments x and y. Assuming the reactions to be parallel to the direction of the wind, we know that $R_1 : R_2 :: y : x$. Hence the load line is divided in the same ratio and R_1 and R_2 are determined, establishing point k.

The polygon of external forces having been completed, the stress diagram is drawn using either of the methods explained earlier. Attention is called to the fact that letters r, s, t, u, v, w, and x occupy the same point in the stress diagram. This indicates that no stresses occur in the web members for the right-hand side of the truss as a result of wind loads on the left. However, these members are stressed when the wind comes from the right, and under this condition zero stresses occur in the web members in the left-hand side of the truss.

Pratt Truss Stress Diagrams

Figure 15.25a represents a Pratt truss subjected to both vertical and wind loads. A roller occurs under the right-hand end of the truss.

The stress diagram for the vertical loads is shown at Figure 15.25b. Note that a roller at a support has no effect on the stresses for vertical loading, each reaction being vertical. The stress diagram presents no difficulties.

Let it be required to draw the stress diagram for wind left, a roller being at the right support. To draw the polygon of external forces, first draw the load line ab, bc, cd, de, and $e-fghij$, Figure 15.25b. Next come the reactions. Assume for the moment that no roller occurs at R_2 and that the reactions are parallel to the direction of the wind. Since the wind loads are symmetrical with respect to the upper chord on the left side of the truss, the resultant wind load is known to act through joint *CDONM* and divides the lower chord into segments x and y. By the method of proportion explained in Section 15.3, the load line is divided into two parts, R_1 and R_2, determining point k'. These are the reactions assuming them to be parallel. However, since a roller occurs at R_2, this reaction must be vertical. That is to say, the actual reaction at R_2 can be only the vertical component of the reaction found for the assumption that the reactions are parallel. Hence, erecting a vertical line from point *fghij* and drawing a horizontal line through point k', their intersection determines the true point k and consequently the actual vertical reaction at R_2. The reaction at the left side of the truss, *KA*, is found by joining the points k and a; thus the polygon of external forces is completed. The stress diagram is now readily drawn. Observe that, for wind left, no stresses occur in the web members at the right-hand side of the truss since letters r, s, t, u, v, w, x, and y fall at the same point.

For the wind right, roller right, the same methods may be employed. The polygon of external forces consists of the load line $abcde-f, fg, gh, hi$, and ij, Figure 15.25d. The point k' is found as for wind left, based on the assumption that the reactions are parallel. Since, however, there is a roller at R_2 the reaction cannot be jk' but will be the vertical component of jk' or jk. The stress diagram is completed without difficulty.

The force polygon of external forces might have been drawn by the following simple method. First draw the load line for wind left $abcde-f, fg, gh, hi$, and ij, Figure

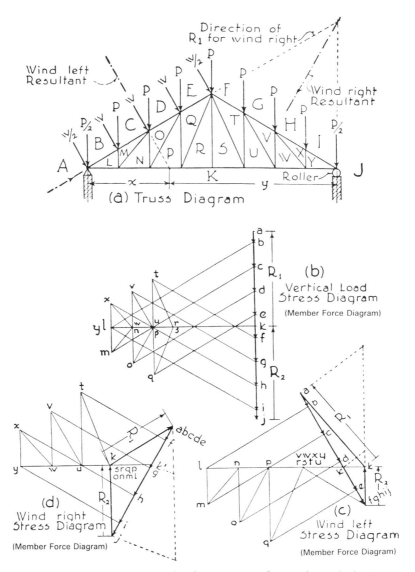

FIGURE 15.25 Graphical analysis of a Pratt truss for gravity and wind loads.

15.25d. Now, since the three external forces, the resultant wind load, R_1 and R_2, are not parallel and are in equilibrium they must have a point in common. Therefore, extend the lines of action of R_2, known to be vertical, and the resultant wind load, until they intersect. A line drawn from this intersection to the point of support at R_1 determines the *direction of R_1 or KA*. Hence in the force polygon, Figure 15.25d, erect a vertical line from *j*. This is the direction of *JK* and point *k* lies at some point on this line. Again draw a line through point *abcde* parallel to the direction, just found, of *KA*. Since *k* is on this line as well as on the vertical line through *j*, it is, of course, at their point of intersection.

By inspecting the truss diagram, Figure 15.25a, it is seen that the left reaction for wind right has the same line of action as the left-side upper chord. This fact is

APPLICATIONS OF GRAPHICAL METHODS   359

verified by the stress diagram and shows that the web members and lower chord on the left side of the truss receive no stresses from the wind loads on the right. The fact that the left reaction and the upper chord are parallel is a mere coincidence and occurs only for a truss of this pitch. To verify this, construct a force polygon for a truss having a $\frac{1}{3}$ pitch.

Cantilever Truss

A truss that is supported at one end only is known as a cantilever truss. Such trusses are used for roofs over shipping platforms or entrances to buildings. Figure 15.26*a* illustrates a truss of this type. It is presented here not because of its common use, but to illustrate the application of fundamental principles of graphic statics in its analysis.

On inspecting the truss diagram we observe that there are three external forces, the load on the truss, the pull from the wall at joint *LEF* and the compressive force on the wall *FA*. In this instance the truss is not symmetrically loaded and therefore it is necessary to find, by some means, the position of the resultant of the four downward loads. This is accomplished by drawing the load line *ab*, *bc*, *cd*, and *de*, Figure 15.26*b*, selecting a pole, drawing the rays, and finally, the funicular polygon. The rays *oa*

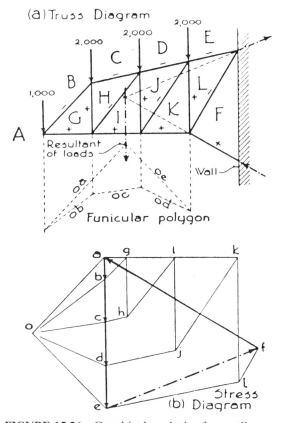

FIGURE 15.26 Graphical analysis of a cantilever truss.

and *oe* hold the resultant *ae* in equilibrium, hence their intersection in the funicular polygon determines a point through which the resultant must pass.

The position of the resultant might readily have been found by mathematics. Although in this truss diagram no dimensions are given showing the position of the loads, it is noted that they are spaced at equal horizontal distances. Then, writing an equation of moments with a point in the face of the wall as an axis,

$$(2{,}000 \times 1) + (2{,}000 \times 2) + (2{,}000 \times 3) + (1{,}000 \times 4) = 7{,}000 \times x$$

or

$$x = 2.28 \text{ spaces}$$

This indicates that the resultant of the loads lies at a point 2.28 spaces from the face of the wall.

The three external forces, being in equilibrium and not being parallel, must have a point in common. The equilibrant at the wall of the force *FA* must have the same line of action as the member *FA*. Therefore, extend the line of action of the member *FA* until it intersects the line of action of the resultant of the loads. A line drawn from this intersection to the joint *LEF* determines the direction of the force *EF*.

To draw the force polygon of external forces, from the point *e* in the load line, Figure 15.26*b* draw a line parallel to *EF* just found. Next draw a line through point *a* parallel to force *FA*. Their intersection establishes the point *f*, thus completing the force polygon of external forces, Figure 15.26*b*.

The stress diagram is begun by considering the forces about joint *ABG* and is completed without any complications.

Compound Fan Truss

A twelve-panel compound fan truss having a roller reaction is shown in Figure 5.27*a*. For the vertical loads, the reactions are vertical and the polygon of external forces is drawn as shown in Figure 15.27*b*. Beginning with the left support, the joints are taken in the following order, *AB10*, *BC21*, *CD32*, and *12340*. It is found that at joints *DE6543* and *4590* there are more than two unknown forces, consequently we cannot proceed in the usual manner. Instead of the members 56 and 67, two other members are substituted temporarily. These substitute members are 5'6' and 6'7' shown by the dotted lines. We may now proceed with the stress diagram by taking the joints in the following sequence, *DE5'43*, *EF6'5'*, and *FG87'6'*, indicating the substitute members as 5'6' and 6'7'. Note that when the substitute member 6'7' is assumed, a zero stress occurs in member 7'8, consequently points 7' and 8 occur at the same position in the stress diagram. In the stress diagram we now see that the stress in *G8* has been determined. This is the actual stress in this member regardless of whether or not the substitute members are introduced. *G8* having been established, we may now discard the substitute members 5'6' and 6'7' and replace the original members 56 and 67. The joints may now be taken in this sequence: *FG87*, *EF76*, *DE6543*, and *4590*. This accounts for all the letters and figures in the left side of the truss and the right side is completed in a similar manner, the stress diagram being symmetrical.

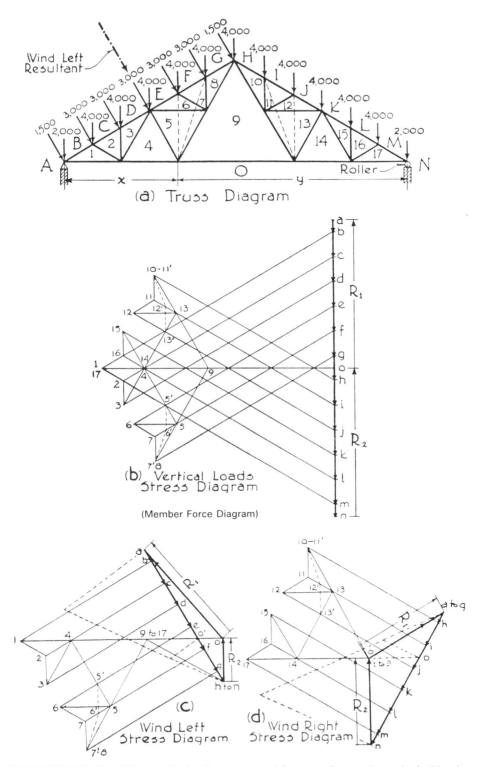

FIGURE 15.27 Graphical analysis of a compound fan truss for gravity and wind loads.

362 GRAPHICAL METHODS FOR INVESTIGATION OF FORCES

The stress diagram for wind left is shown in Figure 15.27c. By the method of the proportional line, explained in Art. 59, the point o' is found, based on the assumption that the reactions are parallel. Since, however, there is a roller at R_2, NO, the right reaction, can resist only the vertical component of the reaction just found. This vertical reaction is shown in Figure 15.27c, and a line drawn from point o to point a determines the left reaction, R_1. In the same manner as described for the vertical load stress diagram, the substitute members are introduced and the stress diagram is developed as has been explained.

The stress diagram for wind right is constructed similarly as shown in Fig. 15.27d.

Twelve-Panel Fink Truss

A number of trusses are solved by the method of adding substitute members as described in the previous article. Figure 15.28a shows a 12-panel Fink truss, The stress diagrams for vertical loads, wind left and wind right are developed by the same methods employed for the compound fan truss.

Crescent Truss—Suspended Loads

A crescent truss, sometimes known as the curved truss, is shown in Figure 15.29a. The total vertical load on the upper chord is 32,000 lb and the suspended load is 14,000 lb, making a total vertical load of 46,000 lb. These loads being uniformly distributed, each reaction is equal to 46,000 ÷ 2 or 23,000 lb. The force polygon for the vertical loads consists first of the downward forces on the upper chord, *ab*, *bc*, *cd*, *de*, *ef*, *fg*, *gh*, *hi*, and *ij*, Figure 15.29b. This is followed by *jk*, an upward force of 23,000 lb, the right reaction. Next come *kl*, *lm*, *mn*, *no*, *op*, *pq*, and *qr*, the suspended loads, and finally the upward force *ra*, 23,000 lb completes the polygon. The stress diagram may now be completed in the usual manner as shown in Figure 15.29b.

The polygon of external forces for wind left is shown separately in Figure 15.29c. There being four different pitches in the upper chord of the left half of the truss, the magnitudes of the wind loads on the four roof surfaces vary as well as their directions. The four forces shown are 4,000 lb, 2,800 lb, 1,800 lb, and 1,000 lb. One-half of each of these forces is transferred to their adjacent panel points. Our first objective is to find the resultant of these different wind loads. Of course we might draw a funicular polygon to find their resultant, but perhaps a simpler method is as follows. Beginning with the wind load *AB*, 2,000 lb, start the force polygon. Between spaces *B* and *C* there are two wind loads, 2,000 lb and 1,400 lb. These two forces are laid off as shown in Figure 15.29c. From space *C* to *D* there are the two forces, 1,400 lb and 900 lb; from *D* to *E* we have forces 900 lb and 500 lb; and from *E* to *F* there is the remaining wind load of 500 lb. There being no wind forces on the right side of the truss, the letters *f*, *g*, *h*, *i*, and *j* fall at one point. The load line for wind left therefore consists of *ab*, *bc*, *cd*, *de*, and *e-f* to *j*, and the resultant of the wind loads is the dotted line in the force polygon drain from point *a* to point *fghij*, Figure 15.29c. Thus we have determined the magnitude and direction of the resultant.

Referring to the truss diagram, Figure 15.29a, we find that the upper chord of the truss has the shape of an arc of a circle and that the center of the circle is designated. Since the wind loads on the different parts of the upper chord are normal to the chord, each separate wind load will pass through this center of the circle. It is a fun-

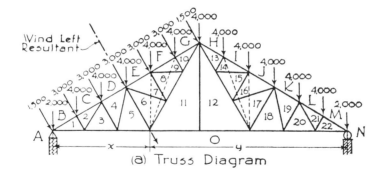
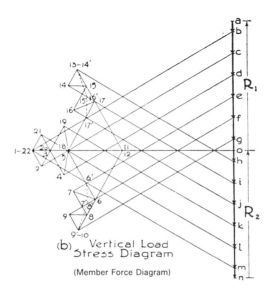
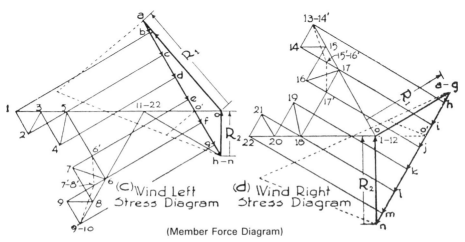

FIGURE 15.28 Graphical analysis of a 12-panel Fink truss for gravity and wind loads.

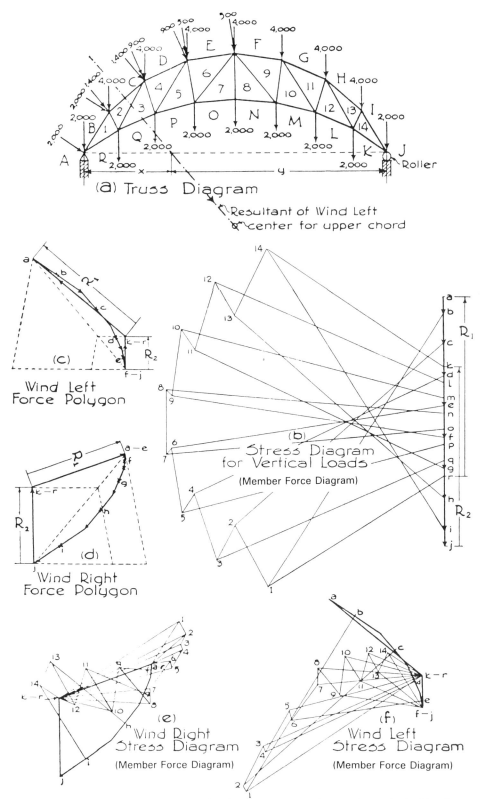

FIGURE 15.29 Graphical analysis of a crescent truss for gravity and wind loads.

damental principle that the resultant of a number of concurrent forces passes through their point of intersection. Therefore, through the center of the circle draw a line parallel to the resultant wind load found in the force polygon, Figure 15.29c. We have now determined the resultant in magnitude, direction and line of action. In drawing the polygon of external forces for wind loads, note that letters k, l, m, n, o, p, q, and r fall at the same point.

By the method described in Section 15.3, and R_1 and R_2 are found assuming the reactions to be parallel. There being a roller at R_2, this reaction must be vertical and the true reactions are consequently found as indicated and identified in Figure 15.29c. This completes the force polygon of external forces for wind left, and the stress diagram is drawn in the usual manner as shown in Fig. 15.29f.

The polygon of external forces for wind right is constructed in a similar manner, as shown in Figure 15.29d, the stress diagram being shown in Figure 15.29e.

Truss for Gambrel Roof

A truss having two different pitches on each half of the truss and having loads suspended from the lower chord is shown in Figure 15.30a. Let it be required to draw the various stress diagrams for this truss and loading, *assuming the horizontal components of the reactions due to the wind loads to be equal.*

A stress diagram for vertical loads is first drawn, Figure 15.30b. The total vertical load is 46,000 lb and since the truss is symmetrically loaded, each reaction, JK and PA, has a magnitude of 23,000 lb. Beginning with point a, the loads on the upper chord are drawn, ab, bc, etc., to point j. Next comes the upward force jk, 23,000 lb, followed by kl, lm, mn, no, and op, the suspended loads. The upward force pa, 23,000 lb, completes the polygon of external forces. In conjunction with this polygon just drawn, the stress diagram is developed, no difficulties being met with, Fig. 15.30b.

A separate force polygon for external forces for wind left is shown in Figure 15.30e. By inspecting the truss diagram we note that there are two wind loads for wind left, their magnitudes being 12,000 lb and 4,000 lb, each having different directions. These loads are apportioned to the adjacent panel joints. Beginning with point a, Figure 15.30e, draw the forces ab, bc, cd, de, and ef, the complete load line for wind left. In drawing this load line note that there are two forces between letters C and D in the truss diagram, so that actually the force CD is the resultant of the forces 3,000 lb and 1,000 lb. To account for all the letters shown on the truss, note that for wind left the letters f, g, h, i, and j fall at one point, and likewise the letters k, l, m, n, o, and p occupy the same position. The resultant of the wind loads in magnitude and direction is the line drawn from point a to point $fghij$. Since the resultant of two concurrent forces passes through their point of intersection, the line of action of the resultant of the wind left forces will pass through point x, see Figure 15.30a. Assuming, temporarily, the reactions to be parallel, R_1 and R_2 are found by the proportional line method as indicated in Figure 15.30e. Data for this problem, however, calls for the horizontal components of the reactions to be equal, therefore R_1 and R_2 are determined as explained in Section 15.3. Referring to Figure 15.30e, note that H_1 and H_2 are the equal horizontal components of R_1 and R_2, respectively. Using the force polygon just completed, a separate stress diagram for wind left is developed in Figure 15.30c.

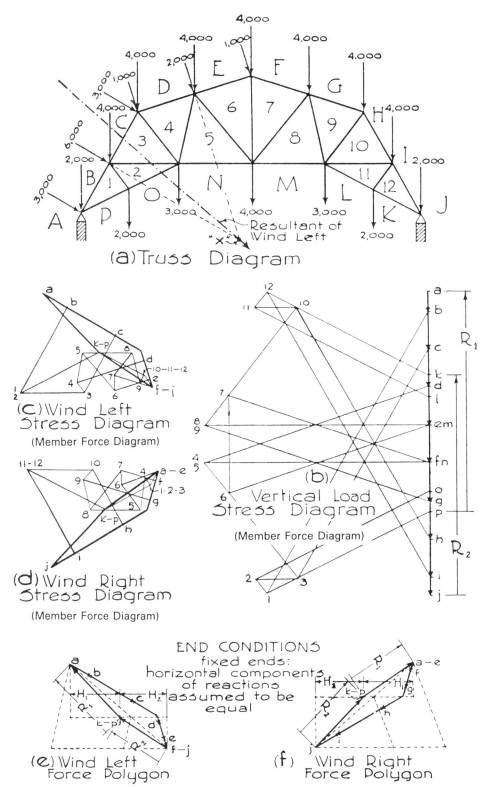

FIGURE 15.30 Graphical analysis of a cambered truss for a gambrel roof (compound slope surface).

In exactly the same manner, the force polygon of external forces for wind right and the complete stress diagram are constructed as shown in Figures 15.30f and d.

Grandstand Truss—Wind Loads

A type of truss used for grandstands, having both vertical and wind loads is shown in Figure 15.31a.

To construct the polygon of external forces for the vertical loads, the load line is first laid off, ab, bc, cd, etc., ending with ij, the total load being 32,000 lb, Figure 15.31b. The reactions due to these loads are, of course, vertical and may be found by means of the funicular polygon, the method shown in Figure 15.21, or by the principle of moments. Although no dimensions are given on the truss diagram, we observe that the vertical loads occur at equal horizontal distances. Since the loading is symmetrical, the resultant of the vertical loads, 32,000 lb, has its line of action at the center of the loading, four spaces from the left support or four spaces from the right end of the truss. Writing an equation of moments about the left support as an axis, we have

$$4 \times 32{,}000 = 5 \times R_2$$

or

$$R_2 = 25{,}600 \text{ lb}$$

Since

$$R_1 + R_2 = 32{,}000 \text{ lb}, R_1 + 25{,}600 = 32{,}000 \text{ lb}$$

or

$$R_1 = 6{,}400 \text{ lb}$$

From point j in the load line measure upward 25,600 lb, thus determining point k, and from k to point a we have 6,400 lb, thus completing the polygon of external forces. The stress diagram is completed without difficulty as shown in Figure 15.31b. Attention is called to the fact that no stress occurs in member KL due to vertical loads, consequently the letters k and l fall at the same point in the stress diagram. It will be seen, however, that this member is stressed when wind loads occur.

In a truss of this type, the column JK is able to resist only vertical forces; hence the horizontal components of the wind loads are resisted by KA, the left reaction. In Figure 15.31a it is seen that framing for the seats joins the left support at joint ALK. These beams or girders resist the horizontal thrust of either wind right or wind left. Consequently, in constructing the polygon of external forces for wind loads, we must assume that JK is vertical; the direction of KA remains to be determined.

Let us first consider wind left. Begin the polygon of external forces by drawing the load line ab, bc, cd, de, ef, and $f\text{-}ghij$, Figure 15.31c. Note that the letters g, h, i, and j fall at the same point for wind left. The three external forces are the resultant wind load, R_1 and R_2. These forces, being in equilibrium and not parallel, must meet in a point. As the wind loads are symmetrically placed on the left upper chord, it is unnecessary

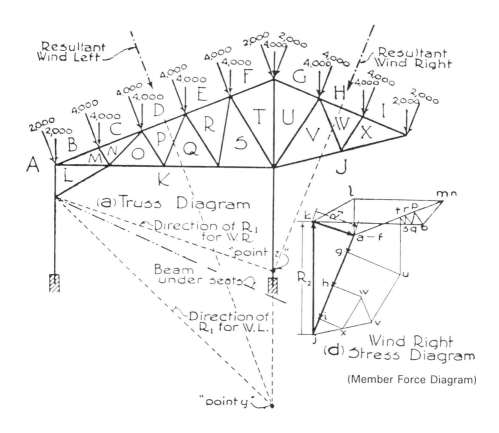
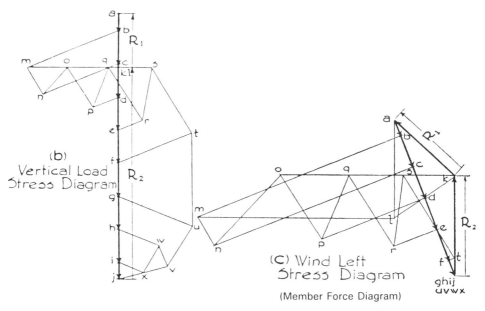

FIGURE 15.31 Graphical analysis of a grandstand trussed bent for gravity and wind loads.

to draw a funicular polygon to find the position of their resultant. We know its line of action is located at the center line of the upper chord. This line of action is extended to intersect the line of action of R_2 which we know to be vertical, the point of intersection being point y. A line drawn from this point to joint ALK determines the direction of R_1, the left reaction, for wind left. Returning to the load line, Figure 15.31c, draw a vertical line, the direction of R_2 through point $ghij$ and also a line through point a parallel to the direction of R_1 just found. Their intersection establishes point k and consequently R_1 and R_2 are established and the polygon of external forces is completed. The development of the stress diagram in conjunction with this polygon presents no difficulties. There being no wind loads on the right side of the truss, no stresses occur in members to the right of the right support and therefore the letters u, v, w, and x occur at the same point as g, h, i, and j.

The polygon of external forces for the wind right stress diagram is drawn in exactly the same manner. The resultant of the wind loads is extended to intersect the line of action of R_2, at point z, and a line from this point to joint ALK establishes the direction of R_1 for wind right, Figure 15.31d. In this diagram observe that $k-abcdef$, which is R_1, reads downward to the right, indicating that the beam under the seats is in tension due to wind coming from the right.

Truss with Monitor

There are several different arrangements of members for the truss with a monitor shown in Figure 15.32a. In the central portion of the truss shown, there are two quadrilateral spaces, each of which is divided by two diagonals. It will be seen on drawing the stress diagrams that these diagonal members are not subjected to stresses when the loading is vertical and uniformly distributed. When, however, wind loads occur, these members are stressed and the diagonal members to be considered are those which are in tension.

The polygon of external forces for vertical loads is readily drawn. LM and MA, the two reactions, are vertical and each is equal in magnitude to 16,000 lb, Figure 15.32c. The stress diagram is drawn in the usual manner, it being observed that for symmetrical vertical loads no stresses occur in the diagonals 89 or 67.

Let us now draw the polygon of external forces for wind left, assuming the horizontal components of the reactions to be equal. Beginning with point a draw ab, bc, cd, de, ef, and fg, Figure 15.32b. Observe that both DE and EF are actually the resultant of two wind loads having different directions. Having completed the load line, first assume the reactions LM and MA to be parallel to the direction of the resultant of the wind loads. The resultant of the wind loads is the line drawn from point a to point $ghijkl$. Therefore, draw the dotted lines through joints $KLM\ 15$ and $AB1M$ Figure 15.32a, parallel to this resultant. These lines represent the lines of action of the reactions based on the assumption that they are parallel to the direction of the wind-load resultant. To find the magnitude of the reactions with this assumption, we construct the funicular polygon. The pole o is selected. Figure 15.32b, the rays drawn and the funicular polygon constructed as shown in Figure 15.32a. In constructing the funicular polygon there are five forces in equilibrium, the wind load on the lower upper chord, the wind load on the vertical side of the monitor, the wind load on the upper part of the monitor and the two reactions. The closing string of the funicular polygon is the dot-and-dash line number 5. Then if we draw, parallel to the closing string, a line from the pole o to the resultant of the wind

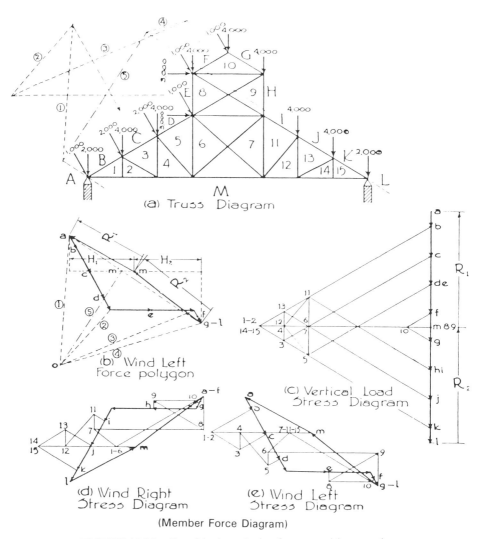

FIGURE 15.32 Graphical analysis of a truss with a monitor.

loads we establish point m', thus determining the reactions based on the assumption that they are parallel. By data, however, we are to consider the horizontal components of the reactions to be equal; hence, using the method explained in Section 15.3 and shown in Figure 15.11, the point m is determined and consequently the two reactions. This completes the polygon of external forces.

To avoid a confusion of lines, this polygon of external forces is redrawn in Figure 15.32e to be used in constructing the stress diagram. In the two quadrangular spaces in the central portion of the truss two diagonals are shown, whereas we know that one diagonal in each space is all that is necessary. For the wind left we will consider only the diagonals that receive tensile stresses. This can readily be determined by trial. First let us draw the stress diagram, assuming the diagonals 89 and 67 to be those shown having directions upward to the right, assuming the remaining pair to be omitted. The stress diagram is now drawn taking the joints in the following sequence, *AB1M, 12M, BC321, 234M, CD543, FG10, EF108, GH9810, DE8965*, etc. By using the method of determining the character of stress explained in Section 6.6, we find that the two diagonals 89 and 67 are each in tension; hence our assumption was correct.

The stress diagram for wind right is shown in Figure 15.32*d*. For wind right, the two diagonals whose directions are upward to the left are considered and these are found to be in tension. The use of two sets of diagonal members is for economy. If only one set were used they would be in tension for wind in one direction and compression for wind in the opposite direction, a reversal of stress. Since, for comparatively long members, less material is required for members in tension, both members are employed, each being designed to resist tension.

Transverse Bent

For certain commercial buildings a common type of construction is a series of trusses supported on columns, the trusses being braced to the columns by sloping members called *knee-braces*. The name given to such a unit is a *transverse bent* as illustrated in Figure 15.33*a*.

The truss shown is an eight-panel Fink truss, the knee braces being members *A3* and *A17*. For vertical loads, the reactions are equal and vertical and no stresses occur in the knee-braces, the stress diagram being constructed as usual and shown in Figure 15.24.

The usual construction is to secure the base of the column to the masonry foundation by means of anchor bolts. Such anchorage is not sufficiently rigid to produce a "fixed end," hence in designing the column it is assumed that no bending moment occurs at the base of the column. Therefore, for this form of connection, we consider the column to be hinged at the base. If, however, the column is deeply embedded in the masonry so as to produce a bending moment at this point when the truss is subjected to wind loads, it is safe to assume the inflection point in the column to occur at half the distance from the base to the knee-brace. For this condition we may assume the column to be hinged at the inflection point and ignore the wind loads acting below this point. Since, for practical considerations, it is quite difficult to construct a column with a base that is actually "fixed," it is generally assumed that the column is hinged at the base and this condition will be considered in the following solution.

In designing the columns for bending due to wind loads, the maximum bending moment occurs in the column on the leeward side of the structure at the point at which the knee-brace joins the column, its magnitude being the horizontal component of the reaction multiplied by the distance from the knee-brace to the base of the column.

The truss diagram dhown in Figure 15.33*a* is actually an incomplete frame if we assume the structure to be hinged at the joints. For this theoretical condition the frame is unstable. Therefore, to enable us to construct a stress diagram for the truss, we assume that the imaginary framing, shown by the dotted lines on the sides of the columns, is substituted. These members are *C1*, *D2*, and *12* on the left side of the truss and the corresponding members on the right side. The stresses in the columns, as found by the stress diagram, are not true stresses, for by construction we know the columns to be in bending. However, the stress diagram gives the true stresses in all the members of the truss including the knee-braces. In the stress diagram the sub-

372 GRAPHICAL METHODS FOR INVESTIGATION OF FORCES

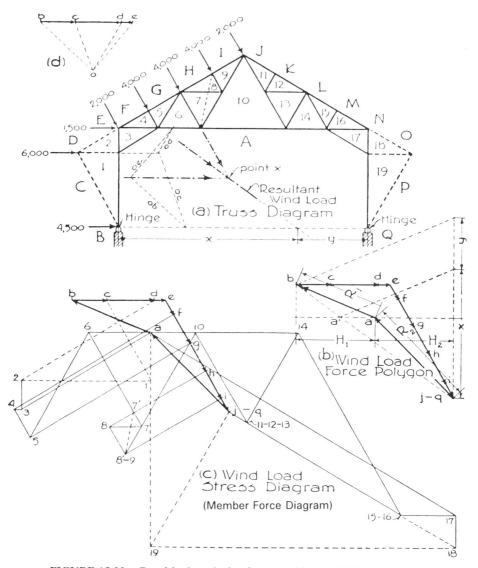

FIGURE 15.33 Graphical analysis of a trussed bent with knee-braces.

stitute framing members are shown by dotted lines and, of course, are ignored in designing the truss.

In the design of a building of this type, the wind load on the vertical sides of the structure must be considered as well as that on the roof. Referring to the truss diagram, Figure 15.33a, the wind load acting horizontally on the side of the building is 12,000 lb and the wind load on the roof is 16,000 lb. These loads are, of course, the loads to be resisted by each truss and are distributed to the panel joints as shown.

The line of action of the resultant of the wind load on the side of the building is found by the force polygon, Figure 15.33d, and its corresponding funicular polygon. The resultant wind load on the roof surface acts at the midpoint of the upper chord, having the same line of action as force GH. The intersection of these two resultants, point x, determines a point through which the resultant of all the wind loads must pass. To find the magnitude and direction of the resultant of all the wind loads, beginning with point b, draw the load line bc, cd, de, ef, gh, hi, and i-$jklmnopq$, Figure 15.33b. The resultant of the wind loads, therefore, is the line drawn from point b to point $jklmnopq$. A line parallel to this resultant is drawn through point x and divides the line drawn from the bases of the columns into the segments x and y. By first assuming the reactions to be parallel, the point a' is found as shown. Assuming the horizontal components of the reactions to be equal, as explained in Section 16.3, the point a is found, thus determining R_1 and R_2.

By use of the polygon of external forces just completed, a separate stress diagram for wind left is drawn as shown in Figure 15.33c. The substitute member $7'8'$ is employed as explained in Art. 76. The auxiliary members shown in the truss diagram have corresponding dotted lines in the stress diagram. Since we know the columns to be in bending, the stresses indicated in the stress diagram for the members $A1$ and $A19$ are, of course, not the true stresses.

Three-Hinged Arch—Vertical Loads

The three-hinged arch is used for long spans such as required by armories and auditoriums. One of the principal advantages in its use is that the crown is permitted, due to the hinges, to rise and fall with temperature variations. In the solution of problems relating to the three-hinged arch the critical problem is the determination of the reactions. Regardless of the loads on the arch, the reactions are not vertical. The horizontal components of the thrusts may be resisted by the masonry foundations or by a tie member connecting the two segments of the truss at the base hinges.

Basically, the three-hinged arch is composed of two separate trusses hinged at the crown and at their points of support. The forces transmitted from one segment of the arch to the other, at the intermediate hinge, must be equal and opposite in direction to comply with the laws of equilibrium.

A three-hinged arch, having the load AB on one segment only, is shown in Figure 15.34a. The three hinges are indicated at points x, y, and z. To find the reactions BC and CA first draw the load line ab, Figure 15.34c. The forces acting on the left segment are CA, AB, and the force transferred from the right segment at hinge y. The right segment is acted upon by two forces, the reaction at hinge z and the force transferred from the left segment at hinge y. Since there are but two forces acting on the right segment, to produce equilibrium these two forces must be equal, opposite in direction and have the same line of action. Therefore, a line drawn through hinge z and hinge y determines the direction of the two forces acting on the right segment and consequently the direction of the reaction BC. There are three external forces acting on the truss, the load AB and the two reactions BC and CA. Since they are in equilibrium and are not parallel they must meet in a common point. Extend the line of action of the right reaction BC until it intersects the force AB. From this point of intersection draw a line to the left support, hinge x, thus determining the direction of CA, the left reaction. In the force polygon, Figure 15.34c, draw from point b a line parallel to the reaction BC, and from point a draw a line parallel to the reaction CA. Their intersection establishes the point c and consequently R_1 and R_2, the two reactions, are now established.

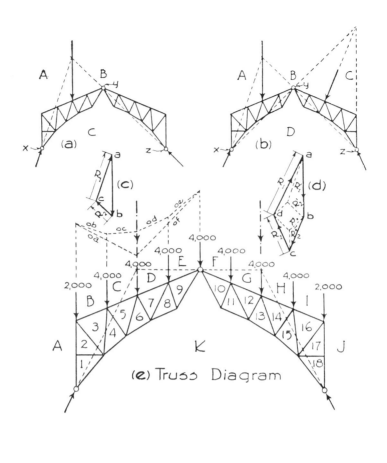
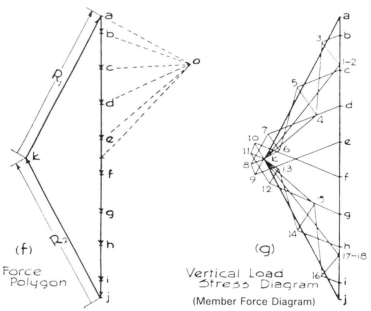

FIGURE 15.34 Graphical analysis of the three-hinged trussed structure for gravity loads.

A three-hinged arch with two unequal loads, AB and BC, on the left and right segments are shown in Figure 15.34b. Let it be required to find the reactions CD and DA.

This is readily accomplished by considering the loads AB and BC separately. The load line ab and bc is first drawn, Figure 15.34d. By the method employed in the previous example, the reactions resulting from the load AB, R_1', and R_2' are found. In the same manner the reactions resulting from the load BC are found. They are shown in Figure 15.34d as R_1'' and R_2''. At the left support we have then the reaction due to the load AB and also the reaction due to the load BC, namely, R' and R_1''. The resultant reaction from these two reactions is found by the triangle of forces and is the true reaction, R_1, due to the two loads AB and BC. Similarly the reaction at the right support, R_2, is determined as shown in Figure 15.34d. The force polygon of external forces then is ab, bc, cd, and da.

A three-hinged arch with vertical loads on the upper chord is shown in the truss diagram, Figure 15.34e. To draw the stress diagram we first draw the polygon of external forces. To accomplish this, first draw the load line ab, bc, cd, . . . ij, Figure 15.34f. As yet the directions of the reactions JK and KA are unknown. The forces acting on the left segment of the truss are three in number, the left reaction, the load on the left segment and the force exerted by the right segment at the hinge at the crown. The resultant of the loads on the left segment is found by the funicular polygon as shown, although this step is unnecessary in this instance since the loads are symmetrically placed and we know the resultant to have the same line of action as the force CD. At the hinge at joint EF 10 K 9, each segment exerts a force which holds the opposite segment in equilibrium. These two forces must be equal, opposite in direction and have the same line of action through the hinge. Since the loads on each segment of the truss are equal, the forces transferred from each segment of the truss must be equal and horizontal. Therefore, a horizontal line drawn through the hinge determines the line of action of these two forces in equilibrium. Considering again the three forces on the left segment, the reaction, the load on the left segment and the force exerted by the right segment, these three forces must meet in a common point since they are in equilibrium and are not parallel. Therefore, extend the horizontal line through the hinge at joint EF 10 K 9 until it meets the resultant of the loads on the left segment. From this point of intersection draw a line to the hinge at the left support, thus determining the direction of the left reaction, R_1. In the same manner determine the direction of the right reaction R_2. From point j in the load line, Figure 15.34f, draw a line parallel to the right reaction JK, and from point a draw a line parallel to the left reaction KA. Their point of intersection establishes the point k and thus the two reactions, JK and KA, are determined.

Using the polygon of external forces just found, a separate diagram, Figure 15.34g, is drawn showing the complete stress diagram for vertical loads. As has been seen, the principal problem in the solution of a three-hinged arch is the determination of the reactions. After this has been accomplished the stress diagram presents no difficulties.

Three-Hinged Arch—Wind Loads

Let it be required to draw the stress diagram of the three-hinged arch with the wind loads as shown in the truss diagram, Figure 15.35a. In order to draw the polygon of external forces, beginning with point a draw the load line ab, bc, cd, de, ef, fg, gh, and

376 GRAPHICAL METHODS FOR INVESTIGATION OF FORCES

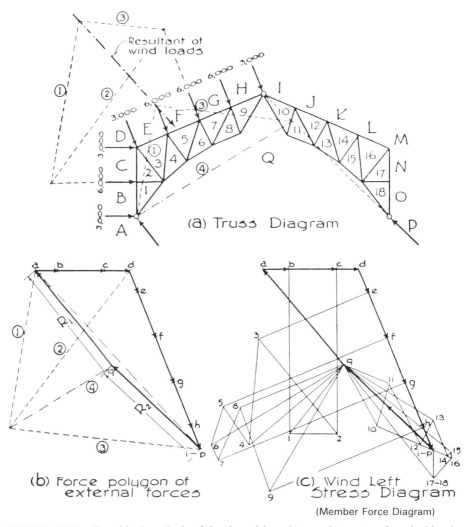

FIGURE 15.35 Graphical analysis of the three-hinged trussed structure for wind loads.

$h-ijklmnop$, Figure 15.35b. The resultant of the wind load, 12,000 lb, on the vertical portion of the truss has the same line of action as the force BC. Likewise the resultant of the wind loads on the sloping part of the upper chord, 24,000 lb, has the same line of action as the force FG. Select a pole and draw the rays numbered 1, 2, and 3. Since the rays 1 and 3 hold the resultant of all the wind loads in equilibrium, the intersection of the sides 1 and 3 of the funicular polygon, shown above the truss diagram, determines a point through which the resultant must pass. Its direction and magnitude are given in the force polygon, Figure 15.35b. Therefore, draw a line parallel to the resultant wind load, $a-ijklmnop$, through the point of intersection of the sides of the funicular polygon 1 and 3.

We may now consider the three external forces, R_1, R_2, and the resultant wind load. Of the left reaction, R_1, we know only a point in its line of action, that is, through the hinge at the left support. Of the resultant wind load we know its magnitude, direction and line of action. As for the right reaction, R_2, we know its line of action is parallel to a line drawn through the hinge at the right support and the hinge at the crown of the arch, as was explained previously. To construct a funicular polygon for those three forces, begin by drawing the string 1, parallel to the ray 1, through the hinge at the left support. From the point where this side of the funicular polygon intersects the resultant of the wind loads draw the string 3 extending it until it meets the line of action of the reaction R_2. The closing string of the funicular polygon is marked number 4, drawn from the intersection of string 3 and the line of action of R_2 to the hinge at the left support. In the force polygon, Figure 15.35b, draw from the pole a line parallel to the closing string 4. Finally, a line drawn from point *ijklmnop*, parallel to the line of action of R_2, *intersects the ray 4 at q*, thus establishing the right-hand reaction R_2. The line from point q to point a is the reaction R_1 and thereby completes the polygon of external forces.

To avoid a confusion of lines, the polygon just completed is redrawn in Figure 15.35c and the stress diagram is developed in the usual manner.

16

FINDING EFFICIENT FORMS FOR TRUSSES*

This chapter shows how to create forms for trusses that have various optimal properties.

16.1 THE NEED FOR EFFICIENT TRUSS FORMS

Among the most valuable skills of the architect or engineer is the ability to shape a structure in such a way that it has the desired structural and aesthetic properties. One might wish, for example, to create a truss in which no member carries more than a given amount of force, in order to be able to build the truss of material that does not exceed a certain size. More significantly, one might shape a truss so that the forces throughout the length of the top or bottom chord are constant, allowing the use of a single size of member from one support to the other. This constant-force truss design has several potential benefits: It can simplify joint design and fabrication. It generally results in a truss that uses a minimum amount of material when compared to other truss forms that carry the same pattern of loads. And it usually produces a truss that is graceful in appearance because its form is derived in direct response to the loads it supports in accordance with natural laws. Truss forms may be optimized by either numerical or graphical methods. The numerical method is demonstrated first.

16.2 A NUMERICAL METHOD FOR FINDING AN OPTIMAL TRUSS FORM

Drawing 1 in Figure 16.1 shows a loading pattern for a top-loaded truss. We would like to find a form for this truss such that the force in the top chord will be a constant

*This chapter was contributed by Edward Allen, who is an architect and educator and the author of *Fundamentals of Building Construction* (Ref. 10).

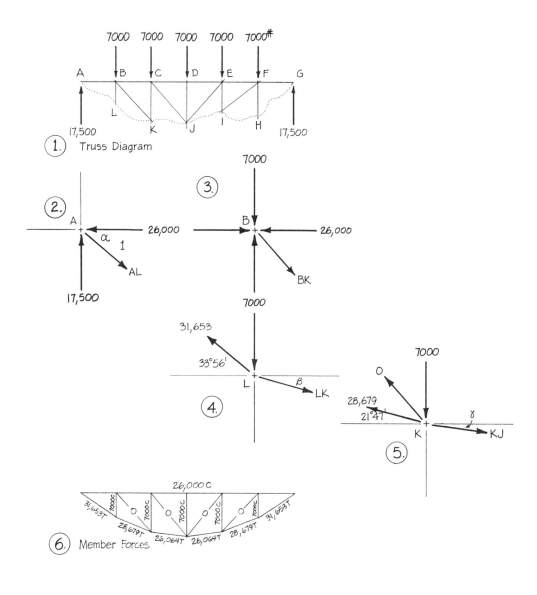

FIGURE 16.1

26,000 lb throughout its length, thus allowing use of a single piece of material of uniform size for this member. The top chord is to be level; the form of the lower chord is unknown, as indicated by the amorphous dotted line.

The numerical solution to this design problem begins at A, the left end joint of the truss, in drawing 2. Two of the three forces that converge on this joint are known. The force AL in the lower chord and the angle α between the upper and lower chords are unknowns. Two simultaneous equations may be written by summing the forces first in the vertical direction, then in the horizontal:

$$AL \sin\alpha = 17,500 \; lb$$
$$AL \cos\alpha = 26,000 \; lb$$

Dividing one equation by the other:

$$\frac{\sin\alpha}{\cos\alpha} = \frac{17,500 \; lb}{26,000 \; lb} = \tan\alpha$$

$$\alpha = 33.94° \quad (33°56')$$

$$AL = \frac{17,500 \; lb}{\sin 33.94°} = 31,653 \; lb$$

Moving next to joint B (drawing 3), three of the five forces are known: the 7000-lb load, and forces of 26,000 lb from the right and the left. The diagonal BK must be a zero-force member; otherwise, its horizontal component would upset the equilibrium between the identical forces in the top chord. Thus, by inspection, we can assign a force of 7000-lb compression to member BK, to balance the only other vertical force on this joint. In similar fashion we deduce that these same relationships must hold true for all the joints in the top chord.

In drawing 4 of Figure 16.1, we diagram the forces on the next joint, L, of the lower chord. The force in member LK and the angle β are unknowns. We again write two equations by summing forces in the vertical and horizontal directions, respectively:

$$7000 \; lb + LK\sin\beta - 31,653\sin 33.94° = 0$$
$$LK\cos\beta - 31,653\cos 33.94° = 0$$

Solving these equations, we find that the angle β is 22.12° (22°7'), and force LK is 28,679 lb. In drawing 5, we diagram the next joint, K, which we solve in similar fashion. By symmetry we now know all the forces and angles in the truss, although it is wise to continue the analysis to the righthand support as a check of accuracy.

Drawing 6 of Figure 16.1 is a summary diagram of the form and forces for the entire truss. Ideally, because the diagonals contain no forces, the truss could be built in the form shown in drawing 7, assuming that the external loading on the truss never varies. The lower chord follows the funicular line of a hanging chain that carries five equal loads at uniform intervals. The vertical struts serve only to carry the superimposed loads to the bottom chord. The top chord acts only to resist the horizontal pull of the bottom chord at each support. If properly sized, the top chord and the verticals are stressed to capacity throughout their length. If the bottom chord is made of a steel rod or cable of constant diameter, it is stressed to 82% of its capacity for a third of its length, 91% for another third, and 100% for the rest. There is little structural material in this truss that is not fully utilized. It is a highly efficient form.

Most trusses, of course, are subjected to loading patterns that vary over time: moving loads, in the case of bridges and floors, and changing combinations of wind and gravity loads, in the case of roofs. Diagonals usually must be added to the ideal form to respond to these conditions. In some cases, however, rigid truss joints or external stiffening devices can be used to deal with changing loads so as to avoid diagonals entirely.

Trusses of this shape have been used in such structures as Eduardo Torroja's Tordera Railway Bridge in Spain, built in the 1930s, and the W.E. Simpson Company's Alamodome roof in San Antonio, Texas (1992). In both these cases, diagonals were added to the ideal form to resist nonuniform loadings. Without diagonals, this form has been used to support column loads of multistory buildings over a column-free ground floor in Gunnar Birkerts's Federal Reserve Bank in Minneapolis (1973) and Datum Engineering's EDS Building in Plano, Texas (1992).

16.3 A GRAPHICAL METHOD FOR FINDING OPTIMUM FORM FOR A TRUSS

Figure 16.2 demonstrates how the preceding truss form may be found graphically rather than numerically. Bow's notation is adopted to facilitate the graphical solution. The load line is constructed in the usual manner. A line is drawn parallel to the load line at a distance of 26,000 lb to the left of it, working to the same scale to which the load line is drawn. The member force diagram (Maxwell diagram) is then constructed in steps 1 through 4 in such a way that all the lines that represent forces in the segments of the top chord end at this line. This determines the inclinations of the various segments of the lower chord, which are added one by one to the truss diagram to the left, generating the form of the truss. At each step, the construction of the lines on the member force diagram precedes and guides the construction of the form of the truss. All the diagonals are zero-force members. The forces in the other members are found by scaling the lengths of the corresponding line segments in the member force diagram.

The form and forces found by this graphical method are identical with those found by the numerical method. The main differences between the two methods are that the graphical method involves no numerical calculations, and it typically takes about one-third as much time as the numerical method to achieve the same result. The accuracy of the graphical method depends on the scale at which the constructions are made and the precision with which they are drafted. At almost any scale, the results are far more accurate than the degree of precision with which the external loadings can be predicted.

An expedient working method for finding an optimum form for a truss is to arrive at the desired form by graphical experimentation, then determine exact construction dimensions and angles by numerical analysis. Given the transparency and rapidity of the graphical method for finding the form, only graphical derivations are presented in the remainder of this chapter.

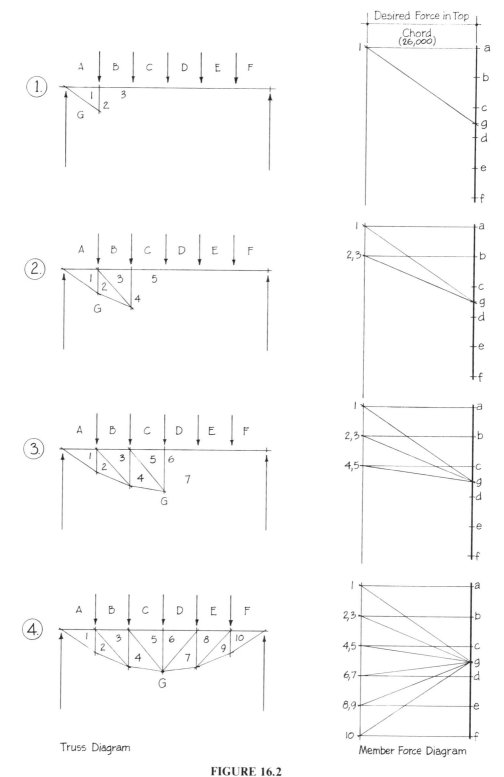

Truss Diagram   Member Force Diagram

FIGURE 16.2

16.4 FINDING A TRUSS FORM THAT HAS UNIFORM FORCE IN THE CURVING CHORD

Examination of the completed member force diagram in Figure 16.2 suggests that one might easily find a form for a truss in which the force is constant throughout the curving lower chord rather than the straight top chord. This condition allows the efficient use of a rod or cable for the bottom chord. This derivation is shown in Figure 16.3. The lines on the member force diagram (c) are constructed first for each pair of members, then the corresponding lines on the truss diagram (a). To begin, working to the same scale as the member force diagram, a compass is set to a radius equal to the desired force in the bottom chord, and an arc is swung about point g on the load line. The intersections of this arc with the horizontal lines that represent the forces in the upper chord segments are the locations of the numbered points in the member force diagram. Except at the centerline of the truss, all the internal mem-

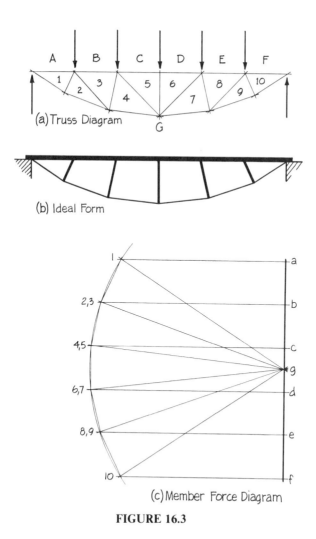

FIGURE 16.3

384 FINDING EFFICIENT FORMS FOR TRUSSES

bers are inclined. Half of them carry zero force under the ideal loading condition and can be eliminated to arrive at the ideal form (*b*).

This form was first discovered by the engineer George Pegram; he used it for the elegantly slender steel roof trusses of the train sheds in the St. Louis Union Station in 1894. Their span, 141 ft, was the longest of any roof truss in the world at that time. These sheds still stand, and impress the viewer with the seemingly impossible delicacy of their trusswork.

Dr. Waclaw Zalewski has developed a similar graphical construction that can be used to derive a truss form that contains true vertical members while still retaining a constant force throughout the length of the bottom chord (Figure 16.4). In this form, the diagonals carry small forces and cannot be eliminated. As in all the examples in this chapter, the construction of the member force diagram leads that of the truss diagram at each step.

In derivations such as these, where a constant force is desired throughout the length of the curving chord of a truss, the graphical method is easy and straightforward, while the numerical method is exceedingly complex.

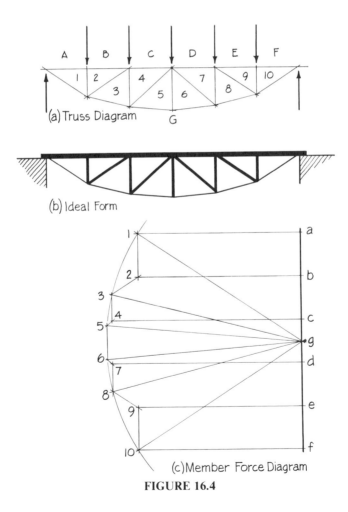

FIGURE 16.4

16.5 A BOTTOM-LOADED CONSTANT FORCE TRUSS

If a similar set of loads is applied along the bottom chord of a truss that has a constant-force top chord, an interesting form emerges (Figure 16.5). All internal members carry zero force, and the ideal form of the resulting truss (*b*) is that of a funicular suspension cable with a compression spreader to resist its horizontal pull. The member force diagram is identical to the diagram that one would use to find the form and forces for a hanging cable. The truss is its own moment diagram: The numerical moment at any point is equal to the height of the truss at that point, measured in feet or inches, times the perpendicular pole distance *g*-1, measured in pounds or kips at the scale of the load line.

Figure 16.6 shows the inversion of this form, which produces an elegantly simple truss whose ideal form (*b*) is that of a tied funicular arch. The member force diagrams for this and the preceding example are identical, although their letters correspond to letters that are differently located on the respective truss diagrams.

A further exploration of this line of investigation is shown in Figure 16.7. The

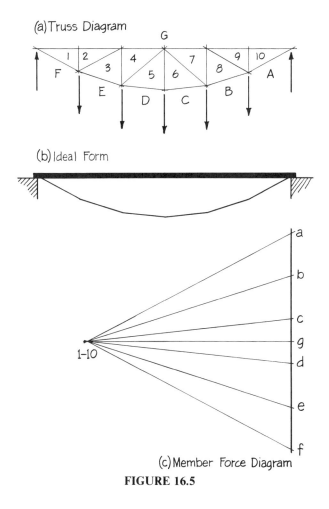

FIGURE 16.5

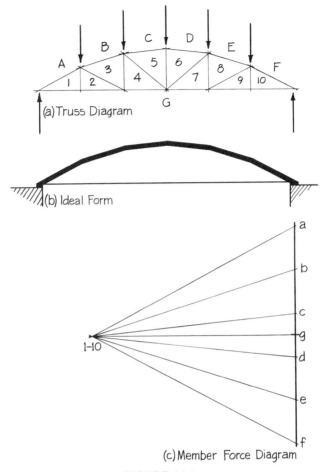

FIGURE 16.6

truss is bottom loaded, but is constructed to have a straight bottom chord with constant force throughout. The resulting form is that of a tied arch with slender tensile verticals to transmit the applied loads to the arch.

16.6 A LENTICULAR TRUSS

A lenticular truss has the unique property that within each panel the forces in the top and bottom chords have the same absolute value, but are opposite in sign. The shape is generated from the member force diagram (c) in Figure 16.8. In each panel of the truss, the inclinations of the top and bottom chord segments are made equal as shown. Lenticular trusses were widely used in the nineteenth century for road and railway bridges, and have been proposed more recently as an efficient means for supporting tension hangers that carry multiple floors from the top of a high-rise building.

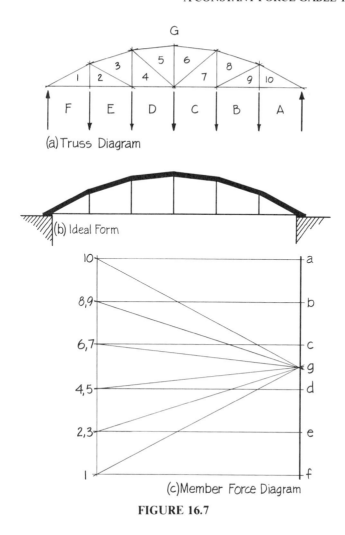

FIGURE 16.7

16.7 A CONSTANT FORCE GABLE TRUSS

Gable trusses may be optimized in the same general manner as flat trusses. Figure 16.9 shows the development of a gable truss in which the force is constant throughout the top chords. The inclination of the top chords is given. A vertical line is constructed to the left of the load line at a distance that yields the desired force in the top chords. This distance is measured along lines $a1$, $b3$, $c5$, $d6$, $e8$, and $f10$, whose inclinations are the same as those of the top chords. The form of the lower chord is determined by the lines that radiate from g on the member force diagram (c). Concrete roof trusses of this form, without diagonals, were used by Robert Maillart in his famous railway warehouse at Chiasso, Italy, designed in 1924.

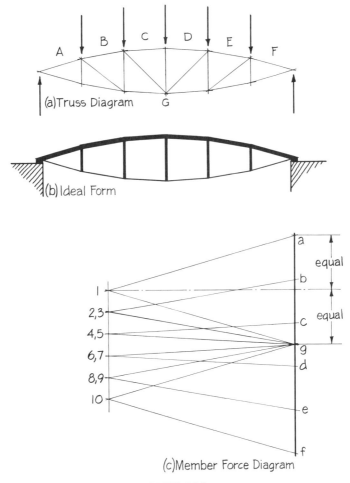

FIGURE 16.8

16.8 CANTILEVER TRUSSES OF OPTIMAL FORM

Figure 16.10 illustrates the development of the form of a cantilever truss that carries five identical top loads while maintaining a constant force in its level bottom chord. As with the other trusses in this chapter, the form of the truss is derived from the form of the member force diagram. The verticals and diagonals in this truss carry no forces under the ideal loading condition. The ideal form of the truss is that of a funicular cable held away from the wall with a spreader that has a constant compressive force throughout.

The truss in Figure 16.11 is the cantilevered equivalent of a lenticular truss. Its form is generated by a member force diagram that has been constructed in such a way that the top and bottom chord forces are identical in each panel. If the page is rotated 90° to the left, the resulting form will be seen to resemble the Eiffel Tower, a structure that was designed using graphical methods. This form of tower is very efficient in resisting wind forces with a minimum of structural material.

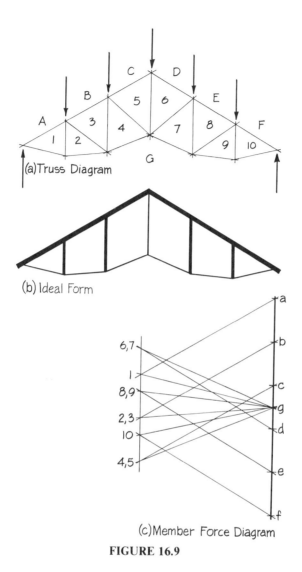

FIGURE 16.9

16.9 OPTIMAL TRUSS FORMS: THE GENERAL SOLUTION

All the examples given to this point have been level, symmetrical trusses with symmetrical gravity loadings. The method that has been used, however, is general, and may be applied to inclined and irregular loading and support conditions. Figure 16.12 shows an inclined truss with varying loads, one of which is nonvertical, applied at irregular intervals. The member force diagram has been manipulated to produce a truss form that has constant force throughout its top chord. Because of the nonvertical load, the fixed end reaction is inclined, and the load line is a quadrilateral. All the lines that represent the forces in the segments of the top chord on the member force diagram have been constructed to the same length, which produces an inclined internal member in the ideal truss.

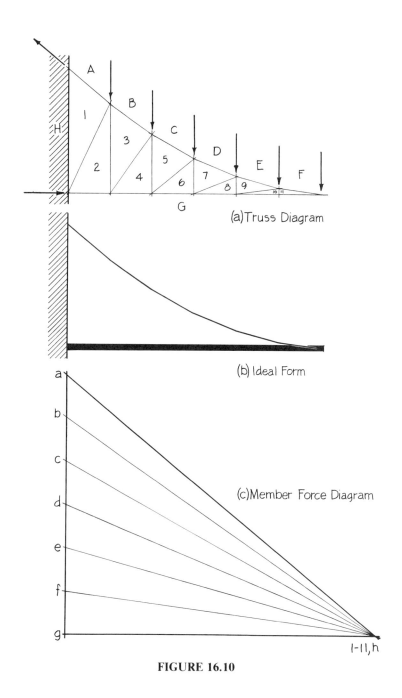

FIGURE 16.10

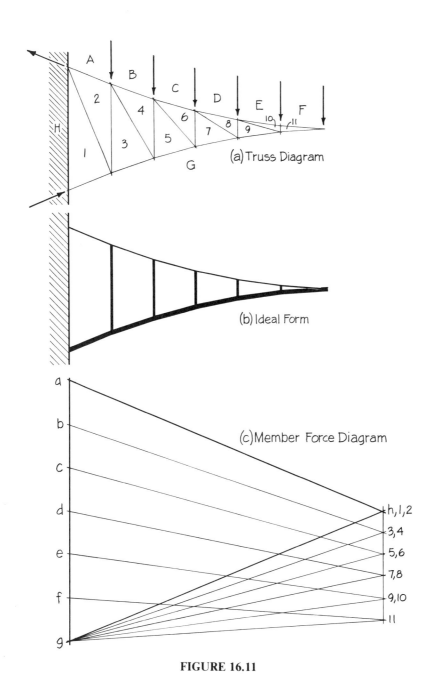

FIGURE 16.11

392 FINDING EFFICIENT FORMS FOR TRUSSES

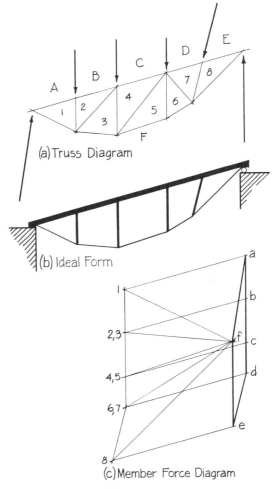

FIGURE 16.12

16.10 SUBOPTIMAL TRUSS FORMS: THE CAMELBACK TRUSS

In many situations, an external constraint is placed upon the form of a truss. Such is the case with a bridge truss whose portals at each end must be tall enough to clear the vehicles on the roadway or track that passes through the bridge. Figure 16.13 demonstrates the derivation of a so-called "camelback" form of truss, which is frequently seen in road and railway bridges. The heights of members 1-2 and 9-10 are fixed by vehicle clearance requirements, thus determining the inclinations of top chord segments G-1 and G-10, and the inclinations of g-1 and g-10 in the member force diagram. For the remainder of the truss, the construction of the member force diagram leads the construction of the truss diagram at each step. The four center segments of the top chord are configured so that they carry equal forces. This allows them to be the same size, which simplifies fabrication and increases overall structural efficiency.

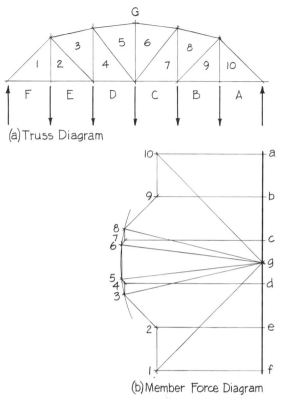

(a) Truss Diagram

(b) Member Force Diagram

FIGURE 16.13

16.11 VISUALIZING FORM IMPROVEMENTS IN TRUSSES OF NONOPTIMAL SHAPE

An important advantage of the graphical method of truss analysis over numerical or computer analysis during early stages of design is that the member force diagram furnishes ample, easily discernable clues as to how the form of an arbitrarily shaped truss might be improved. Consider the scissors truss (1) in Figure 16.14. The scissors form is understandably popular because it supports a gable roof while maintaining a lofty, soaring interior space. But a glance at its accompanying member force diagram reveals that the maximum member forces in this truss are roughly twice as high as the total external load. If this is a welded steel truss these high forces might be of comparatively little consequence, but if the truss is to be made of wood, it is likely that some of the connections will be impossible to make. In examining the member force diagram further, we find that the member forces are so large because the lines in the member force diagram that represent the forces in the ends of the top and bottom chords (*a*-1 and *g*-1, for example) meet at so sharp an angle. If these two lines met at a less acute angle, their lengths would be considerably shorter.

Acting on this clue, we might flatten the bottom chord to increase the angle between the chords, producing a triangular truss (2). Even without measuring, we can see by comparing the two member force diagrams that this strategy has almost

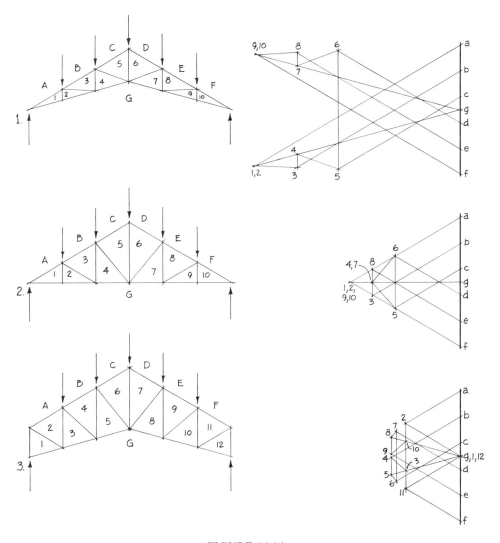

FIGURE 16.14

halved the maximum chord forces. This is good, but the lofty, soaring quality of the enclosed space has been lost. How can we get it back, without returning to a truss form that has excessive member forces?

Realizing that the maximum chord forces occur at the ends of the truss, we could increase still further the angle between the top and bottom chords at the very ends, but pull the bottom chord back up in the middle panels so as to give a more pleasing shape to the enclosed space (3). Comparing member force diagrams once again, we perceive instantly, even without measuring, that the forces in this truss are by far the lowest of any of our three designs. Perhaps this third design contains too much interior volume to be economical or desirable-no matter, we can try other alternatives that retain the basic form of the third truss but decrease its overall height somewhat, because we are now in a position to guess that their member forces will probably not exceed those of the triangular truss option.

This type of experimentation lies at the heart of the structural design process. It is encouraged by the rapidity of the graphical method of analysis, and by the ease with which generalizations may be made about the overall performance of each truss simply by noting the relative compactness of the member force diagram. Numerical analyses tend to take longer, and the bare numbers that they yield are of little help in figuring out how to improve the form of the truss.

APPENDIX A

COEFFICIENTS FOR MEMBER FORCES IN SIMPLE TRUSSES

This section provides quick answers for the internal forces in the members of trusses of common form. The trusses included are displayed in Figure A.1 together with the notation for reference to the table data in Fig. A.2.

For the gable-form trusses, coefficients are given for three different slopes of the top chord: 4 in 12 (18.4°), 6 in 12 (26.6°), and 8 in 12 (33.7°). For the parallel-chorded trusses, coefficients are given for two different ratios of the truss depth to the truss panel module length: 1 to 1 and 3 to 4.

Loadings result from symmetrically placed gravity loads, which are applied to the top chord joints. For the table values, the unit joint load of W in the figure is taken as 1.0. Member force values for other magnitudes of loading may thus be simply multiplied by the actual unit load as determined.

Note that owing to the symmetry of the trusses and the loads, the member forces are the same in each half of the truss. Therefore, we have given the coefficients only for the left half of each truss.

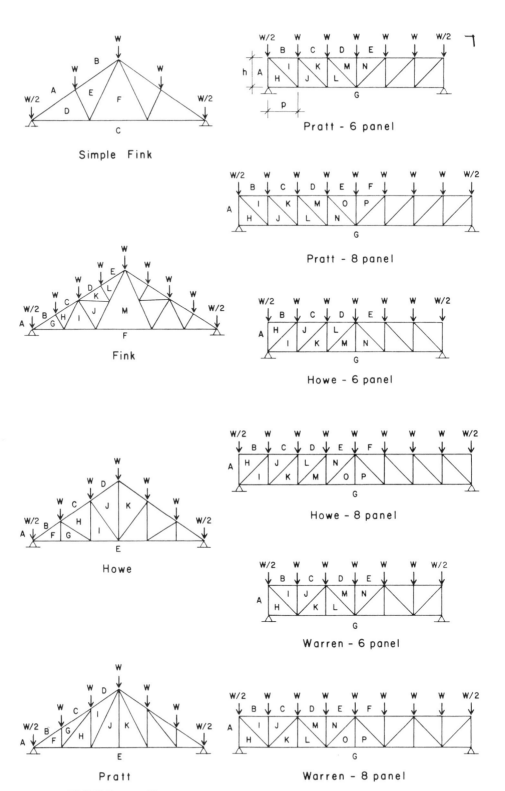

FIGURE A.1 Simple trusses of parallel-chorded and gabled form.

Force in members = (table coefficient) X (panel load, W)

T indicates tension, C indicates compression

Gable Form Trusses					Flat - Chorded Trusses					
Truss Member	Type of Force	Roof Slope			Truss Member	Type of Force	6 Panel Truss		8 Panel Truss	
		4/12	6/12	8/12			$\frac{h}{p}=1$	$\frac{h}{p}=\frac{3}{4}$	$\frac{h}{p}=1$	$\frac{h}{p}=\frac{3}{4}$
Truss 1 - Simple Fink					Truss 5 - Pratt					
					BI	C	2.50	3.33	3.50	4.67
AD	C	4.74	3.35	2.70	CK	C	4.00	5.33	6.00	8.00
BE	C	3.95	2.80	2.26	DM	C	4.50	6.00	7.50	10.00
DC	T	4.50	3.00	2.25	EO	C	—	—	8.00	10.67
FC	T	3.00	2.00	1.50	GH	O	0	0	0	0
DE	C	1.06	0.90	0.84	GJ	T	2.50	3.33	3.50	4.67
EF	T	1.06	0.90	0.84	GL	T	4.00	5.33	6.00	8.00
					GN	T	—	—	7.50	10.00
Truss 2 - Fink					AH	C	3.00	3.00	4.00	4.00
					IJ	C	2.50	2.50	3.50	3.50
BG	C	11.08	7.83	6.31	KL	C	1.50	1.50	2.50	2.50
CH	C	10.76	7.38	5.76	MN	C	1.00	1.00	1.50	1.50
DK	C	10.44	6.93	5.20	OP	C	—	—	1.00	1.00
EL	C	10.12	6.48	4.65	HI	T	3.53	4.17	4.95	5.83
FG	T	10.50	7.00	5.25	JK	T	2.12	2.50	3.54	4.17
FI	T	9.00	6.00	4.50	LM	T	0.71	0.83	2.12	2.50
FM	T	6.00	4.00	3.00	NO	T	—	—	0.71	0.83
GH	C	0.95	0.89	0.83	Truss 6 - Howe					
HI	T	1.50	1.00	0.75	BH	O	0	0	0	0
IJ	C	1.90	1.79	1.66	CJ	C	2.50	3.33	3.50	4.67
JK	T	1.50	1.00	0.75	DL	C	4.00	5.33	6.00	8.00
KL	C	0.95	0.89	0.83	EN	C	—	—	7.50	10.00
JM	T	3.00	2.00	1.50	GI	T	2.50	3.33	3.50	4.67
LM	T	4.50	3.00	2.25	GK	T	4.00	5.33	6.00	8.00
					GM	T	4.50	6.00	7.50	10.00
Truss 3 - Howe					GO	T	—	—	8.00	10.67
BF	C	7.90	5.59	4.51	AH	C	0.50	0.50	0.50	0.50
CH	C	6.32	4.50	3.61	IJ	T	1.50	1.50	2.50	2.50
DJ	C	4.75	3.35	2.70	KL	T	0.50	0.50	1.50	1.50
EF	T	7.50	5.00	3.75	MN	T	0	0	0.50	0.50
EI	T	6.00	4.00	3.00	OP	O	—	—	0	0
GH	C	1.58	1.12	0.90	HI	C	3.53	4.17	4.95	5.83
HI	T	0.50	0.50	0.50	JK	C	2.12	2.50	3.54	4.17
IJ	C	1.81	1.41	1.25	LM	C	0.71	0.83	2.12	2.50
JK	T	2.00	2.00	2.00	NO	C	—	—	0.71	0.83
Truss 4 - Pratt					Truss 7 - Warren					
					BI	C	2.50	3.33	3.50	4.67
BF	C	7.90	5.59	4.51	DM	C	4.50	6.00	7.50	10.00
CG	C	7.90	5.59	4.51	GH	O	0	0	0	0
DI	C	6.32	4.50	3.61	GK	T	4.00	5.33	6.00	8.00
EF	T	7.50	5.00	3.75	GO	T	—	—	8.00	10.67
EH	T	6.00	4.00	3.00	AH	C	3.00	3.00	4.00	4.00
EJ	T	4.50	3.00	2.25	IJ	C	1.00	1.00	1.00	1.00
FG	C	1.00	1.00	1.00	KL	O	0	0	0	0
GH	T	1.81	1.41	1.25	MN	C	1.00	1.00	1.00	1.00
HI	C	1.50	1.50	1.50	OP	O	—	—	0	0
IJ	T	2.12	1.80	1.68	HI	T	3.53	4.17	4.95	5.83
					JK	C	2.12	2.50	3.54	4.17
					LM	T	0.71	0.83	2.12	2.50
					NO	C	—	—	0.71	0.83

FIGURE A.2 Coefficients for member forces in simple trusses.

APPENDIX B

PROPERTIES OF SECTIONS

This section provides properties for the plane cross sections of structural members. Table B.1 provides data for cross sections of simple geometric form. Tables B.2 through B.10 provide data for selected common shapes of steel. Data for the steel shapes is abridged from more extensive tables in the *AISC Manual* (Ref. 2) with permission of the publishers, American Institute for Steel Construction. Table B.11 provides data for cross sections of structural lumber.

TABLE B.1 Properties of Common Geometric Shapes

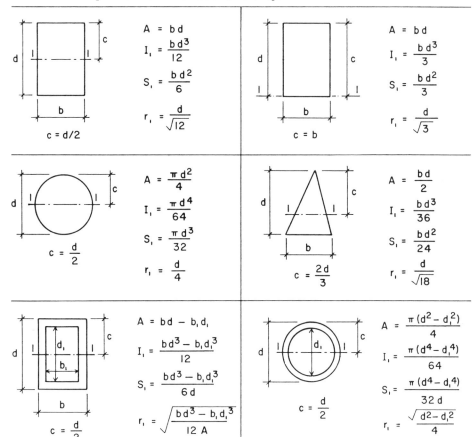

A = Area I = Moment of inertia S = Section modulus = $\frac{I}{c}$ r = Radius of gyration = $\sqrt{\frac{I}{A}}$

TABLE B.2 Properties of Standard Steel Elements: Round Rods

Diameter (in.)	Gross Area (in.^{2})	Net Area at Threaded End (in.^{2})	Radius of Gyration (r) (in.)	Weight per ft (lb)
0.25	0.049	0.032	0.06	0.167
0.375	0.110	0.078	0.09	0.376
0.500	0.196	0.142	0.12	0.668
0.625	0.307	0.226	0.16	1.044
0.750	0.442	0.334	0.19	1.503
0.875	0.601	0.462	0.22	2.046
1.000	0.785	0.606	0.25	2.673
1.125	0.994	0.763	0.28	3.382
1.250	1.227	0.969	0.31	4.176
1.375	1.485	1.16	0.34	5.053
1.500	1.767	1.41	0.37	6.013
1.750	2.405	1.90	0.44	8.185
2.000	3.142	2.50	0.50	10.690
2.250	3.976	3.25	0.56	13.530
2.500	4.909	4.00	0.62	16.703
2.750	5.940	4.93	0.69	20.211
3.000	7.069	5.97	0.75	24.053

TABLE B.3 Properties of Standard Steel Elements: Round Pipe—Standard Weight

Nominal Diameter (in.)	Outside Diameter (in.)	Wall Thickness (in.)	Section Properties				Weight per ft (lb)
			A (in.^{2})	I (in.^{4})	S (in.^{3})	r (in.)	
1.5	1.900	0.145	0.799	0.310	0.326	0.623	2.72
2.0	2.375	0.154	1.07	0.666	0.561	0.787	3.65
2.5	2.875	0.203	1.70	1.53	1.06	0.947	5.79
3.0	3.500	0.216	2.23	3.02	1.72	1.16	7.58
3.5	4.000	0.226	2.68	4.79	2.39	1.34	9.11
4.0	4.500	0.237	3.17	7.23	3.21	1.51	10.79
5.0	5.563	0.258	4.30	15.2	5.45	1.88	14.62
6.0	6.625	0.280	5.58	28.1	8.50	2.25	18.97
8.0	8.625	0.322	8.40	72.5	16.8	2.94	28.55

TABLE B.4 Properties of Standard Steel Elements: Square Structural Tubing

Nominal Size (in.)	Wall Thickness (in.)	Section Properties				Weight per ft (lb)
		A (in.^{2})	I (in.^{4})	S (in.^{3})	r (in.)	
2 × 2	0.1875	1.27	0.668	0.668	0.726	4.32
	0.2500	1.59	0.766	0.766	0.694	5.41
2.5 × 2.5	0.1875	1.64	1.42	1.14	0.930	5.59
	0.2500	2.09	1.69	1.35	0.899	7.11
3 × 3	0.1875	2.02	2.60	1.73	1.13	6.87
	0.2500	2.59	3.16	2.10	1.10	8.81
	0.3125	3.11	3.58	2.39	1.07	10.58
3.5 × 3.5	0.2500	3.09	5.29	3.02	1.31	10.51
	0.3125	3.73	6.09	3.48	1.28	12.70
4 × 4	0.2500	3.59	8.22	4.11	1.51	12.21
	0.3125	4.36	9.58	4.79	1.48	14.83
	0.3750	5.08	10.7	5.35	1.45	17.27
5 × 5	0.2500	4.59	16.9	6.78	1.92	15.62
	0.3125	5.61	20.1	8.02	1.89	19.08
	0.3750	6.58	22.8	9.11	1.86	22.37
6 × 6	0.2500	5.59	30.3	10.1	2.33	19.02
	0.3125	6.86	36.3	12.1	2.30	23.34
	0.3750	8.08	41.6	13.9	2.27	27.48
	0.5000	10.4	50.5	16.8	2.21	35.24
8 × 8	0.2500	7.59	75.1	18.8	3.15	25.82
	0.3125	9.36	90.9	22.7	3.12	31.84
	0.3750	11.1	106	26.4	3.09	37.69
	0.5000	14.4	131	32.9	3.03	48.85

TABLE B.5 Properties of Standard Steel Elements: Rectangular Structural Tubing

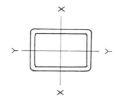

Nominal Size (in.)	Wall Thickness (in.)	Area (in.²)	X-X Axis			Y-Y Axis			Weight per ft (lb)
			I (in.⁴)	S (in.³)	r (in.)	I (in.⁴)	S (in.³)	r (in.)	
3 × 2	0.1875	1.64	1.86	1.24	1.06	0.977	0.977	0.771	5.59
	0.2500	2.09	2.21	1.47	1.03	1.15	1.15	0.742	7.11
4 × 2	0.1875	2.02	3.87	1.93	1.38	1.29	1.29	0.798	6.87
	0.2500	2.59	4.69	2.35	1.35	1.54	1.54	0.770	8.81
	0.3125	3.11	5.32	2.66	1.31	1.71	1.71	0.743	10.58
4 × 3	0.1875	2.39	5.23	2.62	1.48	3.34	2.23	1.18	8.15
	0.2500	3.09	6.45	3.23	1.45	4.10	2.74	1.15	10.51
	0.3125	3.73	7.45	3.72	1.41	4.71	3.14	1.12	12.70
5 × 3	0.1875	2.77	9.06	3.62	1.81	4.08	2.72	1.21	9.42
	0.2500	3.59	11.3	4.52	1.77	5.05	3.37	1.19	12.21
	0.3125	4.36	13.2	5.27	1.74	5.85	3.90	1.16	14.83
	0.3750	5.08	14.7	5.89	1.70	6.48	4.32	1.13	17.27

5 × 4	0.1875	3.14	11.2	4.49	1.89	7.96	3.98	1.59	10.70
	0.2500	4.09	14.1	5.65	1.86	9.98	4.99	1.56	13.91
	0.3125	4.98	16.6	6.65	1.83	11.7	5.85	1.53	16.96
	0.3750	5.83	18.7	7.50	1.79	13.2	6.58	1.50	19.82
6 × 3	0.1875	3.14	14.3	4.76	2.13	4.83	3.22	1.24	10.70
	0.2500	4.09	17.9	5.98	2.09	6.00	4.00	1.21	13.91
	0.3125	4.98	21.1	7.03	2.06	6.98	4.65	1.18	16.96
	0.3750	5.83	23.8	7.92	2.02	7.78	5.19	1.16	19.82
6 × 4	0.2500	4.59	22.1	7.36	2.19	11.7	5.87	1.60	15.62
	0.3125	5.61	26.2	8.72	2.16	13.8	6.92	1.57	19.08
	0.3750	6.58	29.7	9.90	2.13	15.6	7.82	1.54	22.37
8 × 4	0.2500	5.59	45.1	11.3	2.84	15.3	7.63	1.65	19.02
	0.3125	6.86	53.9	13.5	2.80	18.1	9.05	1.62	23.34
	0.3750	8.08	61.9	15.5	2.77	20.6	10.3	1.60	27.48
8 × 6	0.2500	6.59	60.1	15.0	3.02	38.6	12.9	2.42	22.42
	0.3125	8.11	72.4	18.1	2.99	46.4	15.5	2.39	27.59
	0.3750	9.58	83.7	20.9	2.96	53.5	17.8	2.36	32.58

TABLE B.6 Properties of Standard Steel Elements: Structural Tees[a]

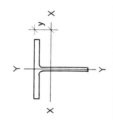

Designation	Area (in.²)	Q_s	Depth (in.)	Stem Thickness (in.)	Flange Width	Flange Thickness	X-X Axis I (in.⁴)	S (in.³)	r (in.)	y (in.)	Y-Y Axis I (in.⁴)	S (in.³)	r (in.)
WT4 × 5	1.48	0.735	3.945	0.170	3.940	0.205	2.15	0.717	1.20	0.953	1.05	0.532	0.841
WT4 × 6.5	1.92	—	3.995	0.230	4.000	0.255	2.89	0.974	1.23	1.03	1.37	0.683	0.843
WT4 × 7.5	2.22	—	4.055	0.245	4.015	0.315	3.28	1.07	1.22	0.998	1.70	0.849	0.876
WT5 × 6	1.77	0.793	4.935	0.190	3.960	0.210	4.35	1.22	1.57	1.36	1.09	0.551	0.785
WT5 × 7.5	2.21	0.977	4.995	0.230	4.000	0.270	5.45	1.50	1.57	1.37	1.45	0.723	0.810
WT5 × 8.5	2.50	—	5.055	0.240	4.010	0.330	6.06	1.62	1.56	1.32	1.78	0.888	0.844
WT5 × 9.5	2.81	—	5.120	0.250	4.020	0.395	6.68	1.74	1.54	1.28	2.15	1.07	0.874
WT5 × 11	3.24	0.999	5.085	0.240	5.750	0.360	6.88	1.72	1.46	1.07	5.71	1.99	1.33
WT5 × 13	3.81	—	5.165	0.260	5.770	0.440	7.86	1.91	1.44	1.06	7.05	2.44	1.36
WT5 × 15	4.42	—	5.235	0.300	5.810	0.510	9.28	2.24	1.45	1.10	8.35	2.87	1.37
WT6 × 7	2.08	0.626	5.955	0.200	3.970	0.225	7.67	1.83	1.92	1.76	1.18	0.594	0.753
WT6 × 8	2.36	0.741	5.995	0.220	3.990	0.265	8.70	2.04	1.92	1.74	1.41	0.706	0.773

WT6 × 9.5	2.79	0.797	6.080	0.235	4.005	0.350	10.1	2.28	1.90	1.65	1.88	0.939	0.822
WT6 × 11	3.24	0.891	6.155	0.260	4.030	0.425	11.7	2.59	1.90	1.63	2.33	1.16	0.847
WT6 × 13	3.82	0.767	6.110	0.230	6.490	0.380	11.7	2.40	1.75	1.25	8.66	2.67	1.51
WT6 × 15	4.40	0.891	6.170	0.260	6.520	0.440	13.5	2.75	1.75	1.27	10.2	3.12	1.52
WT6 × 17.5	5.17	—	6.250	0.300	6.560	0.520	16.0	3.23	1.76	1.30	12.2	3.73	1.54
WT7 × 11	3.25	0.621	6.870	0.230	5.000	0.335	14.8	2.91	2.14	1.76	3.50	1.40	1.04
WT7 × 13	3.85	0.737	6.955	0.255	5.025	0.420	17.3	3.31	2.12	1.72	4.45	1.77	1.08
WT7 × 15	4.42	0.810	6.920	0.270	6.730	0.385	19.0	3.55	2.07	1.58	9.79	2.91	1.49
WT7 × 17	5.00	0.857	6.990	0.285	6.745	0.455	20.9	3.83	2.04	1.53	11.7	3.45	1.53
WT7 × 19	5.58	0.934	7.050	0.310	6.770	0.515	23.3	4.22	2.04	1.54	13.3	3.94	1.55
WT7 × 21.5	6.31	0.947	6.830	0.305	7.995	0.530	21.9	3.98	1.86	1.31	22.6	5.65	1.89
WT7 × 24	7.07	—	6.895	0.340	8.030	0.595	24.9	4.48	1.87	1.35	25.7	6.40	1.91
WT7 × 26.5	7.81	—	6.960	0.370	8.060	0.660	27.6	4.94	1.88	1.38	28.8	7.16	1.92
WT8 × 13	3.84	0.563	7.845	0.250	5.500	0.345	23.5	4.09	2.47	2.09	4.80	1.74	1.12
WT8 × 15.5	4.56	0.668	7.940	0.275	5.525	0.440	27.4	4.64	2.45	2.02	6.20	2.24	1.17
WT8 × 18	5.28	0.754	7.930	0.295	6.985	0.430	30.6	5.05	2.41	1.88	12.2	3.50	1.52
WT8 × 20	5.89	0.784	8.005	0.305	6.995	0.505	33.1	5.35	2.37	1.81	14.4	4.12	1.57
WT8 × 22.5	6.63	0.904	8.065	0.345	7.035	0.565	37.8	6.10	2.39	1.86	16.4	4.67	1.57
WT8 × 25	7.37	0.890	8.130	0.380	7.070	0.630	42.3	6.78	2.40	1.89	18.6	5.26	1.59
WT8 × 28.5	8.38	—	8.215	0.430	7.120	0.715	48.7	7.77	2.41	1.94	21.6	6.06	1.60

[a] Cut from I-shaped elements.

TABLE B.7 Properties of Standard Steel Elements: Single Angles—Equal Legs

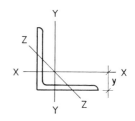

Size and Thickness (in.)	Area (in.²)	Weight per ft (lb)	Q_s	Section Properties— X and Y Axes				Z-Z Axis r (in.)
				I (in.⁴)	S (in.³)	r (in.)	x or y (in.)	
2 × 2 × 3/16	0.715	2.44	—	0.272	0.190	0.617	0.569	0.394
2 × 2 × 1/4	0.938	3.19	—	0.348	0.247	0.609	0.592	0.391
3 × 3 × 1/4	1.44	4.9	—	1.24	0.577	0.930	0.842	0.592
3 × 3 × 5/16	1.78	6.1	—	1.51	0.707	0.922	0.865	0.589
3 × 3 × 3/8	2.11	7.2	—	1.76	0.833	0.913	0.888	0.587
4 × 4 × 1/4	1.94	6.6	0.911	3.04	1.05	1.25	1.09	0.795
4 × 4 × 5/16	2.40	8.2	0.977	3.71	1.29	1.24	1.12	0.791
4 × 4 × 3/8	2.86	9.8	—	4.36	1.52	1.23	1.14	0.788
5 × 5 × 5/16	3.03	10.3	0.911	7.42	2.04	1.57	1.37	0.994
5 × 5 × 3/8	3.61	12.3	0.982	8.74	2.42	1.56	1.39	0.990
5 × 5 × 1/2	4.75	16.2	—	11.3	3.16	1.54	1.43	0.983
6 × 6 × 3/8	4.36	14.9	0.911	15.4	3.53	1.88	1.64	1.19
6 × 6 × 1/2	5.75	19.6	—	19.9	4.61	1.86	1.68	1.18
8 × 8 × 1/2	7.75	26.4	0.911	48.6	8.36	2.50	2.19	1.59
8 × 8 × 5/8	9.61	32.7	0.997	59.4	10.3	2.49	2.23	1.58

TABLE B.8 Properties of Standard Steel Elements: Single Angles—Unequal Legs

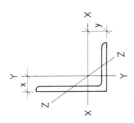

Size and Thickness (in.)	Area (in.²)	Weight per ft (lb)	Q_s	X-X Axis				Y-Y Axis				Z-Z Axis r (in.)
				I (in.⁴)	S (in.³)	r (in.)	y (in.)	I (in.⁴)	S (in.³)	r (in.)	x (in.)	
$2\frac{1}{2} \times 2 \times \frac{3}{16}$	0.809	2.75	—	0.509	0.293	0.793	0.764	0.291	0.196	0.600	0.514	0.427
$2\frac{1}{2} \times 2 \times \frac{1}{4}$	1.06	3.62	—	0.654	0.381	0.784	0.787	0.372	0.254	0.592	0.537	0.424
$3 \times 2 \times \frac{3}{16}$	0.902	3.07	0.911	0.842	0.415	0.966	0.970	0.307	0.200	0.583	0.470	0.439
$3 \times 2 \times \frac{1}{4}$	1.19	4.1	—	1.09	0.542	0.957	0.993	0.392	0.260	0.574	0.493	0.435
$3 \times 2 \times \frac{5}{16}$	1.46	5.0	—	1.32	0.664	0.948	1.02	0.470	0.317	0.567	0.516	0.432
$3 \times 2\frac{1}{2} \times \frac{1}{4}$	1.31	4.5	—	1.17	0.561	0.945	0.911	0.743	0.404	0.753	0.661	0.528
$3\frac{1}{2} \times 2\frac{1}{2} \times \frac{1}{4}$	1.44	4.9	0.965	1.80	0.755	1.12	1.11	0.777	0.412	0.735	0.614	0.544
$3\frac{1}{2} \times 2\frac{1}{2} \times \frac{5}{16}$	1.78	6.1	—	2.19	0.927	1.11	1.14	0.939	0.504	0.727	0.637	0.540
$3\frac{1}{2} \times 2\frac{1}{2} \times \frac{3}{8}$	2.11	7.2	—	2.56	1.09	1.10	1.16	1.09	0.592	0.719	0.660	0.537
$4 \times 3 \times \frac{1}{4}$	1.69	5.8	0.911	2.77	1.00	1.28	1.24	1.36	0.599	0.896	0.736	0.651
$4 \times 3 \times \frac{5}{16}$	2.09	7.2	0.997	3.38	1.23	1.27	1.26	1.65	0.734	0.887	0.759	0.647
$4 \times 3 \times \frac{3}{8}$	2.48	8.5	—	3.96	1.46	1.26	1.28	1.92	0.866	0.879	0.782	0.644
$4 \times 3\frac{1}{2} \times \frac{1}{4}$	1.81	6.2	0.911	2.91	1.03	1.27	1.16	2.09	0.808	1.07	0.909	0.734
$4 \times 3\frac{1}{2} \times \frac{5}{16}$	2.25	7.7	0.997	3.56	1.26	1.26	1.18	2.55	0.994	1.07	0.932	0.730
$4 \times 3\frac{1}{2} \times \frac{3}{8}$	2.67	9.1	—	4.18	1.49	1.25	1.21	2.95	1.17	1.06	0.955	0.727

(*continued*)

TABLE B.8 (*Continued*)

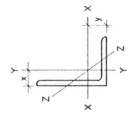

Size and Thickness (in.)	Area (in.²)	Weight per ft (lb)	Q_s	X-X Axis				Y-Y Axis				Z-Z Axis r (in.)
				I (in.⁴)	S (in.³)	r (in.)	y (in.)	I (in.⁴)	S (in.³)	r (in.)	x (in.)	
$5 \times 3 \times \frac{1}{4}$	1.94	6.6	0.804	5.11	1.53	1.62	1.66	1.44	0.614	0.861	0.657	0.663
$5 \times 3 \times \frac{5}{16}$	2.40	8.2	0.911	6.26	1.89	1.61	1.68	1.75	0.753	0.853	0.681	0.658
$5 \times 3 \times \frac{3}{8}$	2.86	9.8	0.982	7.37	2.24	1.61	1.70	2.04	0.888	0.845	0.704	0.654
$5 \times 3\frac{1}{2} \times \frac{5}{16}$	2.56	8.7	0.911	6.60	1.94	1.61	1.59	2.72	1.02	1.03	0.838	0.766
$5 \times 3\frac{1}{2} \times \frac{3}{8}$	3.05	10.4	0.982	7.78	2.29	1.60	1.61	3.18	1.21	1.02	0.861	0.762
$6 \times 3\frac{1}{2} \times \frac{5}{16}$	2.87	9.8	0.825	10.9	2.73	1.95	2.01	2.85	1.04	0.996	0.763	0.772
$6 \times 3\frac{1}{2} \times \frac{3}{8}$	3.42	11.7	0.911	12.9	3.24	1.94	2.04	3.34	1.23	0.988	0.787	0.767
$6 \times 4 \times \frac{3}{8}$	3.61	12.3	0.911	13.5	3.32	1.93	1.94	4.90	1.60	1.17	0.941	0.877
$6 \times 4 \times \frac{1}{2}$	4.75	16.2	—	17.4	4.33	1.91	1.99	6.27	2.08	1.15	0.987	0.870
$7 \times 4 \times \frac{3}{8}$	3.98	13.6	0.839	20.6	4.44	2.27	2.37	5.10	1.63	1.13	0.870	0.880
$7 \times 4 \times \frac{1}{2}$	5.25	17.9	0.965	26.7	5.81	2.25	2.42	6.53	2.12	1.11	0.917	0.872
$8 \times 6 \times \frac{1}{2}$	6.75	23.0	0.811	44.3	8.02	2.56	2.47	21.7	4.79	1.79	1.47	1.30

TABLE B.9 Properties of Standard Steel Elements: Double Angles—Equal Legs

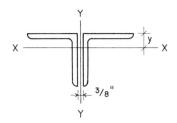

Size and Thickness (in.)	Area (in.²)	Weight per ft (lb)	Q_s	X-X Axis I (in.⁴)	X-X Axis S (in.³)	X-X Axis r (in.)	X-X Axis y (in.)	Y-Y Axis r (in.)
2 × 2 × 3/16	1.43	4.88	—	0.545	0.381	0.617	0.569	0.977
2 × 2 × 1/4	1.88	6.38	—	0.695	0.494	0.609	0.592	0.989
3 × 3 × 1/4	2.88	9.8	—	2.49	1.15	0.930	0.842	1.39
3 × 3 × 5/16	3.55	12.2	—	3.02	1.41	0.922	0.865	1.40
3 × 3 × 3/8	4.22	14.4	—	3.52	1.67	0.913	0.888	1.41
4 × 4 × 1/4	3.88	13.2	0.911	6.08	2.09	1.25	1.09	1.79
4 × 4 × 5/16	4.80	16.4	0.997	7.43	2.58	1.24	1.12	1.80
4 × 4 × 3/8	5.72	19.6	—	8.72	3.05	1.23	1.14	1.81
5 × 5 × 5/16	6.05	20.6	0.911	14.8	4.08	1.57	1.37	2.21
5 × 5 × 3/8	7.22	24.6	0.982	17.5	4.84	1.56	1.39	2.22
5 × 5 × 1/2	9.50	32.4	—	22.5	6.31	1.54	1.43	2.24
6 × 6 × 3/8	8.72	29.8	0.911	30.8	7.06	1.88	1.64	2.62
6 × 6 × 1/2	11.5	39.2	—	39.8	9.23	1.86	1.68	2.64
8 × 8 × 1/2	15.5	52.8	0.911	97.3	16.7	2.50	2.19	3.45
8 × 8 × 5/8	19.2	65.4	0.997	118.0	20.6	2.49	2.23	3.47

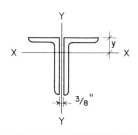

**TABLE B.10 Properties of Standard Steel Elements:
Unequal Double Angles—Long Legs Back-to-Back**

Size and Thickness (in.)	Area (in.²)	Weight per ft (lb)	Q_s	X-X Axis I (in.⁴)	S (in.³)	r (in.)	y (in.)	Y-Y Axis r (in.)
$2\frac{1}{2} \times 2 \times \frac{3}{16}$	1.62	5.5	0.982	1.02	0.586	0.793	0.764	0.923
$2\frac{1}{2} \times 2 \times \frac{1}{4}$	2.13	7.2	—	1.31	0.763	0.784	0.787	0.935
$3 \times 2 \times \frac{3}{16}$	1.80	6.1	0.911	1.68	0.830	0.966	0.970	0.879
$3 \times 2 \times \frac{1}{4}$	2.38	8.2	—	2.17	1.08	0.957	0.993	0.891
$3 \times 2 \times \frac{5}{16}$	2.93	10.0	—	2.63	1.33	0.948	1.02	0.903
$3 \times 2\frac{1}{2} \times \frac{1}{4}$	2.62	9.0	—	2.35	1.12	0.945	0.911	1.13
$3\frac{1}{2} \times 2\frac{1}{2} \times \frac{1}{4}$	2.88	9.8	0.965	3.60	1.51	1.12	1.11	1.09
$3\frac{1}{2} \times 2\frac{1}{2} \times \frac{5}{16}$	3.55	12.2	—	4.38	1.85	1.11	1.14	1.10
$3\frac{1}{2} \times 2\frac{1}{2} \times \frac{3}{8}$	4.22	14.4	—	5.12	2.19	1.10	1.16	1.11
$4 \times 3 \times \frac{1}{4}$	3.38	11.6	0.911	5.54	2.00	1.28	1.24	1.29
$4 \times 3 \times \frac{5}{16}$	4.18	14.4	0.997	6.76	2.47	1.27	1.26	1.30
$4 \times 3 \times \frac{3}{8}$	4.97	17.0	—	7.93	2.92	1.26	1.28	1.31
$4 \times 3\frac{1}{2} \times \frac{1}{4}$	3.63	12.4	0.911	5.83	2.05	1.27	1.16	1.54
$4 \times 3\frac{1}{2} \times \frac{5}{16}$	4.49	15.4	0.997	7.12	2.53	1.26	1.18	1.55
$4 \times 3\frac{1}{2} \times \frac{3}{8}$	5.34	18.2	—	8.35	2.99	1.25	1.21	1.56
$5 \times 3 \times \frac{1}{4}$	3.88	13.2	0.804	10.2	3.06	1.62	1.66	1.21
$5 \times 3 \times \frac{5}{16}$	4.80	16.4	0.911	12.5	3.77	1.61	1.68	1.22
$5 \times 3 \times \frac{3}{8}$	5.72	19.6	0.982	14.7	4.47	1.61	1.70	1.23
$5 \times 3\frac{1}{2} \times \frac{5}{16}$	5.12	17.4	0.911	13.2	3.87	1.61	1.59	1.45
$5 \times 3\frac{1}{2} \times \frac{3}{8}$	6.09	20.8	0.982	15.6	4.59	1.60	1.61	1.46
$6 \times 3\frac{1}{2} \times \frac{5}{16}$	5.74	19.6	0.825	21.8	5.47	1.95	2.01	1.38
$6 \times 3\frac{1}{2} \times \frac{3}{8}$	6.84	23.4	0.911	25.7	6.49	1.94	2.04	1.39
$6 \times 4 \times \frac{3}{8}$	7.22	24.6	0.911	26.9	6.64	1.93	1.94	1.62
$6 \times 4 \times \frac{1}{2}$	9.50	32.4	—	34.8	8.67	1.91	1.99	1.64
$7 \times 4 \times \frac{3}{8}$	7.97	27.2	0.839	41.1	8.88	2.27	2.37	1.55
$7 \times 4 \times \frac{1}{2}$	10.5	35.8	0.965	53.3	11.6	2.25	2.42	1.57
$8 \times 6 \times \frac{1}{2}$	13.5	46.0	0.911	88.6	16.0	2.56	2.47	2.44

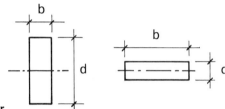

TABLE B.11 Properties of Structural Lumber

Dimensions (in.)		Area (in.²)	Section Modulus[a] (in.³)	Moment of Inertia[a] (in.⁴)	Weight at 35 lb/ft³ (lb/ft)
Nominal	Actual				
2 × 3	1.5 × 2.5	3.75	1.563	1.953	0.9
2 × 4	1.5 × 3.5	5.25	3.063	5.359	1.3
2 × 5	1.5 × 4.5	6.75	5.063	11.391	1.7
2 × 6	1.5 × 5.5	8.25	7.563	20.797	2.0
2 × 8	1.5 × 7.25	10.875	13.141	47.635	2.6
2 × 10	1.5 × 9.25	13.875	21.391	98.932	3.4
2 × 12	1.5 × 11.25	16.875	31.641	177.979	4.1
3 × 4	2.5 × 3.5	8.75	5.104	8.932	2.1
3 × 5	2.5 × 4.5	11.25	8.438	18.984	2.7
3 × 6	2.5 × 5.5	13.75	12.604	34.661	3.3
3 × 8	2.5 × 7.25	18.125	21.901	79.391	4.4
3 × 10	2.5 × 9.25	23.125	35.651	164.886	5.6
3 × 12	2.5 × 11.25	28.125	52.734	296.631	6.8
4 × 4	3.5 × 3.5	12.25	7.146	12.505	3.0
4 × 6	3.5 × 5.5	19.25	17.646	48.526	4.7
4 × 8	3.5 × 7.25	25.375	30.661	111.148	6.2
4 × 10	3.5 × 9.25	32.375	49.911	230.840	7.9
4 × 12	3.5 × 11.25	39.375	73.828	415.283	9.6
6 × 6	5.5 × 5.5	30.25	27.729	76.255	7.4
6 × 8	5.5 × 7.5	41.25	51.563	193.359	10.0
6 × 10	5.5 × 9.5	52.25	82.729	392.963	12.7
6 × 12	5.5 × 11.5	63.25	121.229	697.068	15.4
6 × 14	5.5 × 13.5	74.25	167.063	1127.672	18.0
6 × 16	5.5 × 15.5	85.25	220.229	1706.776	20.7
8 × 8	7.5 × 7.5	56.25	70.313	263.672	13.7
8 × 10	7.5 × 9.5	71.25	112.813	535.859	17.3
8 × 12	7.5 × 11.5	86.25	165.313	950.547	21.0
8 × 14	7.5 × 13.5	101.25	227.813	1537.734	24.6
8 × 16	7.5 × 15.5	116.25	300.313	2327.422	28.3
8 × 18	7.5 × 17.5	131.25	382.813	3349.609	31.9
10 × 10	9.5 × 9.5	90.25	142.896	678.755	21.9
12 × 12	11.5 × 11.5	132.25	253.479	1457.505	32.1
14 × 14	13.5 × 13.5	182.25	410.063	2767.922	44.3

[a] Properties for strongest centroidal axis for rectangular sections.

APPENDIX C

VALUES FOR TYPICAL BEAM LOADINGS

Some of the most common beam loadings are shown in Figure C.1. The forms of the shear and bending moment diagrams are shown for each case, and formulas are given for obtaining values for the reactions, R, the maximum shear, V, the maximum bending moment, M, and the maximum deflection, D. Values for the reactions are not given for symmetrical cases, as each reaction is simply one-half the total load.

Also given in the figure are values designated *ETL*, which stands for *equivalent tabular loading*, also sometimes called *equivalent uniform loading*. These may be used to obtain a hypothetical uniformly distributed load, which, when applied to the beam, will produce the same magnitude of maximum bending moment as that for the given case of loading. One use for the *ETL* is in obtaining approximate values for deflections from graphs or from tabulated values given for uniformly distributed loadings.

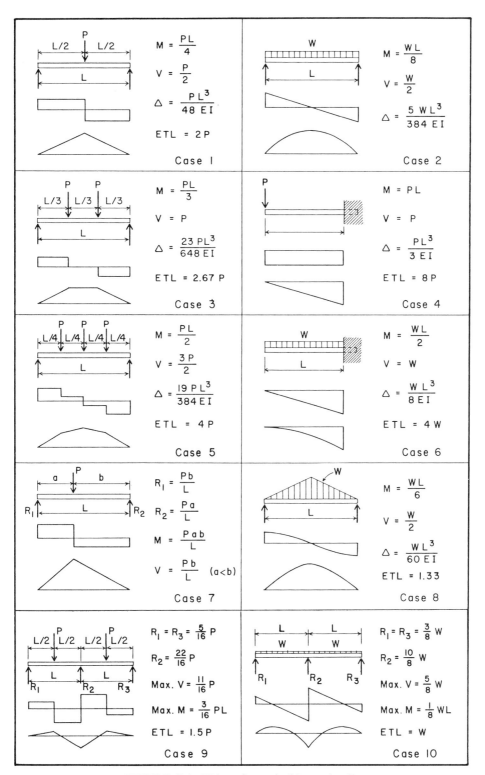

FIGURE C.1 Values for typical beam loadings.

APPENDIX D

USE OF COMPUTERS

Two products of modern technology have had tremendous influence on the working style of building designers: the computer and the pocket calculator. In a short time, these have transformed the level of complexity and the accuracy of mathematical computation possible for even the most routine work. Through the use of prepared routines and computer programs, minimally trained designers can utilize highly sophisticated analyses and design procedures. The materials in this part present a brief discussion of the potentiality for use of current computer hardware and commercially available software in the processes of education and professional design, with particular applications to trussed structures.

Practical utilization of computers requires a layered, sequential learning process. This begins with (1) some fundamental training in basic computer operations and routines, then (2) some simple introduction to engineering and architectural applications, and finally (3) some specific instruction and experience in use of individual programs and software systems for particular tasks.

COMPUTER-AIDED DESIGN

Investigative processes, whether pursued for education or for design application, tend to be more simply adapted to computer operation. Design processes, which typically require many layered stages of judgmental decisions, are generally quite complex; seldom a challenge to the computer equipment, but often overwhelming for the software writer or user. Assuming that the equipment is capable, we consider first the problem of what it takes to use the equipment.

COMPUTER SOFTWARE

Computers and various supportive equipment and materials that surround them have been for a long time in a state of marketing hype. The frenzy has resulted in the development of some useful materials relating to work in various fields. Obviously, the larger the market (in terms of potential buyers), the more that market has been exploited and is now rich with available support equipment and materials.

Building design work does not generally represent a major market for just about anything—textbooks, computer equipment and software, or whatever—and thus the build-up of useful items at competitive prices has been slower. This is sure to change, as upcoming students and young working professionals are increasingly computer conversant.

Some design offices are already heavily involved, equipped, and operative with all that can be done with computers. However, of the total amount of work done in building design, only a small fraction is presently supported directly by computers. This is not true of all areas or stages of design; thus, the engineering work in building design is generally far ahead of many other aspects of the whole building design field. This has significance to the engineering work itself, but the total impact on buildings will not be realized until the whole building design team is computer interactive.

For the field of structural engineering, there is a considerable existing inventory of software available from various sources. Most of the major industry organizations (ACI, PCA, AISC, etc.), professional groups (AIA, ASCE, CSI, etc.), and many individual product manufacturers have either directly developed or are somehow involved with the development of computer data storage and computer-aided design software systems. Inquiry for information about currently available materials can be directed to those organizations. Interest in a specific topic can be pursued by focusing on the individual groups that have special involvement in those topics. (Wood, steel, structural investigation in general, buildings, etc.)

Private producers and marketers of software also have materials, but any limited market (truss design?) precludes a broad coverage. Individuals and small groups (schools, individual offices, small professional groups, etc.) have pet systems that they have developed and use and want to share; maybe for a price. A lot is going on, but it is not easy to learn about as a whole, to assess, or to take full advantage of.

In time, it will probably become relatively easy to find out what can be done in terms of computer-aided design work, to obtain the materials most useful for particular tasks, and to learn how to use the materials productively. If the current level of competition continues, it is also likely that tooling up with the equipment, as well as the software, will become relatively painless.

There is no shortcut to finding out about all of this, and plugging into it requires a commitment to doing whatever you can to get involved—learning about computers and how to use them but more specifically, learning about what exactly is available for application for your particular needs. This leads rapidly to the maze of overblown promotional displays of current commercial products. Dorothy has finally arrived in Oz and badly needs the Good Witch to guide her.

A first acquisition for the newcomer should be any and all summary analyses of the current market and the various potential applications to specific tasks. Different sources of such analyses will produce different opinions, but a general feeling for it all may eventually emerge. Don't buy anything until you kick a lot of tires. Get a lot of opinions that overlap before you commit to anything.

COMPATIBILITY

Marketing in the free enterprise system has of course produced a great deal of incompatibility between competing computer materials. This may go to the bone in terms of fundamental operations, or may be thinly concealed and easily untangled for cross-system use. Don't get trapped in a singular system that only a few people are using, but do expect that cross-system potential will be well-touted if it is there, and not mentioned if it is not (not always true, but something to look for).

Almost all major players in the commercial computer business are backed up by various levels of supportive products. Soon after a new major word-processing or spreadsheet program comes out, a whole array of secondary programs come out to enhance its usage. Some of these may allow the user to cross the non-compatible systems barriers or to use the machine-capacity-eating new programs to be used on more modest, older, cheaper equipment.

Materials relating to specific fields have in some cases centered on particular equipment or basic operating software systems. This occurs naturally when certain equipment or systems favor certain tasks. Thus, the object-oriented system developed by Apple Computers became a leader in graphic applications and still holds strong for architectural applications in general, although the Microsoft applications with Windows now bite strongly into this field.

INVESTIGATION OF TRUSSES

An obvious application of computer support is in the investigation of trusses for internal forces and the determination of the deformed shape of loaded trusses. In terms of existing programs and available software, this is the most fully supported work at present. This is pure physics, mathematics, and graphically expressed manifestations of mathematically expressed relationships.

Use of determined values for internal forces is essential to truss design, but is basically only an interpretive, analytical phase of design; it is not particularly a conceptual, initiating phase nor a conclusive, verification phase. For investigation, a truss form and its dimensions must first be defined. One function of the investigation is thus to verify the logic of the decisions regarding form and dimensions. If other forms or dimensions for the truss (truss profile, member articulation, panel module size, depth-to-span ratios, etc.) can be visualized, then investigations of some range of variables in form and dimension may eventually feed into a judgment about the proper or best form and dimensions.

GLOSSARY

The material in this glossary constitutes a brief dictionary of words and terms frequently encountered in discussions of trussed structures. Many of the words and terms have reasonably well-established meanings; in those cases, we have tried to be consistent with the accepted usage. In some cases, however, words and terms are given different meanings by different authors or by groups that work in different fields. In these situations, we have given the definition as used for this book so that the reader may be clear as to our meaning.

For a fuller explanation of some of the words and terms given here, as well as definitions not given here, the reader should use the index to find related discussion in the text.

Abutment. Originally, the end support of an arch or vault. Now, any support that receives both vertical and lateral load from the supported structure.

Analysis. Separation into constituent parts. In engineering, the investigation for determination of the detail aspects of a particular phenomenon. May be *qualitative*, meaning a general evaluation of the nature of the phenomenon, or *quantitative*, meaning numerical determination of the magnitude of the phenomenon.

Anchorage. Refers to attachment for resistance to movement off a support; usually the result of uplift, overturn, sliding, or horizontal separation. *Positive anchorage* refers to fastening that does not easily loosen, especially with repeated or dynamic loading.

Angle of load to grain. Refers to the orientation of the direction of a load to the direction of the grain in a piece of wood. Grain direction in wood is along the length of the log (tree trunk). Load direction is referred to as *parallel to the grain* (along the grain, or at a zero angle to the grain), *perpendicular to the grain* (90° to the grain), or at some angle to the grain (between zero and 90°).

Beam. A structural element that sustains transverse (perpendicular to its linear axis) loading and develops major internal forces of bending and shear. May be called a *girder* if large in size and providing support for other beams; a *purlin* if a secondary member in a framing system; a *joist* if used in a closely-spaced set and providing direct support for a deck; a *rafter* if used in the manner of a joist, except for a roof structure. Also see *Header* and *Lintel.*

Braced frame. Literally, any framework braced against lateral forces. Building codes use the term to describe a frame braced by triangulation (trussing).

Centroid. The geometric center of an object, usually analogous to the center of gravity. The point at which the entire mass of the object may be considered to be concentrated when considering moment of the mass.

Composite structure. Structural element in which different materials share a load, requiring determination of the distribution of the load between the materials. Example: steel and concrete sharing compression in a reinforced concrete column.

Core bracing. Vertical elements of a lateral-bracing system developed at the location of permanent interior walls for stairs, elevators, duct shafts or rest rooms. Usually used when these elements are placed in a group in the interior of the building (constituting a "core").

Curtain wall. An exterior building wall that is supported entirely by the frame of the building, as opposed to being self-supporting or load-bearing.

Deflected shape. Refers to the profile form of a structure as deformed by the loads.

Deflection. Lateral movement of a member subjected to bending; used mostly to describe the vertical movement (sag) of a spanning member, such as a beam or truss.

Deformation. General description of shape change produced by stress actions in a structure.

Deformed shape. See *Deflected shape.*

Determinate. Having clearly defined limits; definite. In structures, the condition of having the exact sufficiency of stability internally and externally, and thus being capable of analysis by the resolution of forces (static equilibrium) alone. An excess of stability conditions (for example, more than the required number of reactions) constitutes a condition described as *indeterminate* or *statically indeterminate.*

Diaphragm. A surface element (deck, wall) used to resist forces in its own plane by spanning or cantilevering. See also *Horizontal diaphragm, Shear wall,* and *Vertical diaphragm.*

Dual bracing system. Building code term for a lateral bracing system that uses moment-resisting frames (rigid frames) in combination with either shear walls or trussing that works in a load-sharing manner with the moment-resisting frame.

Eccentric bracing. Truss bracing for a frame in which the bracing members do not attach to the joints of the frame at one end or both of the bracing member. Forms include knee-brace, K-brace, chevron brace (also called V-brace). Induces bending and shear in the frame members in addition to the usual truss effects.

Element. A component or constituent part of a whole; a distinct, separate entity; one truss, one column, one nail, and so on.

Equilibrium. A balanced state or condition, usually used to describe a situation in which opposed effects neutralize each other to produce a net effect of zero (no movement, etc.).

Feasible. Capable of being, or likely to be, accomplished.

Force. An effort that tends to change the shape or the state of motion of an object.

Force polygon. Vector plot of a force system; graphical equivalent of an algebraic summation of forces.

Funicular polygon. Graphical plot of a force system that verifies the equilibrium of another force system. Forces in the funicular polygon are components of the forces in the real system and thus add up to the same total system result.

Gable roof. Double-sloping roof formed by joined rafters or rigid frames with a ridge or peak at the top. A *gable* is the upper triangular portion of a wall at the end of a gable roof.

Girder. See *Beam*.

Gusset plate. Intermediate connecting device that allows the attachment of members at a joint by having them all attached to the common plate inserted in the joint.

Header. A horizontal framing element (beam) over an opening in a wall or at the edge of an opening in a floor or roof structure.

Heavy timber. See *Timber*.

Horizontal diaphragm. See *Diaphragm*. Usually a roof or floor deck used as part of a lateral bracing system.

Indeterminate. See *Determinate*.

Internal force diagram. Proper name for the graphical device that is used to determine the values and senses of internal forces in the members of a truss. Also called *stress diagram* or *Maxwell diagram*.

Joist. See *Beam*.

Kern. Zone around the centroid of a cross section within which an applied direct force of tension or compression will not cause net stress reversal on the cross section due to its eccentricity from the centroid.

Knee brace. A form of *eccentric bracing* consisting of a single member that connects to two members at a joint, but does not connect to the joint.

Lateral. Literally means to the side or from the side. Often used in reference to something that is perpendicular to a major axis or direction. With reference to the vertical direction of gravity forces, wind, earthquakes, and horizontally directed soil pressures are called *lateral effects*.

Let-in bracing. Diagonal boards (typically of 1-in.-nominal thickness) nailed to studs to provide trussed bracing in the wall plane. In order not to interfere with installation of surfacing materials, they are usually placed in notches in the stud faces, called let-in.

Light wood frame. Wood structure achieved primarily with framing members of 2-inch-nominal thickness (2 × 4 studs, etc.).

Lintel. A beam built into a wall over an opening in the wall.

Live Load. See *Load*.

Load. The active force (or combination of forces) exerted on a structure. *Dead load* is permanent load due to gravity, which includes the weight of the structure itself plus other permanent elements of the construction. *Live load* is any load component that is not permanent, including effects of wind, earthquakes, temperature changes, shrinkage, and so on; but the term is most often used for gravity loads that are not permanent and generate from the use of the structure. *Service load* is the total load combination that the structure is expected to experience in use. *Factored load* is the service load multiplied by some increase factor for use in strength design. *Ultimate load* usually refers to the limiting load that the structure can resist up to the moment of failure.

Maxwell diagram. See *Internal force diagram*.

Member. One of the distinct elements of an assembled system, such as one beam in a framing system.

Modulus of elasticity. The ratio of unit stress to unit strain in a material subjected to loading. A direct measure of the *relative stiffness* of the material.

Moment of Inertia. The second moment of an area about a fixed line (axis) in the plane of the area. A purely mathematical property, not subject to direct physical measurement. Has significance in that it can be quantified for any geometric shape and is a measurement of certain structural responses (most notably *deflection* due to bending).

Net section. The remaining solid part of a cross section that is reduced by cutouts for holes, notches, and so on; the actual area available for stress development.

Overturn. The toppling, or tipping over, effect of lateral loads.

Panel point. Primarily used to refer to the top or bottom chord joints in a truss.

Parapet. The extension of a wall plane or the roof edge above the adjacent roof level.

P-delta effect. Secondary effect on a linear compression member that occurs when the member is curved by bending, resulting in an added bending due to eccentricity of the compression force.

Perimeter bracing. Vertical bracing elements (shear walls, braced bents, etc.) located at the building perimeter (outer edge). Also called *peripheral bracing*.

Peripheral bracing. See *Perimeter bracing*.

Purlin. See *Beam*.

Radius of gyration. A defined mathematical property; the square root of the product of the moment of inertia divided by the area. Significant in investigation of buckling of compression elements (columns, truss chords, etc.).

Rigid frame. Framed structure in which the joints between members are made to transmit moments between the ends of the members. Called a *bent* when the frame is planar. Called a *moment-resisting space frame* in building codes.

Section. The two-dimensional profile or area obtained by passing a cutting plane through a solid form. *Cross section* implies a section at right angles to another section or to the linear axis of an object. Architectural plans, as commonly drawn, are horizontal cross sections.

Shear wall. A wall used as a *vertical diaphragm*; part of a lateral bracing system; resists force (shear) in the plane of the wall.

Slenderness. Relative thinness. In structures, the quality of flexibility or lack of buckling resistance is inferred by a condition of extreme slenderness.

Space diagram. Drawing of a structure indicating its form and support and loading conditions. May also give dimensions and the magnitudes of the loads.

Stability. Refers to the inherent capability of a structure to develop resistance to forces as a property of its form, orientation, articulation of parts, type of connections, method of support, and so on. Is not directly related to quantified strength or stiffness, except when buckling actions are involved.

Stiffness. In structures, refers to resistance to deformation, as opposed to *strength*, which refers to resistance to force. A lack of stiffness indicates a *flexible* structure. *Relative stiffness* usually refers to the comparative deformation of two or more structural elements that share a load.

Strain. Deformation resulting from stress. It is usually measured as a percentage of deformation, called *unit strain*, or *unit deformation*, and is dimensionless. Total deformation, such as the sag of a beam, is accumulated strain.

Strength. Capacity to resist force (static strength) or to do work (dynamic strength or energy capacity).

Strength design. Design method based on use of *factored loads*, and the estimated ultimate capability (strength) of a structure. Also called *ultimate strength design*. See also *Working stress design*.

Stress. The mechanism for development of force within a material; visualized as pressure on a cut section (tensile or compressive stress) or shearing effect in the plane of the section (shear stress); measured in units of force per unit area. *Allowable*, *permissible*, or *working* stress refers to a modified, reduced stress magnitude used for working stress design. *Ultimate* stress refers to the maximum stress a material can endure before failure.

Stress diagram. Term previously used to describe the *internal force diagram*. The term *internal force* is more proper for identification of tension and compression forces.

Timber. General term for solid-sawn wood elements. Now used in the United States to describe thicker elements, as opposed to other terms used for lighter elements such as those used for the *light wood frame*. Construction produced with the thicker elements is described as *timber construction* or sometimes as *heavy timber construction*.

Truss. A framework of linear elements that achieves stability through triangular formations of the elements.

Ultimate strength. Usually refers to the maximum static force resistance of a structure or of some part of a structure.

Vertical diaphragm. See *Diaphragm*. Usually a *shear wall*.

Working stress. See *Stress*.

Working stress design. Design method based on investigation of stress conditions under the *service loads*. Uses modified stresses that are ultimate stresses reduced by a safety factor to produce *working stresses*.

BIBLIOGRAPHY

The following list contains materials that have been used as references in the development of various portions of the text. Also included are some widely used publications that serve as general references for building design, although no direct use of materials from them has been made in this book. The numbering system is random and merely serves to simplify referencing by text notation.

1. *Uniform Building Code*, 1991 ed., International Conference of Building Officials, 5360 South Workmanmill Road, Whittier, CA 90601. (Called the *UBC*.)
2. *Manual of Steel Construction*, 8th ed., American Institute of Steel Construction, Chicago, IL, 1980. (Called the *AISC Manual*.)
3. *National Design Specification for Wood Construction*, National Forest Products Association, Washington, D.C., 1991.
4. *Timber Construction Manual* 3rd ed., American Institute of Timber Construction, Wiley, New York, 1985.
5. *Standard Specifications, Load Tables, and Weight Tables for Steel Joists and Joist Girders*, Steel Joist Institute, Suite A, 1205 48th Avenue North, Myrtle Beach, SC 29577.
6. Charles G. Ramsey and Harold R. Sleeper, *Architectural Graphic Standards*, 8th ed. Wiley, New York, 1988.
7. Jack C. McCormac, *Structural Analysis*, 4th ed., Harper & Row, New York, 1984.
8. S. W. Crawley and R. M. Dillon, *Steel Buildings: Analysis and Design*, 3rd ed., Wiley, New York, 1984.
9. Donald E. Breyer, *Design of Wood Structures*, McGraw-Hill, New York, 1980.
10. Edward Allen, *Fundamentals of Building Construction: Materials and Methods*, 2nd ed., Wiley, New York, 1990.
11. James Ambrose and Dimitry Vergun, *Simplified Building Design for Wind and Earthquake Forces*, 2nd ed., Wiley, New York, 1990.
12. Harry Parker, *Simplified Design of Roof Trusses for Architects and Builders*, 2nd ed., Wiley, New York, 1953.

13. Harry Parker and James Ambrose, *Simplified Design of Building Trusses for Architects and Builders*, 3rd ed., Wiley, New York, 1982.
14. Frank Kidder and Harry Parker, *Architects and Builders Handbook*, 18th ed., Wiley, New York, 1931.
15. Jerome Sondericker, *Graphic Statics*, Wiley, New York, 1916.
16. Charles Gay and Harry Parker, *Materials and Methods of Architectural Construction*, Wiley, New York, 1932.
17. James Ambrose, *Building Structures*, 2nd ed., Wiley, New York, 1993.
18. James Ambrose, *Construction Revisited*, Wiley, New York, 1993.
19. Walter C. Voss and Edward A. Varney, *Architectural Construction*, Volume One, Book Two, Wood Construction, Wiley, New York, 1926.

INDEX

Allowable stresses for:
   steel, 207
   wood, 148
Angle-to-grain loading for wood, 152, 169

Beam:
   action, 77
   analogy for truss analysis, 96
   with cantilevered ends, 82
   with concentrated loads, 87
   deflection, 79
   determinate, 81
   indeterminate, 88
   with internal pins, 90
   moment, 78
   reaction, 77
   shear, 78
   simple, 77
   values for typical loadings, 415
Bolted joints:
   in steel, 221
   in wood, 165
Braced frame, 3, 35, 135
Bracing of buildings, 19, 35, 135, 200, 247
Bracing of trusses, 3, 129
Bridges, 6

Camelback truss, 392
Cantilever truss, 359, 388
Centering, 7
Combination wood and steel truss, 133, 198
Combined stress in truss members, 114, 162, 217, 219
Components of force, 42
Composition of forces, 42

Connections, 31
Computations, xv
Computers, 416
Crescent truss, 362

Deflection:
   of beam, 79
   of truss, 253
Delta truss, 287
Design forces for truss members, 112, 233
Design values for wood, 148
   modification of, 151
Determinacy of beam, 88

Eccentric bracing, 137
Equilibrant, 44
Equilibrium, 40, 44
Estimation of truss weight, 107

Fabricated trusses, 22, 133
Force:
   analysis, 45
   classification, 41
   components, 42
   composition, 42
   direction, 39
   equilibrium, 40, 44
   graphical representation, 50, 331
   internal, 46
   notation, 45
   polygon, 43
   properties, 39
   resolution, 42
   resultant, 43
   sense, 39

Force (*Continued*)
  systems, 40
  triangle, 42
Funicular polygon, 332

Grandstand truss, 352, 367
Graphical representation of force, 50, 331

Half-timber construction, 10
Hankinson:
  formula, 152
  graph, 169
Heavy timber truss, 192

Indeterminate truss, 257
Industrial products (trusses), 22, 133, 198
Interaction, 162
Internal force, 46
Interstitial space, 28

Joints, method of, 55
Joist:
  combination wood and steel, 133
  girder, 134
  open web steel, 133

Knee brace, 137

Lenticular truss, 386
Let-in bracing, 201
Line of action of force, 39
Loads:
  dead, 99, 317
  duration, 152
  earthquake, 99, 327
  gravity, 99, 104
  lateral, 324
  live, 99, 322
  roof, 111, 321
  seismic, 99, 327
  wind, 99, 111, 249, 324

Magnitude of force, 39
Manufactured trusses, 133
Maxwell diagram, 51
Measurement, units of, xii
Method of:
  beam analogy, 96
  joints, 55
  sections, 93
Moment:
  in beam, 78
  diagram, 81
  of force, 65
Monitor, roof, 369
Movement at supports, 258
Mullions, trussed, 277, 294

Nails, 173
Nonconcurrent forces, 258
Notation, xvi
Notation of:
  angles, 71
  forces, 45

Offset grid, 279
Open web joist, 133
Optimal truss form, 378
Origins of trusses, 3, 15

Ponding, 306
Predesigned trusses, 22, 133

Q_s, 214

Reactions, 77, 100
Resolution of force, 42
Resultant, 43
Roof:
  considerations for, 315
  loads, 111, 321
  pitch, 125
  slope, 125

Sawtooth roof truss, 350
Scissors truss, 11, 12
Secondary stresses, 258
Sections, method of, 93
Sections, properties of, 400
Sense of force, 39
Separated joint diagram, 51
Shear:
  in beams, 77
  diagram, 81
  developers, wood, 175
  plate, 177
Simple beam, 77
Space diagram, 50
Space frame, 34, 261
Spaced column, 160, 196
Split ring connectors, 177
Stability, 100
Static equilibrium, 40, 44
Steel:
  bolted joints, 221
  combined bending and compression, 219
  combined bending and tension, 217
  compression member, 207, 213
  design considerations, 203
  design stresses, 206
  properties of standard elements, 402
  tension member, 206, 208
  truss:
    with bolted joints, 225
    with tubular members, 243

with welded joints, 238
welded joints, 222
Substitute member, 355
Symbols, xvi
Systems of:
  forces, 40
  trusses, 117
Temperature effects, 100
Three-hinged structures, 14, 296, 373
Tied Arch, 10
Tower, trussed, 8, 36, 71, 290
Truss:
  bracing, 3, 129
  camelback, 392
  cantilever, 359, 388
  coefficients for internal force, 397
  column, 290
  combined wood and steel, 198
  deflection, 253
  delta, 287
  depth, 123
  determinacy of, 100
  erection, 313
  form, 118
  height, 123
  indeterminate, 257
  with internal pins, 91
  joints, 129, 303
  lenticular, 386
  loads on, 99, 121
  materials, 125
  open web steel, 133
  optimal form for, 378
  pattern, 120
  predesigned, 22, 133
  reactions, 100
  secondary stresses, 258
  span/height ratio, 123
  stability, 100
  steel, 203
  supports, 100, 258
  systems, 117
  weight, 107
  wood, 145
Trussed arch, 10, 14, 296
Trussed bent, 296, 371
Trussed column, 290
Trussed mullions, 277, 294
Trussed towers, 8, 36, 71, 290
Two-way spanning, 268

Units of measurement, xii

V-bracing, 139
Viewed trusses, 32, 307
Visualization of force sense, 62

Weight of:
  building materials, 318
  trusses, 107
Welded joints, 222
Wind load, 99, 111, 249
Wood:
  bolted joints, 165
  combined bending and compression, 163
  compression members, 155
    safe loads for, 158
  design considerations, 145
  design stresses, 148
  nailed joints, 173
  properties of standard elements, 413
  spaced column, 160
  split ring connectors, 177
  tension member, 153
  truss:
    with heavy timber members, 192
    with multiple-element members, 196
    with single element members, 185

X-bracing, 21, 135